THE SOCIAL AND ECONOMIC ROOTS OF THE SCIENTIFIC REVOLUTION

BOSTON STUDIES IN THE PHILOSOPHY OF SCIENCE

VOLUME 278

THE SOCIAL AND ECONOMIC ROOTS OF THE SCIENTIFIC REVOLUTION

Texts by Boris Hessen and Henryk Grossmann

edited by

GIDEON FREUDENTHAL

PETER MCLAUGHLIN

Editors
Gideon Freudenthal
Tel Aviv University
The Cohn Institute for the History
and Philosophy of Science and Ideas
Ramat Aviv
69 978 Tel Aviv
Israel

Peter McLaughlin
University of Heidelberg
Philosophy Department
Schulgasse 6
69117 Heidelberg
Germany

ISBN 978-1-4020-9603-7 e-ISBN 978-1-4020-9604-4

DOI 10.1007/978-1-4020-9604-4

Library of Congress Control Number: 2008942091

Printed on acid-free paper

9 8 7 6 5 4 3 2 1

springer.com

Preface

The texts of Boris Hessen and Henryk Grossmann assembled in this volume are important contributions to the historiography of the Scientific Revolution and to the methodology of the historiography of science. They are of course also historical documents, not only testifying to Marxist discourse of the time but also illustrating typical European fates in the first half of the twentieth century. Hessen was born a Jewish subject of the Russian Czar in the Ukraine, participated in the October Revolution and was executed in the Soviet Union at the beginning of the purges. Grossmann was born a Jewish subject of the Austro-Hungarian Kaiser in Poland and served as an Austrian officer in the First World War; afterwards he was forced to return to Poland and then because of his revolutionary political activities to emigrate to Germany; with the rise to power of the Nazis he had to flee to France and then America while his family, which remained in Europe, perished in Nazi concentration camps.

Our own acquaintance with the work of these two authors is also indebted to historical context (under incomparably more fortunate circumstances): the revival of Marxist scholarship in Europe in the wake of the student movement and the professionalization of history of science on the Continent. We hope that under the again very different conditions of the early twenty-first century these texts will contribute to the further development of a philosophically informed socio-historical approach to the study of science.

Tel Aviv, Israel
Heidelberg, Germany

Gideon Freudenthal
Peter McLaughlin

Acknowledgements

Over the years in working on this project we have acquired a number of debts of gratitude. We are supremely indebted to Jürgen Renn, Director at the Max Planck Institute for the History of Science, who has shared our interest in the work of Hessen and Grossmann for many years. He provided us with the workplace for the project and has constantly encouraged and unstintingly supported our cooperative work.

Many scholars have assisted us in various ways. We would like to thank the following scholars for generously sharing sources, information and resources: Serge Guerout, Pablo Huerga Melcon, Anna K. Mayer, Klaus Schlüpmann.

Rick Kuhn gave us hundreds of pages of photocopies of Grossmann's manuscripts. Jürgen Scheele lent us his personal photocopy of Grossmann's MS which became the basis of our edition. Rose-Luise Winkler gave us early drafts of her German translation of Hessen's paper and has constantly shared her knowledge of Russian-language sources along with a wealth of other information about Hessen. Tatiana Karachentsev assisted us at different stages with Russian sources.

Back in 1987 Gabriella Shalit translated Grossmann's essay of 1935 for *Science in Context*, Phillipa Shimrat has now newly translated Hessen's paper of 1931 for this volume. We thank them both for cordial cooperation.

For help in tracking down sources and for access to manuscript materials we would like to thank the the librarians of the Max Planck Institute for the History of Science (Berlin), the Archives for Scientific Philosophy at the University of Constance, the Archives of the Polish Academy of Sciences (NAUK), Warsaw, and the Universitäts- und Stadtarchiv, Frankfurt am Main.

For help with transcription and proofreading we would like to thank Krishna Pathak, Oliver Schlaudt, Elke Siegfried and Helena Mastel.

A grant from the German Science Foundation (DFG) allowed McLaughlin to examine the Grossmann manuscripts in the archives in Warsaw. A grant from Tel Aviv University financed the new translation of Hessen.

Berlin, September 2008

Contents

Contributors

Gideon Freudenthal Tel Aviv University, Tel Aviv, Israel

Rick Kuhn Australian National University, Canberra, Australia

Peter McLaughlin University of Heidelberg, Heidelberg, Germany

Classical Marxist Historiography of Science: The Hessen-Grossmann-Thesis

Gideon Freudenthal and Peter McLaughlin

Boris Hessen's "The Social and Economic Roots of Newton's 'Principia'" (1931) and Henryk Grossmann's "The Social Foundation of Mechanistic Philosophy and Manufacture" (1935) are the classic programmatic examples of Marxist historiography of science. The two works were produced completely independent of one another, but both scholars were working within the same intellectual tradition with the same conceptual tools on the same topic.[1] The positions they develop overlap and complement one another. They have enough in common that the enlarged thesis that emerges from their work may be called the "Hessen-Grossmann-Thesis."[2]

While many Marxists have contributed to the historiography of science, Hessen's and Grossmann's work displays a specifically Marxist approach: they conceptualize science as one kind of labor within the system of social production. Their discussions of the social context and the cognitive content of science are modeled on Marx's analysis of the labor process. Thus, whatever the merits of other Marxists' contributions to the history of science, from Friedrich Engels' *Dialectics of Nature* to various contemporary forms of social constructivism, Hessen's and Grossmann's work is integrally linked to their specific intellectual tradition and could only have been made by a scholar from that tradition.

Hessen's paper immediately caused a stir and was quickly applauded by enthusiastic supporters and held up as a negative paradigm of externalism by detractors who warned against "crude Marxist" explanations of science. Grossmann's paper, in many ways similar in thrust, has remained almost unknown to historians of science, published as it was in German in French exile. Around 1946, now in American exile, Grossmann completed a monographic study with the title (later changed) *Descartes' New Ideal of Science. Universal Science vs. Science of an Elite*. This manuscript is published here for the first time, along with some shorter materials

G. Freudenthal
Tel Aviv University, Tel Aviv, Israel
e-mail: frgidon@post.tau.ac.il

[1] Grossmann (spelled Grossman in Poland) became aware of Hessen's paper somewhat later and mentions Hessen in in a 1938 book review of Georg Sarton's *The History of Science and the New Humanism* (1931) and G. N. Clark's *Science and Social Welfare in the Age of Newton* (1937).
[2] Freudenthal, 1984/1988.

G. Freudenthal, P. McLaughlin (eds.), *The Social and Economic Roots of the Scientific Revolution,* Boston Studies in the Philosophy of Science 278,
DOI 10.1007/978-1-4020-9604-4_1,

on related subjects also in English translation. Together with the two classic papers, these provide an introduction to the basic approach of Marxist historiography of science.

Boris Hessen's paper was originally published in an English translation in London at the Congress at which it was to have been read. Two Russian versions (with only minor differences) appeared later in the Soviet Union.[3] The original English version was done in a great hurry by the staff of the Soviet Embassy on the eve of the Congress[4] and thus, unsurprisingly, left something to be desired. Many of the mistakes are obvious to anyone with a serious knowledge of the subject matter, but the text has placed unreasonable demands on a general readership. Hessen's essay is here published in a new translation.

The purpose of the following introduction is to facilitate a fresh appraisal of the position argued for by Hessen and Grossmann. This reappraisal is necessary because what has come to be known to historians and philosophers of science as the "Hessen-Thesis" has little to do with the theses that Hessen and Grossmann actually propound, but is rather a projection based on misunderstandings and preconceptions of what a Marxist thesis ought to be. The reader will be able to ascertain what a Marxist analysis of science is by reading the texts themselves and can then judge it on its merits not its reputation.

Hessen's "The Social and Economic Roots of Newton's *Principia*" formulates three theses, the first of which was independently proposed by Grossmann and the second of which Grossmann also later assented to.[5]

– The first thesis concerns the relation of economic and technological developments in the early modern period and the relation of these two to the emergence of modern science: Theoretical mechanics developed in the study of machine technology.

---

[3] Hessen 1933, 1934.

[4] See J. G. Crowther, *Fifty Years With Science*, London (Barrie & Jankins) 1970, pp. 76–88, for the details.

[5] As Grossmann's original title "Universal Science versus Science of an Elite" suggests, he refers to two respects in which modern science is universal in contrast to previous forms of knowledge. In the first place the universality of its method, modeled on mathematics, makes it applicable to all subject matter and thus undermines the guild-like knowledge of specialists. Secondly, there is no secret knowledge in science, no skills that are handed down only in personal contact between master and apprentice. Universal method is accessible to all and thus gives everyone the key to participation in the scientific endeavor. The universal and democratic aspects of science are hence intertwined. At the end of his *Universal Science* Grossman quotes Descartes' address to the reader in his *Discours de la Méthode* in which he explains why he wrote the essay in French rather than in Latin: "The last sentences of the *Discours* constitute an open challenge to the specialists. Descartes addresses his work not to them but to the broad intelligent public, to every man with good sense, and is convinced that these men are better able to appraise his work than the specialists" (127–128). However, Grossmann supports these ideas mainly by references to declarations of intention by Descartes and others, whereas his own research, documented in this work, concentrates on the relation between technological practice and concept formation in science. Grossmann seems to have realized this discrepancy himself since he changed the name of the work in later manuscript versions to "Descartes and the Social Origins of the Mechanistic Concept of the World."

- The second thesis draws the converse conclusion: In those areas where seventeenth-century scientists could not draw on an existing technology (heat engines, electric motors and generators) the corresponding disciplines of physics (thermodynamics, electrodynamics) did not develop.
- The third thesis concerns the ideological constraints placed on science in England at the time of the "class compromise" or "Glorious Revolution" (1688): Because of this compromise Newton drew back from fully endorsing the mechanization of the world picture and adapted his concept of matter so as to be able to introduce God into the material world.

1 Economics, Technology and Science

The titles of the essays by Hessen and Grossmann published in this volume refer to different topics. Hessen's paper of 1931 addresses a specific book: Newton's *Principia*.[6] Grossmann's essay of 1935 refers to the "Mechanistic Philosophy" in general and his essay of 1946 names a different person in its title: Descartes. Nonetheless all three papers have one shared topic and cover much the same historical ground. Hessen's and Grossmann's topic is the Scientific Revolution that culminated in the seventeenth century, which in their view had been prepared by developments since the thirteenth or fourteenth century. The great scientists and philosophers, Galileo, Descartes, Huygens and Newton (and many others, of course), represent its peak of achievement. Grossmann therefore stresses that the Scientific Revolution was *completed* in the period of Galileo and Newton but that it had begun much earlier. Both Hessen and Grossmann view mechanics and not cosmology (e.g. the Copernican Revolution) to be the core of the scientific revolution. This is in itself significant inasmuch as they thus focus not on the conflict between a geocentric and a heliocentric worldview, but rather on the mechanization of the world picture, in which natural phenomena are explained, like machines, by mechanical laws of motion only.

1.1 Economic Needs and Technical Problems

The point of departure of Hessen's argument is the correlation between problems in economics, technology and science in the time up to Newton. Certain economic demands or needs are correlated with certain technical problems or developments, which in turn are correlated with fields of scientific study:

> Consequently, we shall first investigate the historical demands presented by the emergence and development of merchant capital. Then we shall consider what technical problems were posed by the newly developing economy and what complex of physical problems and knowledge, essential for solving these technical problems, they generated (p. 5).

---

6 The Russian title is: "The Socio-Economic Roots of Newton's Mechanics."

Economics is said to *present* demands, which *pose* technical problems, which *generate* scientific problems. Each of these steps must be explained. Considering three major social areas: industry or general production, transportation and war, Hessen presents lists of examples of correlated technological and scientific endeavors.

Let us examine some examples: Hessen notes that the further development of trade ("merchant capital") depended on improved transport. The favorite, that is, most efficient means of transport for goods was naturally by water. Economic development, he says, "set transport the following technical problems":

1. to increase the tonnage capacity of vessels and their speed,
2. to improve the floating qualities of ships,
3. to develop means for better navigation,
4. to improve the construction of canals and docks.

Now the technical problems 1, 2 and 4 based on this economic need correspond to the scientific fields of study, hydrostatics and hydrodynamics; technical problem 3, improving navigation (the determination of longitude), involved the development of chronometers and was hence also correlated with studies in mechanics.[7]

Or take industry: mining in particular involved raising the ore from down in the mines. This technical task was tackled with the aid of various complex machines compounded out of the simple machines – which are studied by statics. Ventilating and draining the mines was accomplished by air and water pumps, which are studied by aerostatics and hydrostatics. The use of artillery in war involved determining the trajectory of projectiles and can hence be related to some of the most celebrated work by Galileo and Newton.

Such correlations do not yet present a thesis on the emergence of modern science. The correlations have to be explicated and explained. There would seem to be two alternatives to explain the correlation. The first takes technology to be the *goal* of science and perhaps the *motive* for pursuing science in the first place. The second takes technology to be the *precondition* of science and conjectures nothing about motives:

A. Technology was developed *in order to* facilitate economic development, and science studied the particular problems that it studied *in order to* improve technology.
B. Technology was developed *in order to* facilitate economic development, and science developed *by means of* the study of the technology that was being applied or developed.

Both share the first, but not the second, proposition. These two positions (A and B) are significantly different in their conceptualizations of the relation of science

[7] Hessen is not asserting that our distinct disciplines existed at the time but rather that these disciplines are what arose out of the study of these problems.

to technology. Is technology the goal of seventeenth-century science or rather its subject matter? The first expresses the position usually attributed to Marxist historiography of science – and emphatically rejected by Koyré, Hall and other traditional internalists in the history of science. The second alternative (B) is a formulation of the Hessen-Grossmann-Thesis. Let us briefly explicate both views. The first view (A) involves four steps of argument which develop a more or less strong form of economic determinism:

1. a (causal) connection is established between economic interests and technical projects;
2. it is shown that the technical projects involve technological problems;
3. it is shown that these technological problems correspond to fields of study in science;
4. it is asserted that scientists were *motivated* by economic (or technical) interests to solve the technological problems and therefore also to study the corresponding fields of science.

Proposition (1) seems problematic since it seems to imply that economic or, more broadly conceived, social interests or needs determine technological solutions to the problems of society. Yet it is fairly easy to see that many needs go unfulfilled and many demands call forth no corresponding supply. This is a point made by R.K. Merton with reference to explaining technological invention on the basis of economic needs and scientific discovery on the basis of technological needs. While technological inventions often responded to needs,

> it is equally true (wrote Merton) that a multitude of human 'needs' have gone unsatisfied throughout the ages. Moreover, countries which are generally considered to be the most needy of inventions, such as Amazonia and India, have, in fact, relatively little invention. In the technical domain, needs, far from being exceptional, are so general that they explain little. Each invention *de facto* satisfies a need or is an attempt to achieve such satisfaction.[8]

Merton here also points to a significant asymmetry in discussions of needs and their fulfillment: Basic needs tend to be more general, the means to their fulfillment more particular. Some needs (e.g. nourishment) are common to all societies, but each fulfills these needs in a different way. And needs can go unfulfilled, while means (inventions) always satisfy some need or other, or we go back to the drawing board. Thus before a perceived need can generate an action calculated to satisfy it, it must be made more specific with regard to the means available. A key element of Marx's analysis of the labor process was that the *will* of the producer must be subordinated to a concrete *purpose* before anything gets done.[9] And this concrete purpose

---

[8] Merton 1938, 157. Moreover, "where an observer from a culture which has an established tradition of attempts to improve material welfare and to control nature may often detect a need in another society, that need *may not exist* for the members of the society under observation, precisely because of a difference in values and aims."

[9] *Capital*, Chapter 7, Section 1, The Labour-Process or the Production of Use-Values. "At the end of the labour-process, we get a result that at its commencement already existed ideally in the representation of the workman. He not only effects a change of form in the natural material on

is itself formulated with the help of the available means. In fact, means developed for one purpose (and need) may also allow other needs to become purposes or make previously unrealistic purposes realistic.

The simple straightforward view, that needs determine their fulfillment abstracts from the question of whether the means to satisfy the needs are available, but it also makes a second mistake, in that it takes the needs or interests to be just given, thus overlooking the extent to which the needs themselves are concretized with respect to the means of their possible satisfaction. While some needs may be formulated fairly generally, say transportation, any particular need must be formulated somewhat more precisely before any action can be taken to satisfy it: The means available progressively concretize the wish to an ever more concrete purpose (e.g., first to improve transport, improve shipping, improve the hydrodynamic properties of the ships, improve the shape of the bow).

The need or desire to expand commerce, ascertained by Hessen, is not of itself a need to improve shipping. Merchants could have turned to transport by land or adapted to merely local commerce; they could have switched to handling smaller merchandize, developed new kinds of commodities, or substituted local products for imported ones. The decision to improve shipping for the purpose of expanding commerce, which constitutes the concretized form of the need for the appropriate technology, presupposes the consideration of possible concrete projects characterized by the means for their realization. Furthermore, the fact that an economic need is conceptualized as a *technological* problem is not self-evident. Again Merton reflected on this problem:

> Economic and military needs, then, may be satisfied by other than technologic means. But given the routine of fulfilling these wants by technologic invention, a pattern which was becoming established in the seventeenth century; given the prerequisite accumulation of technical and scientific knowledge which provides a basic fund from which to derive *means of meeting the felt need*; and it may be said that, in a limited sense, necessity is the (foster) mother of invention.[10]

Grossmann deepens these reflections by considering the difference between the economies of Roman antiquity and the late medieval European town. Only in the latter did a need to expand production involve the need to devise a machine or technical device to do something. Grossmann suggests that technology was marginal as long as production could depend on a social perpetuum mobile, slavery. With the rise of the medieval town the situation changed: urban labor costs money and therefore the search for an artificial perpetuum mobile begins: Although experience shows that a perpetuum mobile in the strict sense cannot be found (Grossmann, 1935b, 67–68; 1946, 106), machines can indeed replace human labor. (Grossmann, 1946, 69) Whereas a shortage of labor in Roman antiquity would have been formulated

---

which he works, but he also realizes in that material a purpose, that he knows, that determines as a law the way he acts, and to which he must subordinate his will." MEW 23, 193 (our translation).

[10] Merton 1938, 158. Cf. also pp.155–59. See also "Science and Economy of seventeenth Century England," pp. 6–7. The peculiar notion that needs cause their own fulfillment is so strongly embedded in everyday culture that even Merton slips into it at one point (1938, 148).

as a need for more slaves, in early modern Europe such a shortage was formulated as a need for more and better machines. This latter need of course presupposes that machines are already used in production, so that the experience made with them can be used in conceptualizing such a need. However, whereas machines were at first used primarily to do things that human labor power could not do or to apply force beyond the capacity of a human to provide (stamping mills; iron-production) (Grossmann, 1946, 89–90), they later also replaced regular human labor. Descartes' announcement in the *Discourse on the Method* that science and technology would diminish human labor could look back on a long tradition (Grossmann 1946, 78).

The thesis may be generalized: the means available are decisive in conceptualizing a need. Now this notion seems to turn the widespread understanding of Marxist "externalism" upside down. Means are not developed in order to satisfy existing needs (or interests), but the concrete conception of needs, purposes which may explain action, depends on the means available, that are then used to satisfy them. To a certain extent then, the means available can determine the possibility or at least the reasonableness of certain needs, interests and desires. Thus when Hessen speaks of the "needs of the rising bourgeoisie" or the "demands of the economy and technology" these are of course mediated by the available means to their fulfillment.

These qualifications do not mean that economic developments may not be used as a factor in the explanation of technical developments. On the contrary, they indicate that to explain an action, we should refer to a concrete *purpose,* not to an abstract wish or need. The purpose of an action presupposes a need and plausible means for its satisfaction. The synthesis of both forms a purpose. Similarly, a wish (e.g., to improve transport, or even the more concrete wish to build better boats) does not account for the course of action taken. This depends on the circumstances involved, especially on the available means for action.

1.2 Technical Problems and Science

Now that we have a basic idea of the relation of economic needs to technological problems that both of the above-mentioned interpretations of Hessen's first thesis seemed to share, we can take up the second step, the relation of science to technology, in which the two versions openly differ. The first version (A) maintained that science studied the particular problems that it studied *in order to* improve technology. This was formulated as the proposition:

4. that scientists were *motivated* by economic (or technical) interests to solve the technological problems and therefore also to study the corresponding fields of science.

If we abstract from the special case of explicitly biographical studies in which the individual scientist is the focus of attention, it is questionable whether the personal motives of scientists are at all relevant to the historical understanding of science.

Just as economic needs must be further specified with regard to the technical possibilities of their realization before they can be acted upon, so too must scientific goals be further specified in terms of the means available (methods, techniques, instruments). Thus what it means to pursue this or that concrete scientific problem depends on the means available in the arsenal of science of the time. Given a certain state of knowledge with its material and symbolic means, with its instruments, experimental systems and theories, its open questions and common methods, scientists on the whole will engage in similar activities irrespective of their personal motivations: they will look for solutions to the open questions, search for salient points and innovative ideas etc., whether their motivation is ambition, greed or the quest for truth. A motivation to engage in science is required, but the particular nature of the motives would seem to be of no great importance since it does not determine the particular course of action taken. This specific activity is determined by the state of science, the methods and means available. Neither Hessen nor Grossmann addressed the motivations of scientists since they did not consider them to be relevant to understanding the development of science.

From the considerations above, some important conclusions follow. First, personal motivations of scientists are irrelevant to the project of explaining scientific development on a social scale. Scientific developments depend on the material and symbolic means which determine both the concrete problems and their possible solutions, not on the personal motivations of the scientists. This of course does not exclude the possibility that on a social scale social and economic interests may directly and indirectly exert pressure on institutions and individuals to favor among possible projects those which are socially desired or of immediate economic utility. The history of science in the seventeenth century is full of such examples,[11] as is contemporary funding policy. Second, the actual relation of science and technology in the seventeenth century can be reformulated in the light of the considerations above. Two questions should be considered: First whether the technology of the seventeenth century belonged to the "means" of scientific inquiry and thus made some scientific endeavors possible (and excluded others); and second, whether it was not precisely the specific difference between science and technology that made science possible. This difference is the fact stressed by critics of the "Hessen-Thesis", namely that scientific research was not subordinated to the service of practical ends. In the framework of the Hessen-Grossmann thesis this can be formulated in the following way: Scientists could explore the possibilities contained in the scientific means at hand (whatever their provenience) and had no need to subordinate their inquiry to finding solutions to pressing social or economic problems. The suggestion that science was dependent on a distance from practical pressures that made room for activities that can be viewed as the disinterested quest for truth, does not presuppose that science is an endeavor *sui generis*, but rather attempts to achieve

---

[11] For a short discussion see Robert K. Merton, "Science and the Economy of Seventeenth Century England."

a deeper understanding of science through its specific difference to technological invention. We shall discuss these two questions in the next section.

The view formulated in version (A), often referred to as Baconian Utilitarianism, explicates a widespread misunderstanding of the first Hessen thesis as a claim about the personal motives of individual scientists: that they pursued science to advance technology, production and economic gain. Most critics therefore have believed they must deny the argument concerning the motivation of the scientists involved and that this denial refutes Hessen. This position is just as flawed as the position it criticizes. It presupposes that what is at issue is the *real* motivation of the scientists involved: economic interest, a disinterested quest for truth, or the glorification of God through the study of his works. Critics of Hessen, arguing that the motives of scientists were not in fact utilitarian, have implicitly accepted the presupposition that if the motives had been utilitarian, this would bear significantly on the explanation of the Scientific Revolution. And they seem to presuppose that the determination of technology by economic needs, which they do not deny, is similarly to be explained by the motives of the economic actors involved. In technology as opposed to science, economic motives are not considered implausible or disreputable. But it is not just Marxists of the 1930s who are accused of questioning the motives of seventeenth century scientists; Francis Bacon in particular was accused of this by Alexandre Koyré, who distinguished a "propagandist" of science like Bacon or a craftsman interested in building something from real scientists like Galileo and Descartes "who seldom built or made anything more than a theory."[12] Koyré argued further that the wish to create technology cannot have been the motive of scientists to pursue science because key areas of technology were already in place before and independent of science; thus technology cannot have determined science in any way at all.[13]

However, even Bacon's famous dictum that "Nature to be commanded must be obeyed"[14] need not be read as saying merely that if we want to dominate nature better, we should learn more about its laws. It can just as well be read to assert that in those places where we have in fact succeeded in commanding nature (technology), we must have been obeying nature's laws. Thus, studying successful technology is the key to scientific knowledge of nature. Hessen on more than one occasion appeals to this alternative form of Baconianism, for instance, when he reports that Galileo began his *Discorsi* "with an address to the Venetians praising the activity of the arsenal at Venice and pointing out that the work of that arsenal provided *a wealth of material for scientific study*."[15] It is the critic's assumption that the only conceivable relation of technology to science is that of a motive in the mind of a scientist that blinds them to the possibility that practice with technology might influence ideas about nature. Although, as we shall see below, the disciplinary reception of Hessen

---

[12] Koyré 1943 ("Galileo and Plato") 400–401; A.R. Hall (1952, 163) joins the critique of Bacon saying (without citing evidence or explaining what a seventeenth-century scientist is) "Few of the public apologists for science were themselves scientists."

[13] Koyré [1948] 1961, 308.

[14] *Novum Organum* I, §2.

[15] Hessen 165 (our italics); see also the quotation from Galileo in Hessen's Appendix (no. 1).

is particularly influenced by postwar anti-communism, there is in much of it also a deeper-lying basic inability to consider any materialistic explanation at all.

2 The First Thesis: Technology Opens Horizons for Science

The first thesis, advanced by both Hessen and Grossmann, asserted that science developed by means of the study of existing technology. This means first that the concept of *nature* changed. As the feudal mode of production was gradually replaced by the capitalist mode of production, as the towns became increasingly more important and the country increasingly less, as artisan production and manufacture increased in importance vis à vis agriculture, the concept of nature changed as well. Traditional agricultural labor was supportive of natural processes. Nature turned seed into grain on its own but could do this better when supported by human labor. But machines are not products of nature but man made. Once machines, which were traditionally seen as ways to outwit nature, began to be conceived as natural agents, two consequences ensued. Machines were understood to obey natural laws not to abrogate them. Second, the world was conceived as an ideal machine and natural phenomena as its operations. This has been called the "mechanization of the world picture." Thus both the concepts of *nature* and of *machine* or *mechanics* change. Nature is no longer conceived as an organism governed by teleology, and mechanics no longer as a collection of contrivances to outmaneuver nature. Rather nature and mechanics coincide. A consequence of this unification for the theory of motion is first that the Aristotelian distinction between "natural" and "forced" motion loses its basis and its sense: natural motion is conceived as if produced by a machine, the laws governing the exertion of power of a machine are the laws of nature. To study nature hence means to study man-made machines, not nature untouched. The *machina mundi* is now conceived as an automatic machine (a clock) which functions according to the laws of nature.[16] The science of mechanics, which investigates the laws governing the functioning of machines thus became the science par excellence, a universal science exploring the function of all machines, natural and artificial alike.

The increase of the economic importance of technology and its associated activities was followed by an improved social position of those involved. As remarked by Hessen and Grossmann in passing and discussed by Merton in detail, economic, technical and scientific occupations improved their social positions in the seventeenth century, such that the social elite, which earlier went into other fields of occupation, now went in significantly larger numbers into science.[17] Of course, the establishment of scientific societies, the financial support of scientific endeavors etc. should also be considered the effects of this economic and technical development.

[16] See on this point Freudenthal 1986, 59–66 and McLaughlin 1994. More specifically, the machine studied in scientific mechanics until and including Newton and the generation following him were "transmission machines." The importance of this issue will be discussed below.

[17] Merton 1938, Chapters II–III.

Second, the rise in social prestige of technology (due to its increased economic significance) made the progressive merging of two traditions possible which, for social reasons, were earlier segregated from one another: the mechanical and the liberal arts, the knowledge of craftsmen and the knowledge of the learned.[18] On the one hand, the new group of sophisticated craftsmen (architects, instrument makers etc.), so-called *virtuosi*, were better educated and occupied higher social positions than ordinary craftsmen. On the other hand, the improved social locus of the mechanical arts made it possible for the learned now to engage in their study. The learned profited from the knowledge of the craftsmen (either directly or from the new technical literature) and also gained a field for their own observation and experimentation (Grossmann 1935a, 187–88). It seems reasonable to conjecture that the new experimental-mathematical science was born out of the fusion of the experimental tradition of the craftsmen with the systematic and mathematical tradition of the learned.[19]

The main thesis common to Hessen and Grossmann builds on these considerations. It says that the science of mechanics (so-called "theoretical" mechanics) developed in the study of contemporary technology, of "practical" mechanics. This thesis is diametrically opposed to the wide-spread view, which is also regularly attributed to Hessen, that practical mechanics was guided by the science of mechanics and that theoretical mechanics was pursued in order to apply it in practice. More or less the opposite is the case. Both Hessen and Grossmann maintained that the primary occupation of scientific mechanics in the early modern period was to study already *existing* technology and understand how it functions, not to improve it – however much the one or other scientist personally may in fact have wanted to do just this.

Third, since scientific mechanics developed through the study of practical mechanics and its tradition, it owed much of its theoretical structure and conceptual character to practical mechanics. The Hessen-Grossmann-Thesis addresses the determination of the cognitive content of science, which was traditionally shunned by sociologists of science (Merton included). The thesis attempts to explore the horizon of cognitive possibilities on the basis of the material and symbolic means employed. One corollary concerning early modern science is Grossmann's contention that the origin of essential conceptual presuppositions of mechanics is to be found in practical mechanics. This will be discussed in the next section.

[18] At one point Grossmann (1946, 70) dates the first beginning of modern science with this merger, the subsumption of mechanics under geometry as part of the liberal arts in the *De divisione* of Gundisalvus (12th century).

[19] This thesis is usually associated with the work of Edgar Zilsel, "The Sociological Roots of Science," *American Journal of Sociology*, *47* (1942), 245–279, but it is also present in Hessen's and Grossmann's papers.

2.1 Grossmann on the "Real Abstraction" in Transmission Machines

Grossmann scrutinized the genesis of the general, abstract and quantitative concept of motion. Simple observation does not offer us pure motion. In everyday life and in technical practice, motion always occurs together with other phenomena: friction, heat, force etc.; and it is always a qualitatively specific motion: straight, curved, upwards, downwards etc. The traditional Aristotelian conceptualization of motion as "natural" or "forced" shows that a process of abstraction can go in different ways from the modern direction. Grossmann's studies of the genesis of the abstract concept of motion, in which all these concrete forms of motion are left out of consideration, took up the question of what recommended or favored one kind of abstraction over another: What made it possible to replace the quite intuitive and traditionally sanctioned concepts of motion with entirely different concepts, which had earlier seemed abstruse?

From the perspective of everyday human practice, scientific concepts of motion are extremely counter-intuitive. In our experience bodies do not move in uniform inertial motion. This does not preclude the possibility of our forming laws for counter-factual cases, but it may very well render them implausible and raise philosophical doubts as to whether they are merely *entia rationis* or have a *fundamentum in re* and an application in experience. Grossmann's thesis, which will be elaborated in more detail below, claims that the new concept of motion was acquired in the study of technical, mechanical practice.

Grossmann's thesis may be read as a cognitive-psychological or as a philosophical thesis. The psychological thesis attempts to explain the plausibility or credibility of a particular conceptualization in spite of everyday experience and in spite of traditional learning by pointing out a sphere in which such concepts could seem plausible. The philosophical thesis attempts to explain why such concepts can be taken to have reference and where the referents are to be found. Together, Grossmann's considerations may be applied to explain why the rise of technology also gave rise to a new conceptualization of natural phenomena, why these new concepts did indeed find reference in the real world by way of technology, and finally also why this conceptualization of nature seemed plausible within certain strata of society.

Grossmann refers in his paper to some ideas of Marx concerning the introduction of machines into production.[20] Marx pointed out that the introduction of engines into production presupposed that the function of a motive power had been separated practically from the various specific operations performed on the object worked upon. Once an automated tool, a "machine" in Marx's terminology, is introduced instead of a tool guided by the skilled hand of the craftsman, human labor is reduced to the function of a motive force of this machine. Only then can it be replaced by an animal or some other natural power (wind, water, gravity).

Consider a grain mill. It may be seen to consist of three parts: the engine or "motor mechanism" on the one end and the grinding device ("tool or working machine") on the other. These two are connected by a transmission mechanism

[20] *Capital* I, MEW 23: 401–407 CW 35, 384–89.

("transmitting machine"), which transmits and sometimes transforms the motion produced by the engine to the grinding device. Sometimes the transmission machine transforms circular motion into rectilinear or vice versa. One power source can be replaced by another which fulfils the same function. It may be a water wheel or a windmill, a human or an animal. Similarly, the same engine may be attached to different devices: It can drive a mill, a lathe, or some other device. These instruments, which directly form the working piece, may be automatic or guided by hand. Marx called "working machine" such an automatic instrument whether moved by an engine or a human worker (in contradistinction to an artisan working with a tool). He insisted that the crucial step in industrialization was the invention of such machines, that is, automatic instruments. He argued that the introduction of motor mechanisms presupposes that the labor process has been emancipated from its artisan form, in which the functions of the instrument and of the engine are inseparably intertwined. Only when the movement of the hand, which both drives and guides a tool, has been broken up into the function of an automatic instrument (which needs no guidance) and a power source producing a standard motion, can human power be replaced by an engine. Only when what once was skilled labor is performed by a machine, can an engine also replace human physical power. Once different kinds of labor are performed by different machines driven by the same kind of motion, these different machines could be attached to the same kind of engine. Only then could such engines be introduced into the process of production, and indeed they were so introduced.

Grossmann takes this idea a step further into the cognitive realm and asks what the origin of an abstract concept of *motion* or *work* produced by *force* was.[21] Evidently it makes no sense to form a general concept of motion if instances of this motion cannot be transformed into various specific motions known from experience. "Motion in general" does not exist aside from its different individual forms. In light of the discussion of Marx above, the question can be put this way: Under what circumstances does the concept of *motion* or *work* (a homogeneous form of motion against resistance) make sense? Evidently, such a concept does not make sense when we study the work of a craftsman: here the aspects of a purposeful modification of the object by a special form of motion (dependent on the nature of the material and the purpose of the craftsman and on his instruments), cannot be separated from the application of physical force that is moving the instrument. Motive force, skill, purpose and instrument form a unity. It does, however, make sense to *distinguish* between these aspects as soon as they are in fact *separated* or when it seems possible to separate them. This separation of motive force from the purposeful guidance of the hand in the process of labor is the same that was conceived by Marx as a presupposition for the introduction of motor mechanisms into the labor process. Grossmann conceived it to be also the starting point for conceptualizing "motion"

[21] Marx, too, extended this idea to the cognitive realm, but did not develop it further. He suggested that difficulties in the use of transmission machines connecting the motive force and the mechanical tool led to the study of friction and the flywheel. "In this way, during the manufacturing period, were developed the first scientific and technical elements of modern mechanical industry" (*Capital* Chapter 15, CW 35, 80; MEW 23: 397).

in an abstract and quantitative fashion. When (1) various different kinds of labor have been separated from the motive power applied in performing them, then motive power could also be conceptualized separately, and when (2) various kinds of the motion (circular, straight) produced by various motive powers (e.g. water, animal, man) could also be transformed one into the other by appropriate transmission machines, then it also made sense to form concepts of abstract motion and force, referring exclusively to the faculty of performing labor as such, i.e., moving heavy bodies against resistance, especially raising heavy bodies in the gravitational field of the Earth.

> It is evident that man, in all these technological upheavals, acquired new, important material for observing and contemplating the actions of forces. In the machines, in the turning of the water wheels of a mill or of an iron mine, in the movement of the arms of a bellows, in the lifting of the stamps of an iron works, we see the simplest mechanical operations; those simple quantitative relations between the homogeneous power of water-driven machines and their output, viz. those relations from which modern mechanics derived its basic concepts. Leonardo da Vinci's mechanical conceptions and views are only the result and reflection of the experiences and the machine technology of his time, when one new technical invention follows the other or the previous inventions are improved and rationalized. (Grossmann, 1935a: 193–194)[22]

Grossmann's use of "reflection" in this context should not be taken to mean that scientific mechanics derived its concept of homogeneous motion simply from observation of working machines. In the footnote to this passage, Grossmann refers to technical literature, which emerged in the middle of the fifteenth century; and Hessen points to a new kind of expert: scientific engineers who had been working in the mines since the 15th century (Hessen, 161, 169).

Hence the process of concept formation in scientific mechanics refers back to practical mechanics in two ways: *first,* by the direct study of machines, and *second,* by appropriating the knowledge contained in practical mechanics, whether through personal contact with practitioners, or through the technical literature. Indeed, it is easy to show that some concepts of practical mechanics were adopted by scientists and were still used even after scientific knowledge has superseded them (e.g. *force* for *momentum*).[23]

---

[22] Hessen reasons in a similar way: "Since the time of the Crusades industry had developed enormously and had a mass of new achievements to its credit (metallurgy, mining, the war industry, dyeing), which supplied not only fresh material for observation but also new means of experimentation and enabled the construction of new instruments" (169).

Hessen adds an important consideration. The concept of "abstract work" abstracts not only from its specific form of motion in space, but also from the transformation of work from one form (mechanical kinetic and potential energy) into others (e.g. thermal and electric energy).This even more demands an explanation as to the direction abstraction took.

[23] In the Preface to the *Principia* Newton also extensively referred to practical mechanics in order to demonstrate the wide range of applicability of the third law of motion (see Scholium to the laws of motion; Newton 1999, 424–430); there, too, we find a typical expression adopted from the technical literature can be found: "The effectiveness and usefulness of all machines or devices consists wholly in our being able to increase the force by decreasing the velocity and vice versa." (Newton 1999, 429) Compare this with John Wallis' pre-Newtonian concept of force, which recurs

2.2 The Differing "Purposes" of Science and Technology

The question of the origin of concepts of classical mechanics can be also differently formulated: We can ask why the concepts used in scientific mechanics did in fact have reference and why they were *believed* to apply to the world in general and to technology in particular. Why did the laws of statics and dynamics, as developed by using mathematical representations apply to inclined planes, pulleys etc., or why did the laws of motion of bodies (later: "point masses") refer to real bodies and real machines? And why did no scientist in the seventeenth century doubt that statics and dynamics refer to the real world and have application to machines and projectiles in spite of empirical evidence that they did not? The discovery of the parabolic trajectory of projectiles, which is explicitly and recurrently heralded as the solution to an essential problem of ballistics, is far from an adequate description of the trajectory of a shell shot from a cannon. In fact, without some previous reason to believe that theories are about the real world, an experimental test with artillery would more likely refute than confirm such a theory. This question deserves some elaboration because the discrepancy between practical and theoretical mechanics was put forward in what is arguably the most serious criticism of Hessen, by A.R. Hall,[24] but it seems that his arguments prove the exact opposite of what he believes they prove.

The discussion above showed a very close connection between science and technology: technology was not only presented as the direct object of study of mechanics, but also as a determinant of its concepts in significant ways. What could and could not be conceived was discussed in reference to what could and could not be learned from contemporary technology. These considerations have to be followed by an examination of the specific differences between the points of view of science and technology. If indeed both technicians and scientists studied the same field, often the same machines, in what does their knowledge differ?[25]

This difference is the main concern of A.R. Hall's study *Ballistics in the Seventeenth Century*. Hall studied ballistics – one of the techniques to which Hessen

---

to the five "common" machines: "And this is the foundation of all machines for facilitating motion. For in whatever ratio the weight is increased, the speed is diminished in the same ratio; whence it is that the product of the weight and the speed for any moving force is the same." (Wallis, letter to Oldenburg, November 15, 1668; *Oldenburg Correspondence*, 1966ff: V, 168)

[24] Hall 1952 (*Ballistics*). Hall in fact mentions Hessen only once in the book and doesn't include his name in the index or bibliography, but the text is a sustained argument against the utility thesis.

[25] Newton clearly saw that his subject matter was the same as that of practical mechanics. He just as clearly underestimated the difference between scientific and practical knowledge seeing it merely in the degree of precision: "Practical mechanics is the subject that comprises all the manual arts, from which the subject of mechanics as a whole has adopted its name. But since those who practice an art do not generally work with a high degree of exactness, the whole subject of mechanics is distinguished from geometry by the attribution of exactness to geometry and of anything less than exactness to mechanics. Yet the errors do not come from the art but from those who practice the art. Anyone who works with less exactness is a more imperfect mechanic, and if anyone could work with the greatness exactness, he would be the most perfect mechanics of all" (Newton 1999, 381).

referred (161–164) – but he generalized his conclusions to science and technology in general.[26]

A general formula for the curve of a projectile was found by Evangelista Torricelli (1608–1647), but it was not for the use of gunners. For these, Hall tells us, "Torricelli published tables of ranges and altitudes from which, the range at any one angle having been measured, the rest could be found by the rule of three."[27] By presenting both his theory and practical rules for guns side by side, Torricelli acknowledged the existing gap between theory and practice. Confronted with the fact that the trajectory of real projectiles was not a parabola, Torricelli claimed that his study *De Motu Projectorum* was purely theoretical.

According to Hall, in presenting his formula Torricelli demanded to be treated as a philosopher and a mathematician, not as someone talking about application in the real world: "the tables and instruments he had described were not for measuring the ranges of cannon balls, but certain geometric lines associated with geometric parabolas."[28] However, Hall admits that Torricelli explicitly spoke in his essay of guns shooting shells at walls of cities, that he printed tables giving measurements in paces, and that his readers might easily have supposed that "when he talked of guns he meant real guns."[29]

Hall distinguishes three different stages in the relation between gunnery and scientific ballistics since the Renaissance: With Leonardo practical and theoretical mechanics were not yet separated; they separated with the development of scientific mechanics (as in the work of Galileo and Torricelli); and they met again under different conditions after the work of Newton and Jean Bernoulli, through which science could much better explain and describe the real trajectory of projectiles.[30]

We can now suggest answers to the questions formulated above.

Idealizations and counter-factual assumptions are indispensable in science. Inertial motion, point masses, uniform acceleration in free fall and the parabolic trajectory of projectiles – all these blatantly contradict experience. Real canon balls are not point masses and do not move like them, they are not shot in a vacuum but in the resisting medium of the air. The inertial motion of the projectile cannot be observed nor can a parabolic trajectory. This difference between idealization and reality immediately implies that the results of science cannot be directly applied to experience. And the other way around: The purposes of

[26] Hall 1952, 163.
[27] Hall 1952, 95.
[28] Hall 1952, 97.
[29] Hall 1952, 99.
[30] "By the third quarter of the century everyone in the van of the scientific movement admitted that the primary principles of dynamics laid down by Galileo were fundamental to all future work, but it was also apparent that in their simple form they were not true for the world of experience.... It was necessary to discover the complex mathematical rules which link the world of scientific abstraction to the world of nature, if it was to be proved that the one was indeed appropriate to the other..." (Hall 1952, 128).

science (in the sense discussed above) were not adopted from technology but rather formulated within science. Scientific knowledge developed only when it was not required to give immediate solutions to existing problems. It required on its part freedom to concretize its own problems and develop its impractical solutions in order to develop concepts that transcend immediate technical knowledge and necessities.

In Torricelli's time, the gap between gunnery and science was indeed sensible. This is another way of saying that specific scientific knowledge had already developed considerably and independently of practical technical knowledge.

And yet, there was no doubt in the mind of its practitioners that both kinds of knowledge, the technical and the scientific, were about the same empirical matters of fact. The Hessen-Grossman thesis explains why: because science studied – albeit in its specific way – real technology. This thesis was formulated and substantiated for one historical case by Hessen's critic, A.R. Hall. Only the terminology and the ideological prejudices are different:

> Philosophers from Galileo to Newton . . . used the problems of ballistics as a gymnasium in which to develop their powers for larger and more important researches (Hall, 1952, 158).

The gap between science and technology opened because science developed in a different direction from its starting point in practical knowledge (this was the foundation of its subsequent strength), it progressively closed as science advanced enough to explain and predict existing technology much better than the most experienced practitioners. The gradual rapprochement was achieved by a superposition of additional laws (motion in a resistant medium applied to the motion as determined by force, inertia and gravity) and by the improvement of technology that rendered its functioning more standardized.

The growing success demonstrated not only the success of the science of mechanics in solving this particular problem, but also that science's analytical procedure is adequate. This procedure rests on the presupposition that the innumerable observed phenomena consist of a limited and small number of basic law-governed processes independent of each other and their composition. On this assumption, the first task of science was to analyze the phenomenon in question into its constituents. In a second step the laws governing these single processes were to be determined, and finally the initial phenomenon had to be explained as resulting from a composition of the different processes. Newton's success in determining the trajectory of a projectile under the complex conditions on the surface of the Earth proved not only that his theory was adequate, but also that the analytical approach of science (the "analytic-synthetic method") was adequate. Thus were the initial idealizations and counter-factual assumptions justified in retrospect. But we should also note that scientists in this age were aware of the technical origin of their problems and therefore never doubted the reference of their concepts or the applicability of theoretical mechanics to practical mechanics. The Hessen-Grossman thesis that science developed in the study of contemporary technology does not mean that it served technology, but on the contrary, that it put technology in the service of its own enquiry – and it did not itself contribute to technology for decades, if not for centuries.[31] And nevertheless, there was never any doubt that scientific mechanics is *about* real machines and that *in principle* it will finally explain how they function better than the practitioners can.

2.3 Grossmann on the Mechanization of Mathematics

Grossmann's ideas about the development of mathematics and of universal method as found in Descartes are expressed in the original subtitle of his monograph, *Universal Science vs. Science of an Elite. Descartes New Ideal of Science* (renamed after 1946 as *Descartes and the Social Origins of the Mechanistic Concept of the World*). The "New Ideal of Science" refers on the one hand to a science that becomes active, replacing the contemplative ideal of science in antiquity and promising together with technology to dominate nature in the service of mankind. On the other hand the "new ideal" refers to a universal science, a universal method, appropriate for investigation in all areas of inquiry. The original title of the book expresses a complimentary concern: the new science is not the science of an elite because the means developed and employed in it do not require any specialized knowledge, being powerful enough that even non-specialists could achieve significant results with them. On these questions, Grossmann pursues his ideas concerning the cognitive import of technology for contemporary science in general. The introduction of machines makes the virtuoso superfluous; specialized, highly trained craftsmen are not required any more to ensure the quality of the product. Similarly, the mechanization of mathematics makes mathematics a universal method and hence makes science in general accessible to all.

The mechanization of mathematics refers to three different factors: (1) the use of mechanical devices in mathematics (slide rules, logarithms etc.); (2) the use

[31] Hall believed that the gap between technology and science refuted Hessen: Science did not in fact contribute much to technology in the seventeenth century, and this proves, Hall believed, that the improvement of technology cannot have been the motivation for scientists to engage in research (Hall 1952, 163–64).

This argument is mistaken in two regards. First, it misunderstands the Hessen-Thesis and construes it as referring to the motivations of individual scientists. Second, it presupposes that because an expectation was not fulfilled, it cannot have been a motivation.

Hall's ideological commitments are obvious. He believes that sociological history of science as such reduces great scientific discoveries to "no more than the response of a quick mind to the most pressing need of the moment" (162) wrenched from science "by the strong hand of economic necessity" (165). In contrast, Newton's work on the trajectory of projectiles in resisting media in the second book of the *Principia* does not mean that he "was *guilty*... of allowing himself to be dominated by the technological problems of his own day" (Hall 1952, 129, our italics) but rather that he solved an ancient philosophical problem. The philosopher studies science "in order to satisfy his intellectual curiosity" (p. 4). In short: scientists had "higher" motives than economic gain. This motivation is evident in many writings on the topic. See again Hall 1963, 15: The social studies of science has created says Hall "a certain revulsion from the treatment of scientists as puppets."

of propositions of mechanics in mathematics (as in determining a tangent by the inertial component of the curved trajectory of a body); and (3) the "mechanical" performance of a mathematical algorithm without reflection, just as a machine can be operated by a worker who does not understand its structure. This latter characteristic of machines and algorithms alike bears on Grossmann's view that modern science is not the science of an elite. A machine can guarantee invariably high quality that is independent of the virtuosity of the workman. Grossmann emphasized the analogous democratic aspect of this development in science (1946, 13), namely that science becomes accessible to all, not the secret of a few virtuosi (1946, 121–2). He believed that this was the reason why Descartes did not conceal his method from the masses, but rather propagated it, even including women in the intellectual endeavor (1946, 125–126). Grossmann enthusiastically celebrated Descartes as someone who a century and a half before the French Revolution proclaimed the fundamental equality of humans in respect to reason (1946, 126) and saw in his decision to write in the vernacular instead of scholarly Latin, another manifestation of this democratic stance (1946, 128).[32]

In our context it is important to emphasize the similarity of Grossmann's conception of labor to his concept of scientific work and the decisive role he ascribes to the means employed. As is well known, Marxism ascribes the means of production a decisive role in social life, but Grossmann focuses on the claim that a worker does not himself have to possess the knowledge embodied in the means he uses. The qualification of the worker in mechanical production may vary inversely as the quality of the means. With the universal language of ideas conceived by Descartes, a peasant might do better than a contemporary philosopher (1946, 19, 22).

The idea of the universality of the means refers to both sides of the labor process: to the subject and to the object. On the side of the subject, it means that anyone can operate them; on the side of the object, it means that they are applicable to every object. The reason for this is that science examines the simple *relations* and *proportions* between things, not the multifarious *natures* of the things themselves (1946, 25–27). Before the universal method can be applied, the various objects must be first reduced to common dimensions, analogous to spatial dimensions, which are common to all material objects and are studied by geometry (1946, 26–27, 30–34).

Now, this comparison of Cartesian universal method to machines is not an arbitrary analogy suggested by Grossmann. In fact, Descartes himself praises his geometry with the same words with which he praises his machines for grinding lenses. The universal method should be an algorithm that can be operated mechanically, without mathematical thought. An external mechanism should likewise be able to perform these operations. Thus Descartes' project loses its eccentric flair and appears as a step in a long development which worked on the mechanization of mathematics by means of mechanical calculation devices such as sliding rules

[32] Grossmann also believed that Descartes may also assume contemporary significance as he "foresaw the great intellectual crisis of today," that is, specialization because of which nobody can understand "social and intellectual life as a whole" (1946, 21).

and logarithms and finally led to Descartes' attempt to automate the intellectual processes themselves (1946, 49–52).

We discussed above Grossman's idea that the abstract and mathematical concept of motion resulted from the study of machines. We concentrated on the transformation of one form of motion into another which enabled the formation of the concept of motion as such, abstract motion. We can now enrich the picture and address the mathematical character of the new concept of motion, resulting from the application of new mathematical methods to abstract motion, and vice versa, the formation of new mathematical concepts resulting from the study of new forms of motion in machines. Not only did mathematics study motion and use mechanical devices, but mathematical teachings were also conceptualized in mechanical terms. The conceptualization of the infinitesimal calculus in terms of motion ("fluxions") and the analysis of motion by means of the infinitesimal calculus is an obvious example. It can be shown at least for some cases that the conceptualization of the infinitesimal in mathematics and of the mathematical concept of motion in mechanics were developed in one and the same argument and were dependent on the same experience with mechanical devices.[33]

3 The Second Hessen-Thesis: The Limited Horizon of Science

Whereas Grossmann concentrated on the positive contribution of practical mechanics to science in the form of prerequisites and fundamental concepts, Hessen also pointed to the limits to theoretical mechanics drawn by practical mechanics. Hessen's second thesis is the converse of the first: if theoretical mechanics was made possible by mechanical technology, then other fields of physics, that did not figure prominently in the 17th century may not have developed because the requisite technology had also not yet been developed. Specifically, Hessen maintains that the primitive state of steam-engine technology did not permit a science of heat and its relations to mechanical forms of energy.[34] Thus, for instance, the conservation of energy could only take the form of the conservation of *mechanical* energy. Other forms of energy such as heat and electromagnetism as well as the transformation of one form into another could only be fully integrated into experimental science after their practical transformation in the steam engine and generator. This argument has been almost universally ignored. Hessen (rightly) gives some credit for the idea to Friedrich Engels, and this reference has been interpreted as another sign

[33] Grossman did not present specific cases to substantiate his claim that the formation of the general and abstract conception of motion was dependent on the study of machines, nor did he analyze specific mathematical examples. For a case study that shows on the example of Giambattista Benedetti (1530–1590) the dependence of conceptualization of mathematics and the concept of motion on experience with treadles, which transform rectilinear into circular motion, see Freudenthal 2005.

[34] Hessen, 193–203.

of dogmatism.[35] One of the few exceptions was Grossmann, who in a review of a book by G.N. Clark, Hessen's most prominent early critic, pointed out that Hessen attempted, "to understand the general character of classical mechanics and physics in distinction to the later development of these sciences."[36]

Hessen's second thesis, in fact, addresses the same issue as the first thesis shared by Hessen and Grossmann but from another perspective. The first thesis was an answer to the question, under what conditions certain abstractions are possible and why the resulting concepts are believed to have reference of some kind. Hessen's second thesis is an answer to the question, under what circumstances certain abstractions are not possible. The first thesis was that the study of machines transforming kinetic energy into potential energy and vice versa, or rectilinear into circular motion and vice versa provided the basis for the abstractions of mechanics. The second thesis is that precisely because only these machines could be studied, other branches of physics (thermodynamics and electrodynamics) did not develop at the time, nor was a general law of the conservation of energy formulated (Hessen 1931, 188, 193). Conversely, this substantiation of the second thesis strengthens the argument for the first thesis: Once steam engines were invented and applied in industry, thermodynamics could (and did) develop through the study of this motion (194, 199). Hessen writes:

> As soon as the thermal form of motion appeared on the scene, and precisely because it appeared on the scene when it was indissolubly bound up with the problem of its conversion into mechanical motion, the problem of energy came to the forefront. The very way in which the problem of the steam engine was formulated ("to raise water by fire") clearly points to its connection with the problem of the conversion of one form of motion into another. It is not by chance that Carnot's classic work is entitled: *On the Motive Power of Heat* (202).

It should be stressed here that the Hessen-Grossmann thesis attempts to explain the determination of the horizon of empirical scientific inquiry, not the horizon of the scientific imagination or speculation. It is easy to object to Hessen's thesis by pointing to almost eternal ideas and to projects that were conceived centuries before they were realized. Human flight for instance, seems to be one such timeless dream documented in human culture at least since Greek mythology but technically realized only in the late eighteenth century. Here and elsewhere, it is the lack of concreteness of the "wish to fly" which gives it the appearance of a timeless idea. As soon as we concretize this abstract wish and consider the different projects designed to realize it (imitating a bird's wings by Daedalus, combining balloons and baskets by Montgolfier, or using a propeller and an internal combustion engine by

[35] I.B. Cohen (1990, 56) sums up: "There follows a lengthy discussion of the post-Newtonian economic and technical history of steam power . . . during which Hessen lauds the thought of Engels." H.F. Cohen (1994, 329) understands only that "because of his lack of familarity with the steam engine, the principle of the conservation of energy in the definitive form given to it by Marx's collaborator, Friedrich Engels, had eluded Newton." Whereas the misunderstanding of Hessen's first thesis is probably due for the most part to an unreflected, dogmatic idealism, the inability even to see what the second thesis asserts is less ethereal in origin and is probably just due to anti-Marxism. On this see below Section 5.

[36] Grossmann 1938. This text is printed in full below on pp. 231–235.

the Wright brothers), we immediately realize that "wishing" to fly (like a bird) and having the concrete purpose to fly (in a plane) are somewhat different: the concrete purpose (to construct a particular flying device) is indeed dependent on existing technology.

The same holds true in the case considered here: Long before the development of thermodynamics many philosophers and scientists expressed the conviction that work can be transformed into heat, perhaps also heat into work. Bacon is famous for such ideas. Leibniz was even quite specific in explaining the apparent non-conservation of kinetic energy ("motion") in inelastic collisions or wherever friction occurs and speculated that it was converted into the motion of very fine particles of matter and appeared as heat. Such ideas may have played a role in preparing the ground for future concepts, but they did not establish an area of empirical scientific inquiry into heat and work. Leibniz used this speculation as an ad-hoc hypothesis to explain away the apparent loss of motion in empirical bodies so as to uphold his thesis that the overall quantity of *vis viva* was conserved. But his conceptualization of kinetic energy as *vis viva* was confined to mechanical forms of motion and did not envision transformation of motion into a qualitatively different form, heat.[37] According to the Hessen-Grossmann thesis, such notions demanded that the transformations be technically controlled and thus be subject to repeatable experiment and quantitative investigation.

Hessen's second thesis explaining what was not possible in science in the 17th century has a further consequence for the later development of physics in the 18th and 19th centuries. Hessen suggested an explanation for the historical order of the development of physics such that the study of the forms of energy followed the development of technology: mechanics (simple machines and their compounds), thermodynamics (steam engine), electrodynamics (electric motor) (200–201). As Grossmann remarked, it "allows us to comprehend the historical order of the individual stages of this development" (1938, 234).

The first and the second thesis are thus two aspects of the same thesis on the determination of the horizon of scientific study by the perspective on nature from the point of view of the dominant manner and means of its appropriation. The technology studied, whether the technology of general production or of the laboratory determined which phenomena of nature (*sub specie machinae*) could be studied and also which could not (yet) be scientifically studied, what aspects of phenomena were in the focus of inquiry and what aspects were peripheral, as it also determined and guaranteed the reference of the theories developed.

4 The Third Hessen-Thesis: Mechanistic World Picture and Newtonian Ideology

Hessen's third thesis addresses a different kind of determinant of science, ideology. The first and second theses specify the horizons of science, but do not address such

[37] On Leibniz' usage of this ad hoc hypothesis which in fact follows a similar maneuver of Denis Papin, see Freudenthal 2002.

other relevant factors as the "political, juridical, and philosophical theories, religious beliefs and their subsequent development into dogmatic systems," (177) that might shape the character of science. Hessen also offers a first sketch of a more detailed consideration of one aspect of Newton's work and its world-view consequences. Here, too, he goes a step beyond Grossmann.

Grossmann stressed the generalization of mechanical models in scientific research, taking his examples mainly from Descartes, whose research was guided by mechanical analogies and who is the paradigmatic representative of the mechanistic world picture for Hessen as well. Thus Descartes used the analogy to the deflection of a shell shot by a cannon on the surface of a river to illustrate problems of reflection and refraction (1935a: 203–204), he also studied organisms as if they were clocks or other automats (1935a, 208), and generalized the fundamental principles of mechanics to apply to the formation of the universe out of matter and motion according to mechanical laws (1946, 110–112). Finally, mechanical models of the universe driven by clockworks, naturally suggested that the world as a whole is to be studied and explained in the same way as machines are. Mechanization does not stop here: It is not only the natural universe, which is mechanized in this way but also the social world. This is how Grossmann interprets Hobbes's method of analysis and synthesis: Hobbes conceives of the state as a machine made up of parts. In order to be understood, the state has to be taken apart (in thought) and the parts studied, just as the working of a machine is comprehended when the properties of its parts are made known and their assemblage is understood (1935a, 209–210).

Grossmann wrote about the dependency of the mechanization of the world picture and its generalization on the study of machines and on the state of mechanics without dealing with the social interests of the protagonists and their ideological articulations. This was probably also due to the fact that his essay was written in criticism of Franz Borkenau, who had stressed the role of ideology and the struggle between social groups in the formation of world pictures. Although Grossmann did not criticize the relevance of these factors, he severely criticized the inadequacy of Borkenau's analysis and the notion that the conceptualization of basic processes in natural philosophy was primarily due to local social interactions, or that the mechanistic world picture was a projection of the social structure of the workshop in manufacture onto the cosmos (Grossmann 1935a, 162–164).

Hessen uses the discussion of the mechanistic world picture to make a few first steps towards showing what other social factors might be relevant to concept formation in science. This is a subtle discussion couched in an occasionally strident Marxist vocabulary that can be somewhat off-putting outside the school and has led generations of historians without a sufficient background in Marxism to underestimate Hessen's argument and miss the level of discourse. Hessen here presents his (third) thesis on the different forms in which materialism, idealism and dualism are related to the science of the day and to the ideological positions of their protagonists and adherents. Hessen discusses Newton's world picture in contrast to Descartes, John Toland, and modern physics. Each contrast enables him to discuss not only a different aspect of Newton's dualism, but also a different aspect of the formation of ideological positions. Let's take them in turn. Since Newton and Descartes basically share a comparable scientific background, Hessen ascribes the significant *differences* in their philosophical views to differences in their social reality and in their stances towards that reality.

Hessen maintained that Descartes' mechanistic world picture was an adequate generalization of mechanics. His question is why Newton was a dualist *within* science. Although Descartes is usually seen a representative of dualism, if not its inventor (mind and body: *res cogitans* and *res extensa*), Hessen sees Descartes as taking the materialist-mechanistic view as far as was possible on the basis of science of his day. He developed a materialistic conception of the entire physical world and turned to idealism only where his science was powerless – in understanding thought. Newton, on the other hand, introduced dualism into the physical world itself and allowed God and other "active principles" to interfere with mechanical causality in the world. Hessen believed that Newton's dualistic world picture was not necessitated by limitations inherent to the science of the day since Descartes had already gone farther. He traced the shortcomings of Newton's worldview in this regard back to the "class compromise" of the Glorious Revolution, of which Newton was a typical advocate. This is the third thesis that Hessen presents in his paper.

> This ideological cast of mind of Newton, who was a child of his class, explains why the latent materialistic germs of the *Principia* did not grow to become a consistent system of mechanical materialism, like the physics of Descartes, but were interwoven with his idealistic and theological views, to which, on philosophical questions, even the materialistic elements of Newton's physics were subordinated (183).

Hessen specified the points in which he considered Descartes' philosophy to be superior to Newton's: First, Descartes introduced a conservation law for motion and thus forbade God to interfere in the physical world. Conceiving of the world as a dynamically isolated system excludes the possibility that a causal agent outside this world may interfere with its internal law-governed causal processes. Newton did not consider the world to be such a closed system although the law of the conservation of (kinetic and potential) energy can be easily inferred from his laws of motion. Second, Descartes introduced a historical element into physics, describing the evolution of the cosmos from material bodies in motion, whereas Newton used the argument from design to introduce God.

The contrast between the dualism of Newton and the speculative materialism of Toland focuses on Newton's severing of matter from motion. The conceptualization of matter as essentially inert, lacking activity, was the basis on which Newton could plausibly argue for the introduction of active principles, especially God, to account for phenomena of motion in the world. Toland did not share Newton's reductionist concept of matter. However, his non-mechanistic concept of matter could not be based on contemporary science. Nevertheless, the difference between these philosophical positions shows that science alone does not determine philosophical generalizations. There are other factors, such as the ideological positions and the social movements with which they are associated. Toland, according to Hessen, perspicuously criticized the deficiencies of Newtonian mechanics, the dualism of inert matter and active principles (182); and Richard Overton (1599–1664) advocated the unity of form and matter. But this materialistic worldview could not be put on a scientific footing at the time.

Hessen believed that "modern physics" (general relativity theory) opened a new possibility to establish a materialist, non-mechanistic monism. The argument runs as follows (187–190): If motion, as in Newton, is not an essential but merely an accidental property of matter (a *mode*, not an *attribute*),[38] then not only does the law of inertia hold, but matter is also "absolutely inert" ("inert in the full meaning of the word"). This means that a state of absolute rest is essential to matter. From this two consequences follow: first, that "such a conception of modality of motion must inevitably lead to the introduction of an external motive force, and in Newton this role is performed by God" (187); and second that a frame of reference for absolute inertia must be introduced – and this function is fulfilled by Newton's absolute space. Absolute space, again, was conceived by Newton as the *sensorium Dei*, thus again ascribing to God a role within physics. Hessen therefore sums up: "Thus the idealistic views of Newton are not incidental, but organically bound up with his conception of the universe" (190).

The scientific deficiencies of these two idealistic consequences disappear with General Relativity: "In modern physics, the view of the inseparability of motion from matter is being more and more accepted" thus "modern physics rejects absolute rest." (189), consequently neither immaterial forces nor an immaterial absolute space are called for in modern physics.

General relativity overcomes the dualistic conceptions of inert matter and active, immaterial entities and allows a monistic conception to be based on scientific grounds, whereas in Newton's time, the monist materialism of Overton and Toland could not be based on science. And yet, Newton did not simply draw necessary conclusions from the state of physics. Descartes' philosophy proves that on the basis of scientific mechanics a mechanistic-materialist philosophy could be developed. Newton actively engaged in propagating religion – as is evident also in his involvement in Bentley's "Boyle Lectures" and other occasions (184–186).

Although Hessen did not elaborate his own methodological assumptions, it seems that he held the following view: Given a certain physical theory, a horizon for a general conception of the universe is opened up. Thus mechanics in the age of Descartes and Newton opened up a horizon with a mechanistic world picture in its centre. Depending on the ideological struggles of the time and the positions of the persons involved, this horizon spanned a great distance: from radical mechanical materialism based on science (Descartes and in some sense Spinoza) to the compromise position (politically and philosophically) of dualism partially based on science (Newton) and finally to the monistic, politically radical, non-mechanical materialism of Toland, which, however, could not be based on science. Einstein's relativity opened up a new horizon of possibilities of basing a non-mechanical materialism on science. Seen from the perspective elaborated in the study of Newton, we may

[38] Hessen uses Cartesian/Spinozist terminology such as "attributes" and "modes" instead of "essential properties" and "accidents" in the discussion of matter and motion in Newtonian physics.

say that General Relativity does not conceive matter as "absolutely inert". Hence it also overcomes the dubious status of *force* and *space* (ab)used by Newton to introduce God into the physical universe. The conclusion, which was not spelled out by Hessen, is nevertheless easy to draw: from the point of view of (Hessen's) materialism, General Relativity is superior to Newtonian mechanics. A second conclusion was spelled out for Newton but not for the discussion about relativity theory (considered idealistic by some Soviet theorists) in which Hessen was involved at the time in the Soviet Union: A theory of physics allows for more than one philosophical interpretation and which position is endorsed depends on the ideological stance of the persons involved.

5 The Reception of Hessen, Grossmann and Merton

Grossmann's "The Social Foundation of Mechanistic Philosophy and Manufacture" (1935a) was hardly noticed at all at the time since it was published in the German-language *Zeitschrift für Sozialforschung* that appeared in French exile at a time when the institution sponsoring it was on its way into American exile. The essay was written as a critique and correction of Franz Borkenau's *The Transition from the Feudal to the Bourgeois World picture,* a nominally Marxist work that the Institute for Social Research had commissioned (and even published) but whose results were considered superficial and embarrassing to the Institute. Grossmann was charged with the task of saving face for the Institute.[39] If Grossmann's work was noticed at all, then as a refutation of Borkenau's book, not as a contribution in its own right. Alexandre Koyré, who knew Grossmann personally and professionally, remarked that Grossmann's critique of Borkenau was "far more instructive than this work itself," but this remark was used merely to dismiss all attempts to understand the Scientific Revolution on the background of technical practice – without actually dealing with any of them.[40] Grossmann's essay, though quite polemical against Borkenau's superficial scholarship, was otherwise measured in its claims and unassailable in its scholarship; it presented its few readers with the paradox of an obviously orthodox Marxist refuting all the positions they associated with Marxism: Grossmann does not explain science as pursued for the sake of production. Grossmann's paper fell into oblivion like many other excellent works of the 1930s. The only impact of his work on mainstream history of science was Koyré's footnote against Borkenau, which has been taken up by other historians in the same form and for the same purpose.[41] With the exception of occasional friendly citations by Lynn

[39] Kuhn 2007, 165.

[40] See Koyré 1978 (*Galileo Studies*) 39, note 8.

[41] Koyré's remark is repeated almost verbatim by Canguilhem (1948, 308); Dijksterhuis (1961, III, ii, p. 241) follows suit; and H.F. Cohen (1994, 582) in turn follows him. Not even R.K. Merton, for whom Hessen's paper was so important and who was also well acquainted with German scholarship, knew of Grossmann's essay in 1938 when his *Science, Technology and Society* was

White Jr.,[42] this has constituted Grossmann's entire reception in official history of science. When socio-historical approaches revived in the 1970s, Grossmann was rediscovered by the German-speaking Left,[43] but his influence went no farther. An English translation appeared in 1987 in the first volume of the journal of *Science in Context* – without altering the reception significantly.

The reception of Hessen's essay, "The Social and Economic Roots of Newton's *Principia,*" is quite a different story: this paper was notorious from the day it was published. It had an immediate and lasting impact on the participants of the conference at which it was presented, it was widely read in the 1930s and is still widely known – at least in facile misrepresentations. The story of the immediate and later reception is worth telling since this reception had a constitutive role in the development of history of science as a discipline.

Hessen's paper was presented at the Second International Congress on the History of Science held in London in 1931. The Soviet Union sent a delegation of eight speakers headed by N.I. Bukharin. Bukharin had already lost the power struggle, and his positions in state and party were largely ceremonial. However, he was a prominent enough communist for the British tabloid press to go wild criticizing the Labour government for issuing a visa and to put pressure on the organizers not to let the Russians use the Congress as a forum for their political propaganda. The summer of 1931 was at the height of the Great Depression, there had already been some significant unrest encouraged by communists (and others) both domestic and foreign. Thus, an otherwise uneventful academic gathering was turned into a high-profile political affair. Anti-communism in both the tabloid and the professorial form has always been an important factor in the reception of Hessen.[44]

However the event was not political just because of the Russians and the reaction to them. The "Comité Internationale d'Histoire des Sciences" had been founded at a conference in Oslo in 1928, and a first public conference had been held in Paris in 1929, but the 1931 Congress in London was planned to be more than just another academic meeting of historians of science. The organizers, a group of British educators, historians and scientists, hoped to use the congress as an occasion to drum up support for the history of science as an instrument for improving science education and for increasing public support for science itself. They hoped to increase the part played by intellectual history at the expense of traditional political history and to

---

published, nor did he cite it at the time of its reprinting in 1970. This is especially surprising since Merton did cite Borkenau's book in 1938. (See Merton 1970: 155n, 156n, 191n, 228.)

[42] For instance, "Natural Science and Naturalistic Art in the Middle Ages," *American Historical Review 52*, 1947, 422.

[43] When Borkenau's book was reprinted by a scholarly publisher in Germany in 1971, a pirate edition soon appeared with Grossmann's paper as an appendix. One example of this positive reception of Grossmann is Wolfgang Lefèvre, *Naturtheorie und Produktionsweise*, Neuwied: Luchterhand, 1978.

[44] I. B. Cohen (1990, 55–56) reported, "Far from being a study of the interpretation of science 'as a product of the life and tendencies of a society' or even a general example of 'the dependence of science on social factors,' Hessen's analysis was couched in the narrow doctrinaire canons of rigid Marxist dialectical materialism." If there is a logic to this sentence, then it means that an argument formulated in Marxist terminology need not on principle be taken seriously.

raise the social status of science and the scientist. This was a basically progressive group with fairly realistic goals and means.[45] Whatever the organizers reservations towards a personalizing form of historiography may have been, they did consider the personal motivations of individuals to be a significant explanatory factor, and some of their prominent public pronouncements were fairly extreme. Hessen could rightly attack them for endorsing A.N. Whitehead's ludicrous assertion that "Our modern civilization is due to the fact that in the year when Galileo died, Newton was born."[46] Whatever the original plan for the Congress, the arrival of the eight Soviet delegates changed its nature fundamentally. The press was unconcerned with the status of science education and very concerned with the Soviet guests. The Russians had their own agenda and wanted to speak from the floor in all the sessions on all the issues. Since they had registered only their names in advance but not submitted their contributions, none of them were on the program, and a special session at the end of the Congress had to be arranged for their papers. The Russian Embassy had them all translated during the week.[47]

At the time of the Congress Hessen was Director of the Physics Institute at Moscow University and was strongly involved in the defense the theory of relativity in the Soviet Union. (As we have seen above (p. 25) Hessen also argued in his paper that relativity was better compatible with dialectical materialism than was Newtonian physics.) The Soviet debate on relativity was carried out in philosophical, historical, and physical controversies. Science studies in the Soviet Union had become a serious professional discipline,[48] and Hessen had already done many years of research before he joined the delegation to London. When he was arrested in August 1936, he had a 700-page textbook on the social conditions of the rise of classical physics in proof.[49]

Although many of the Congress' organizers and attending members had excellent professional credentials in science or in history, with the exception of Charles and Dorothy Singer almost all the British participants were basically dilettantes in the history of science. The science journalist J.C. Crowther, perhaps a somewhat biased witness, characterized the regularly scheduled papers as "reminiscences from

[45] See especially A.K. Mayer (2002, 2003), who points out that the organizers were in many respects actually opposed to such a personalizing form of historiography. See also C.A.J. Chilvers 2003, 2006.

[46] Hessen 151. Those attacked were F.S. Marvin and importantly G.N. Clark.

[47] In spite of the additional session, an excursion to Oxford planned for the same time was not cancelled. Only registered members but not the general public were admitted to the session. As in the other sessions the speakers were given only ten minutes to present their ideas. The ten-minute limit at this session was enforced by the organizer, Charles Singer, ringing a ship's bell (Werskey 1971, Needham 1971, vii).

[48] See Winkler (forthcoming) for a selection of papers from the period.

[49] The incomplete proof sheets of this approximately 700-page collection of translated sources and expositions by Hessen were found in 2004 by the late V.S. Kirsanov in the papers of the historian of mathematics A.S. Youschkevitsch, whose father-in-law, V.S. Gochman, may be presumed to have been the translator at least of the Latin texts.

the elderly, and trivia from obscure amateurs."[50] At the Congress they were all outflanked, out-argued and just plain out-talked by the highly professional, well-prepared Soviet delegates, who had a particular viewpoint and argued for the social determination of the development of science.[51] As J.D. Bernal reported after the Congress: The Russians "had a point of view, right or wrong; the others had never thought it necessary to acquire one." Nonetheless, their presentations had little effect: "The Russians came in a phalanx uniformly armed with Marxian dialectic, but they met no ordered opposition, but instead an undisciplined host, unprepared and armed with ill-assorted individual philosophies. There was no defense, but the victory was unreal."[52] In a newspaper interview given by the President of the Congress, Charles Singer, summing up its results, no mention at all is made of the conflict that dominated discussion, of the extra session with the Soviet delegates, or even of the presence of such a delegation.[53] The polarization that began at this Congress and the strategy of dealing with it has plagued historical studies of science ever since.

After attacking the cult of genius (unintentionally cultivated by his British hosts) on the first page of his paper, Hessen asserted that there is an alternative view, historical materialism, and proceeded to give three pages of remedial Marxism for the professors, to whom he thought the information would be new. How right he was may be gathered from the report filed on his return by the Soviet delegate Modest Rubinstein, in which he mentions that "the editor of the leading philosophical journal *Mind* in Cambridge, after a long conversation with Comrade Bukharin, asked him: 'Who was that Engels you kept referring to?'"[54]

The great impact of the Soviets, in particular Hessen, is well documented. J.G. Crowther reported later that Hessen's paper "gave the first concrete example of how science should be interpreted as a product of life and tendencies of society," since it demonstrated "the depth and range of Newton's dependence on the ideas promulgated by the epoch." He concluded that Hessen's paper "transformed the study of the history of science, and out-moded the former conceptions of the subject, which treated it as governed only by the laws of its internal logical development."[55] Two eminent British historians of science, J.D. Bernal and Joseph Needham, both of whom had been present at the Congress, testified repeatedly to the significance of the event for their own work and for the development of British science studies. Needham, who at the time had just completed a history of embryology, believed that further historiography should do "for the great embryologists what has been so well done by Hessen for Isaac Newton." In a magazine article shortly after the Congress

---

[50] "Social Crisis and Scientific Inspiration," [1931] *Fifty Years* p. 77.

[51] According to the Congress documentation in the first session, besides the scheduled speakers, six presentations were made from the floor – five of them by Soviets. See *Archeion 14* (No. 2) 1932, 277–288.

[52] Bernal 1949, 336.

[53] Singer 1931.

[54] Rubinstein 1931, 95. The editor of *Mind* in 1931 was G.E. Moore.

[55] J.G. Crowther, *The Social Relations of Science*, London: Macmillan [1941] 1967, pp. 431.

Bernal summarized the view presented by Hessen and the others: "The development of pure science is dependent on that of economics and technics both for the problems they present it with and for the *means provided for their experimental study.*"[56]

The reaction of the audience in 1931 was divided according to political sympathies. Scholars with socialist inclinations were impressed with the Soviet contributions in general and with Hessen's in particular. The reaction of those not basically sympathetic to socialism was reserved if not hostile. But a significant asymmetry between the two predispositions should not be overlooked. The sympathizers, familiar with some of the background, could actually understand the arguments of the Russians. But due to the Marxist terminology of the presentations, the more traditionally inclined in the audience did not comprehend the assertions being made, much less the arguments supporting them. In Bernal's words they were disposed "not to listen to the arguments which followed, with the feeling that anything so ungentlemanly and doctrinaire had best be politely ignored."[57]

The first real counterattack came several years later in a paper by G.N. Clark, the opening speaker of the first session and thus the direct victim of the first wave of the Soviet assault, but also the indirect victim of Hessen's criticism of a scholar who had praised Clark. This has remained the only substantial explicit response to Hessen.[58] Although the two best qualified contemporary judges, Robert K. Merton and Henryk Grossmann, almost immediately demonstrated that Clark had not only misunderstood Hessen's position and ignored his arguments, but also championed an approach that was patently untenable, Clark's attempted refutation remains the model for standard critiques of Hessen: they ignore the substance of the argumentation, insist on purported mistakes in historical details, characterize the dispute as being about the utilitarian motives of the scientists, and use the word "crude".

The correlations between technological enterprise and scientific research in the seventeenth century emphasized by Hessen were not denied by Clark nor by his followers. There were at first reservations as to whether *scientists* themselves were really engaged in the technological projects. But as pointed out by a sympathetic reader in 1939 and by an unsympathetic reader in 1993, most seventeenth-century scientists were indeed also personally engaged in technological projects.[59] What remains of Clark's critique is the rejection of a thesis, which he erroneously ascribes

---

[56] Bernal 1949, 337, our emphasis.

[57] Bernal 1949, 338.

[58] Hessen's foil was a position attributed to Whitehead and Clark by Marvin in an enthusiastic review of Clark's *The Seventeenth Century*. Clark's paper appeared in *Economic History 3*, 1937 and was reissued the same year (with incorrect source data) as a chapter of his *Science and Social Welfare in the Age of Newton*.

[59] R.K. Merton 1939, 5: "It is neither an idle nor unguarded generalization that *every English scientist of this time* who was of sufficient distinction to merit mention in general histories of science at one point or another explicitly related at least some of his scientific research to immediate practical problems." Richard Westfall (1993, 65) wrote: "That is, about three-quarters of the scientific community did participate in some technological enterprise. To be frank, I must say that this is a considerably higher proportion than I would have predicted when I began."

to Hessen, namely that the real personal motive of all the scientists in seventeenth-century England was utilitarian. He mentions that optical research may really have been pursued in order to correct eyesight defects, that Robert Boyle was genuinely pious in his motives, and that many people probably had motives for pursuing science far different from improving production:

> In surveying the social background of the scientific movement we have now distinguished five different groups of influences, which have worked upon science from the outside; those from economic life, from war, from medicine, from the arts, and from religion.... There still remains a motive which we have not considered.... The disinterested desire to know, the impulse of the mind to exercise itself methodically and without any practical purpose, is an independent and unique motive.[60]

Had Hessen believed that improving economic production was the personal motive of all seventeenth-century scientists, he might have been dealt a severe blow by Clark's critique. But as Joseph Needham remarked in a book review: "it is surely not essential to Hessen's case to assume that the actions of a scientist have a *consciously* economic motive."[61] R.K. Merton, too, was less than impressed by Clark's arguments:

> Clark's recent critique of Hessen's essay may be taken to illustrate the confusion which derives from loose conceptualization concerning the relations between the motivation and the structural determinants of scientists' behavior.... Clark's criticism of Hessen narrows down to a repudiation of the thesis that economic factors are alone determinant of the development of science. In company with Hessen I hasten to assent to this undisturbing renunciation.[62]

Clark's misunderstanding is clear: he presumes that an argument about the determinants of scientific thought must recur to the personal motivations of the scientists – whereas it was just this personalizing tendency of the British organizers that Hessen had attacked at the outset. (See the discussion above p. 28)

This fundamental misunderstanding, shared by most later readers, cannot be due merely to the provocative form in which Hessen presented his position, since R.K. Merton's *Science, Technology and Society in Seventeenth-Century England* suffered basically the same fate without having engaged in any provocative language. Merton's book dealt with two aspects of seventeenth-century English science: the role of Puritanism in fostering scientific activity and the role of technology in determining the focus of research. In the first half of the book he saw the role of Puritanism as important though limited:

> But if this congeniality of the Puritan and the scientific temper partly explains the increased tempo of scientific activity during the later seventeenth century, by no means does it account

---

[60] Clark 1937, 376–77.

[61] Needham, "Capitalism and Science" (review of Clark's *Science and Social Welfare*) *Economic History Review 8* (1938) p. 198.

[62] Merton 1939, 6. Unaware of Merton's rejection of Clark's critique of Hessen, I.B. Cohen in 1990 praised Clark and attacked Hessen – in a laudatio on Merton (Introduction to: *Puritanism and the Rise of Modern Science*): "Hessen's analysis was couched in the narrow dotrinaire canons of rigid Marxist dialectical materialism" (55–56), whereas "Clark's book is so well balanced and devoid of any overriding thesis (or theses), that it did not have the impact of Hessen's paper" (59).

> for the particular foci of scientific and technologic investigation. Which forces guided the interests of scientists and inventors into particular channels?[63]

The answer to this question came in the second half of the book, which dealt with technology and openly acknowledged a debt to Hessen's "provocative thesis."[64] Merton's work has become synonymous with his thesis on the Puritan context of early modern science while his second thesis on the economic and technological context has been ignored. In 1970, when the book was reissued, Merton remarked on the "puzzle" that although he thought his argument of 1938 was set out clearly enough to be understood, nevertheless the responses during more than three decades made him doubt this. He did the numbers: Ninety percent of the responses dealt only with the thesis on the interrelations between Puritanism and the institutionalisation of science and neglected the second thesis on the economic and military influences on the spectrum of scientific work, although he had dedicated significantly more space to the second thesis than to the first. He traced this back to the idealistic proclivities of historians.[65]

Back to Hessen: G.N. Clark's response established a pattern that declined over time. Clark in 1937 is politely condescending: "If Professor Hessen, who is a physicist, had found as a collaborator a trained historian, he would have been able to eliminate some crudities from his article." I.B. Cohen, looking back on the early 1950s and his own first reading of Hessen (and seeing no connection to the Cold-War climate), reported in 1990: "We must remember, however, that many historians of science who were practicing a traditional historiography were repelled by Hessen's crude Marxian dialectical materialism." And H.F. Cohen in 1994 gives up all semblance of argument when confronted with "... this narrow-minded piece of bigoted dogmatism at its Stalinist crudest ..."[66]

Another variant on this theme in more recent literature has been the attempt to contextualize the alleged "utilitarianism" of Hessen and thus to relativize its repugnance. Loren R. Graham has done "for Boris Hessen what Boris Hessen tried to do for Isaac Newton."[67] Noting Hessen's involvement in debates over the theory of relativity in the Soviet Union prior to the conference in London, Graham suggests that at least that part of Hessen's essay which relates the emergence of classical

---

[63] Merton 1938, 137.

[64] Merton 1938, 142–3, note 24.

[65] Merton 1970: xi–xii. A.R. Hall in "Merton Revisited" (1963, 10) involuntarily confirms this judgment, concluding his discussion of Merton with the words: "social forms do not dominate mind; rather, in the long run, mind determines social forms." As Steven Shapin noted in his enlightening essay "Understanding the Merton Thesis" (1988): "On the evidence of some of those historians who have endeavoured to refute what they represent to be his thesis, Merton's 1938 monograph and related texts can scarcely have been read at all" (594).

[66] Clark 1937, 363; I.B. Cohen 1990, 59 and H.F. Cohen 1994, 332. Because he assumes that the motivation and thus the knowledge of the scientists is at issue, H.F. Cohen accuses Hessen of crass presentism.

[67] Graham 1985, 706; for a similar interpetation see Schäfer 1988. Graham (1993, 143–151) offers a similar narrative but tones down the externalism.

mechanics from economics and technology may just have been written to pacify the party bosses:

> The overwhelming impression I gain from the London paper is that Hessen had decided to 'do a Marxist job' on Newton in terms of relating physics to economic trends, while imbedding in the paper a separate, more subtle message about the relationship of science to ideology.[68]

It seems clear that Graham wishes to explain how an apparently sophisticated Marxist, who defended relativity and quantum mechanics when they were under attack in the Soviet Union, could write the blunt and worthless essay of 1931. But in the end he attributes to Hessen a crude, vulgar sociological argument about Newton and then makes the same kind of argument about Hessen himself: Hessen had ulterior motives for writing that Newton had ulterior motives for writing the *Principia*.

The externalist appeal to Hessen's commitment to defending relativity obviates the need for a serious consideration of the arguments for his position. In the essay of 1931 Hessen presented an extensive argument concerning the ideological implications of "modern physics" and of Newtonianism and argued that relativity is more congenial to Marxist materialism than is Newtonianism. But this argument is not even mentioned by Graham and those who have followed his interpretation. Here, as in the discussion of the Hessen-Thesis itself, the critics seem to believe that to give an externalist explanation of science means to uncover the *ulterior motives* of the scientists involved and that this ulterior motivation somehow discredits the scientific work done. This is of course an entirely different kind of externalism than that of Hessen, who – however one judges him – offered an externalist account of the cognitive content of seventeenth-century science not of its purported lack of content or lack of truth.[69]

Hessen asserts repeatedly that he sees the relation of technology to science as one of providing material for science to study or means with which science can pursue the study of nature; and sympathetic readers like Bernal, Needham, Grossmann, and Merton recognized and insisted on just this fact. There are some later brief, presentations of Hessen's position from scholars sympathetic to social-historical approaches that are basically accurate.[70] Nonetheless more traditional historians

---

[68] Graham 1985, 716. Graham could not have known of the extensive historical work done by Hessen and his group on the history of the Scientific Revolution (see above p. 28). But an unprejudiced reading of Hessen's essay must recognize the erudition of the work and the close argumentation that excludes a quick "Marxist job."

[69] Graham sums up: "My conclusion is that Hessen's paper is better understood as a result of his peculiar and threatened situation in the Soviet Union than as a model of Marxist analysis of science, either vulgar or sophisticated" (1985, 707).

[70] See especially Steven Shapin's short entry "Hessen thesis" in the *Dictionary of the History of Science*; see also the remarks by A. P. Youschkevitch in the bibliography entry on Soviet literature in the DSB article on Newton; and Wersky 1970 and Schaffer 1984. Wittich and Poldrack 1990 place Hessen's work in the context of Soviet discussions on science policy and the possible contribution of science to the industrialization of the Soviet Union. For a recent discussion of the role of artifacts in the conceptualization of nature, see H.S. Davis Jr 2001. For a rich but not entirely consistent discussion of Marxist analysis of science see 38–113; for Hessen, especially 83–86 and 97–99.

of science have always seen him as engaging in completely unsubstantiated speculations about the motives of scientists. Even such a serious scholar as Richard Westfall, who scrupulously avoided cheap polemics, forced Hessen into the mould forged by Clark: "Although Hessen did not mention Baconian utilitarianism explicitly, he concerned himself entirely with something identical to it."[71] The reader who takes the trouble to study Hessen's paper will discover that Hessen is quite capable of mentioning the things he wants to deal with.

6 Conclusion

More than seven decades have passed since the first publication of Hessen's and Grossmann's original essays. During this time Grossmann's essay has been ignored and Hessen's essay misinterpreted to fit the preconceptions of internalist historians. The three important theses presented in these essays on the horizons opened for science by contemporary technology, on the limitations of these horizons and finally on the ideological impregnation of science were neither comprehended nor seriously discussed. A real discussion of these theses and a controversy in which other interpretations of the historical record might try to compete with the Hessen-Grossmann-thesis would certainly significantly contribute to our understanding of the Scientific Revolution and of the origins of the modern world in general.

The aspects discussed above certainly do not exhaust the substance of Hessen's and Grossmann's ideas. There can be little doubt that these essays contain seminal work of the highest order: original, daring and well argued. Many individual observations have already been taken up by later scholarship (e.g. issues of the Newtonian ideology). However the Hessen-Grossmann-Theses are still as novel as they were seven decades ago when they were first proposed. This applies to explaining the horizon of cognitive possibilities on the basis of the means employed, both the horizon *opened* by specific means (this is the first Hessen thesis as elaborated by Grossmann) and also to the *limits* set by these possibilities (this is the second Hessen thesis). But this applies also to the encompassing view in which the broad perspective of the emergence of science within the transition from feudal to capitalist society proves relevant to issues of cognitive development and vice versa. This comprehensive approach indebted to Marx, and especially to his theory of labor in its material and social determination, is the basis of the theses

- on the formation of social needs on the basis of potential means available to their fulfillment and therefore,
- on the demands that rising capitalism put on technology,

[71] Westfall 1983, 107. Against Clark Merton remarks: "Motives may range from the desire for personal aggrandisement to a wholly 'disinterested desire to know' without necessarily impugning the demonstrable fact that the thematics of science in seventeenth century England were in large part determined by the social structure of the time" (1939, 4).

- on the ensuing change in the social locus of technology and the entry of a knowledge of practical mechanics into theoretical study,
- on the correlation between the fields of rising technology and the disciplines developed in science (mechanics, hydrostatics and hydrodynamics), and also
- on the disciplines that did not develop because no corresponding technology existed which could be studied (thermodynamics, electrodynamics), finally also
- on the ideological influence on the conception of one's own practice and the objects involved: in Newton's case the "absolute inertness" of matter and the role of God and active principles in his world picture.

Hessen's and Grossmann's essays can be read as explorations of the implications of Marx's theory of labor as applied to science.[72] Their debt to Marx's historical theory and especially to his theory of the transition from one "socio-economic formation" to another is evident in Hessen's paper, which begins with a short and strident summary of these views based primarily on the "Introduction to Critique of Political Economy." In Grossmann this view is applied in his sketch of the socio-economic development from the fifteenth to the seventeenth century. This general background theory is the source of both Hessen's and Grossmann's peculiar perspective on science from the point of view of the transition from the feudal socio-economic formation to the capitalist mode of production.

This common origin and point of reference for both Hessen and Grossmann explains the extensive agreement between their views, in spite of the fact that their original essays were independently written. Marx's general theory was for both of them the frame of reference in their work, and Marx's concept of labor served them as an analytic tool in their detailed analysis of the cognitive content of physics and thus of the Scientific Revolution.

Bibliography

Archeion. Archivio di storia della scienzia 14, 1932.

Bacon, Francis ([1620] 1858) *Novum Organum, The Works of Francis Bacon*, vol. 4 (ed. J. Spedding, R. Ellis, D.D. Heath) London.

Bernal, John Desmond ([1931] 1949) "Science and Society," *The Spectator*, July 1931 (Reprint in J.D. Bernal, *The Freedom of Necessity*, London: Routledge and Kegan Paul, 1949).

Borkenau, Franz (1934) *Der Übergang vom feudalen zum bürgerlichen Weltbild*, Paris: Felix Alcan (Reprint Darmstadt: Wissenschaftliche Buchgesellschaft, 1971; pirate edition with reprint of Grossmann's critique: Berlin: Junius, 1971).

Canguilhem, Georges (1948) "Les philosophes et la machine" *Critique* 1948, in *Études de la pensée philosophique*, Paris: Armand Colin, 1961, 279–309.

Chilvers, C.A.J. (2003) "The Dilemmas of seditious men: the Crowther-Hessen correspondence in the 1930s," *British Journal for the History of Science 36*, 417–435.

Chilvers, C.A.J. (2006) Afterward to B. Hessen, *Les racines sociales et économiques des Principia de Newton* (ed. and transl. by S. Guérout) Paris: Vuibert.

[72] *Capital I*, esp. Chapters 7 and 15.

Clark, G.N. (1937) Social and Economic Aspects of Science, *Economic History 3*, (reprint in: G.N. Clark, *Science and Social Welfare in the Age of Newton*, Oxford: Clarendon Press, 1937).
Cohen, H.F. (1994) *The Scientific Revolution. A Historiographical Inquiry*, Chicago: University of Chicago Press.
Cohen, I.B. (ed.) (1990) *Puritanism and the Rise of modern Science. The Merton Thesis*, New Brunswick, NJ & London: Rutgers University Press.
Crowther, J. G. (1970) *Fifty Years With Science*, London: Barrie & Jankins.
Crowther, J.G. ([1941] 1967) *The Social Relations of Science*, London: Macmillan.
Davis, Henry F., Jr. (2001) *Technology and the Origin of Nature Concepts: The Impact of the built World on Aristotle and Galileo's Concepts of Motions*, (dissertation Duquesne UMI).
Dijksterhuis, E.J. (1961) *The Mechanization of the World Picture*, Oxford: Oxford University Press.
Freudenthal, Gideon (1986) *Atom and Individual in the Age of Newton*, Dordrecht: Reidel (Boston Studies in the Philosophy of Science, vol. 88).
Freudenthal, Gideon ([1984] 1988) "Towards a Social History of Newtonian Mechanics. Boris Hessen and Henryk Grossmann Revisited," in: *Scientific Knowledge Sozialized, Selected Proceedings of the 5th Joint International Conference on the History and Philosophy of Science Organized by the IUHPS Veszprém, 1984*. ed. by I. Hronsky, M. Fehér und B. Dajka, Dordrecht: Kluwer (Boston Studies in the Philosophy of Science, vol. 108) 193–212.
Freudenthal, Gideon (2002) "*Perpetuum mobile*: the Leibniz-Papin Controversy," *Studies in History and Philosophy of Science 33*, 573–637.
Freudenthal, Gideon (2005) "The Hessen-Grossman Thesis. An Attempt at Rehabilitation," *Perspectives on Science 13*, 166–193.
Graham, Loren (1985) "The Socio-political Roots of Boris Hessen: Soviet Marxism and the History of Science," *Social Studies of Science 15*, 705–722.
Graham, Loren (1993) *Science in Russsia and the Soviet Union. A Short History*, Cambridge: Cambridge University Press.
Grossmann, Henryk (1935a) "Die gesellschaftlichen Grundlagen der mechanistischen Philosophie und die Manufaktur," *Zeitschrift für Sozialforschung 4* (1935) 161–231. "The Social Foundations of the Mechanistic Philosophy and Manufacture." This volume (103–156).
Grossmann, Henryk (1935b) letter to Friedrich Pollock and Max Horkheimer of August 23, 1935. This volume, 229–231.
Grossmann, Henryk (1938) Review of G.N. Clark, *Science and Social Welfare in the Age of Newton* (1937) and George Sarton, *The History of Science and the New Humanism* (1937) *Zeitschrift für Sozialforschung* 7 (1938) 233–237. This volume, 231–235.
Grossmann, Henryk (1946) Descartes and the Social Origins of the Mechanistic Concept of the World, unpublished monograph, originally titled: "Universal Science versus Science of an Elite. Descartes' New Ideal of Science." This volume 157–229.
Hall, A. Rupert (1952) *Ballistics in the Seventeenth Century: a Study in the Relations of Science and War with Reference Principally to England*, Cambridge: Cambridge University Press.
Hall, A. Rupert (1963) "Merton Revisited, or Science and Society in the Seventeenth Century," *History of Science 2*, 1–16.
Hessen Boris (1931) "The Social and Economic Roots of Newton's 'Principia'," in: *Science at the Cross Roads*, Kniga, London 1931, 149–212. New translation, this volume, 41–101.
Huerga Melcón, Pablo (1999) *La Ciencia en la encrucijada*, Oviedo: Fundación Gustavo Bueno, Pentalfa Ediciones.
Huerga Melcón, Pablo (2001) "Raíces filosóficas de Boris Mijailovich Hessen: critíca al mito del externalismo de Boris Hessen," *ILUIL 24*, 347–395 (unpublished English manuscript: "Philosophical Roots of Boris Hessen: A Criticism of the Myth of Hessen's Externalism").
Koyré, Alexandre (1948) "Les philosophes et la machine," *Critique*, (reprinted in *Études de la pensée philosophique*, Paris: Armand Colin, 1961, 279–309).
Koyré, Alexandre (1943) "Galileo and Plato," *Journal of the History of Ideas 4*, 400–428.
Koyré, Alexandre (1978) *Galileo Studies* (transl. by John Mepham) Sussex: Harvester .

Kuhn, Rick (2007) *Henryk Grossman and the Recovery of Marxism*, Urbana, IL: University of Illinois Press.

Lefèvre, Wolfgang, *Naturtheorie und Produktionsweise*, Neuwied: Luchterhand, 1978.

Mayer, A.K. (2002) "Fatal Mutilations: educationism and the British Background to the 1931 International Congress for the History of Science and Technology," *History of Science 40*, 445–472.

Mayer, A.K. (2004) "Setting up a Discipline, II: British History of Science and the End of Ideology," *Studies in History and Philosophy of Science 35*, 1–72.

McLaughlin, Peter (1994) "Die Welt als Maschine. Zur Genese des neuzeitlichen Naturbegriffs," in A. Grote (ed.) *Macrocosmos in Microcosmo. Die Welt in der Stube. Zur Geschichte des Sammelns*, Opladen: Leske & Budrich, 439–451.

Merton, Robert K. (1938) *Science, Technology and Society in Seventeenth-century England*, [Osiris, 1938] 2nd ed. New York: Harper, 1970.

Merton, Robert K. (1939) "Science and the Economy of Seventeenth Century England," Science and Society 3, 3–27.

Needham, Joseph (1938) "Capitalism and Science" (review of Clark's *Science and Social Welfare*) *Economic History Review 8*, 198–99.

Needham, Joseph (1971) Forward to P.G. Werskey (ed.) *Science at the Cross Roads* 2nd ed. London: Cass 1971

Rubinstein, Modest (1931) "O poezdke na mezhdynarodnyy kongress v London po istorii nauki i tekhniki" [On the trip to the international congress on the history of science and technology], Report to the Communist Academy, August 1, 1931, *Vestnik Komakademii, 93–100.*

Sarton, George (1931) *The History of Science and the New Humanism*, New York: Holt.

Schäfer, Wolf (1988) "Äussere Umstände des Externalismus. Über Boris Hessen und das Projekt einer Geschichte der Wisssenschaftsforschungs-Geschichte" in Hans Poser and Clemens Burrichter (eds.) *Die geschichtliche Perspektive in den Disziplinen der Wissenschaftsforschung*, Berlin: Universitätsbibliothek der Technischen Universität, 7–46.

Schaffer, Simon (1984) "Newton at the Crossroads," *Radical Philosophy*, No. 37, 23–28.

Science at the Cross Roads (1931) (papers presented to the International Congress of the History of Science and Technology, held in London from June 29th to July 3rd, 1931 by the delegates of the U.S.S.R.) London: Kniga.

Shapin, Steven (1982) "Hessen thesis" *Dictionary of the History of Science* (ed. by W. F. Bynum) London: Macmillan.

Shapin, Steven (1988) "Understanding the Merton Thesis," *Isis 79*, 594–605.

Singer, Charles (1931) Interview: "Science & History. The Congress and its work. False Perspectives." *The Observer*, July 5, 1931, 13.

Wallis, John, letter to Oldenburg, November 15, 1668; *Oldenburg Correspondence, vol. 5*, Madison, WI: Univ. of Wisconsin Press 1968.

Werskey, G. (1978) *The Visible College: a Collective Biography of British Scientists and Socialists of the 1930s*, London: Lane.

Westfall, Richard S. (1983) "Robert Hooke, Mechanical Technology, and Scientific Investigation," in: John G. Burke (ed.) *The Uses of Science in the Age of Newton*, University of California Press, 85–110.

Westfall, Richard S. (1993) "Science and Technology during the Scientific Revolution: An empirical Approach," in J.V. Field/Frank James (eds.) *Renaissance & Revolution*, Cambridge: Cambridge University Press, 63–72.

White, Lynn Jr. (1947) "Natural Science and Naturalistic Art in the Middle Ages," *American Historical Review 52*, 1947.

Wittich, Dieter and Horst Poldrack (1990) "Der Londoner Kongress zur Wissenschaftsgeschichte 1931 und das Problem der Determination von Erkenntnisentwicklung," in: *Sitzungsberichte der Sächsischen Akademie der Wissenschaften zu Leipzig*, Philologisch-historische Klasse, vol. 130, No. 5.

Winkler, Rose-Luise (forthcoming) An den Ursprüngen der Entstehung wissenschaftssoziologischen Denkens im ersten Drittel des XX. Jahrhunderts. (Rußland und Sowjetunion).
Zilsel, Edgar (1942) "The Sociological Roots of Science," *American Journal of Sociology 47*, 245–279.

A Note on the texts

Hessen: "The Social and Economic Roots of Newton's *Principia*"

Hessen's text has been newly translated from the 1933 Russian edition by Philippa Schimrat and compared by her and the editors with the 1931 English publication to avoid merely stylistic differences. Spelling is left British style. The editors have checked the technical terminology from economics and physics. Sources quoted by Hessen in Russian translation have been replaced by the English originals or by English translations directly from the original languages except in the few cases where the source could not be located. Where the text closely paraphrases Marx or Engels the passage is indicated or quoted in a footnote. In the original English publication many proper names were garbled due to transliteration into Cyrillic letters and then back into Latin letters; the names are spelled here as in the *Dictionary of Scientific Biography*. In the original English publication there were also many typographical mistakes in dates: Where the Russian edition contains the correct date (often the date given in F. Rosenberger's *Geschichte der Physik*), it has been adopted. Around a dozen sentences were omitted in the original English (or perhaps added for the Russian publication); these are indicated by footnotes.

There have been a number of editions of Hessen's text. We have been able to trace the following editions in western languages:

"The Social and Economic Roots of Newton's 'Principia'," in: *Science at the Cross Roads*, Kniga, London, 1931, 149–212, also as separatum paginated 1–62.

"The Social and Economic Roots of Newton's 'Principia'," in: *Science at the Cross Roads*, with a preface by Joseph Needham and an introduction by P.G. Werskey, London: Cass, 1971.

The Social and Economic Roots of Newton's 'Principia', preface by J. P. Callaghan. Sydney: Current Book Distributers, 1946.

"The Social and Economic Roots of Newton's 'Principia'," exerpt in G. Basalla (ed.) *The Rise of Modern Science. External or Internal Factors?* Lexington MA: Raytheon, 1968, pp. 31–38.

The Social and Economic Roots of Newton's 'Principia' with a preface by R.S. Cohen, New York: Fertig, 1971.

Социально-экономические корни механики Ньютона (Russian version) М.-Л-д. 1933. (2nd ed. 1934. II-ое изд.).

Excerpt under the title *Отдельная глава: Классовая борьба эпохи английской революции и мировоззрение Ньютона in*: *Природа* 3–4 (1933): pp. 16–30.

"De sociala och ekonomiska förutsättningarna för Newton 'Principia'," (Swedish translation by Maria Eckman) in: Ronny R. Ambjörnsson (ed.) *Idé och klass: texter kring den kommersiella revolutionens England*, Stockholm: PAN/Nordstedts, 1972, 90–145.

"Die sozialen und ökonomischen Wurzeln von Newtons 'Principia'," (German translation) in: Peter Weingart (ed.) *Wissenschaftssoziologie* vol. 2: 262–365. Frankfurt/Main: Fischer Athenäum, 1974.

Les Fondements sociaux et économiques des Principia *de Newton* (French translation) in: Serge Guérout (ed.) Paris: Bibliothèque Interuniversitaire Scientifique Jussieu, 1978.
Les racines sociales et économiques des Principia *de Newton*, (translated and edited by Serge Guérout with an afterword by Christopher Chilvers) Paris: Vuibert 2006.
"Le radici sociali ed economiche dei Principia di Newton," (Italian translation) in: N. Bukharin et al. *Scienza al bivio*, Bari: De Donato, 1977, 183–244.
Las Raíces Socioeconómicas de la Mecánica de Newton, (Spanish translation) in P.M. Pruna (ed.) Havana: Editorial Academia, 1985.
"Raices sociales y economicas de los *Principia de Newton*," (Spanish translation by H. Valanzano) in: A Cheroni (ed.) *Newton, el hombre y su obra*. Montevideo: Universidad Montevideo, 1988: 1–60.
"Las raíces socioeconómicas de la mecánica de Newton," (Spanish translation) in Pablo Huerga Melcón (ed.) *La ciencia en la encrucijada*, Oviedo: Fundación Gustavo Bueno, 1999, 563–630.
Οι κοινωνικές και οικονομικές ρίζες των Αρχών Φυσικής Φιλοσοφίας του Νεύτωνα (Greek translation and Postface by Dimitris Dialetis) Athens: Nefeli, 2009.

Grossmann: "The Social Foundations of the Mechanistic Philosophy and Manufacture"

Grossmann's text was translated from "Die gesellschaftlichen Grundlagen der mechanistischen Philosophie und die Manufaktur," *Zeitschrift für Sozialforschung 4* (1935) 161–231, by Gabriella Shalit and originally published in *Science in Context 1* (1987) 129–180. It appears here with minor revisions by the editors.

Grossmann: Descartes and the Social Origins of the Mechanistic Concept of the World (first publication)

There are basically six manuscripts relevant to this work: It originated as a German paper on "machinism" begun after the "Social Foundations" paper of 1935. After this first ms a typoscript was made (ca. 25 pp.), then a handwritten expanded version, then an improved copy. This is a longhand German manuscript (A) with the footnotes interrupting the flow of the text. The title is: "Der Maschinismus als Modell beim Aufbau der cartesischen Algebra oder der Science Universelle." This seems to have been the basis for the first English typescript (B) – which is presumably the work of a translator, since this was Grossmann's procedure with his economics papers (Kuhn 2007, 193); the first English version is professionally typed, not handwritten. Very many additions and corrections in longhand were made to this typescript, which was then retyped in 1946 (128 pp.). The resulting second typescript (C) *Universal Science versus Science of an Elite. Descartes' New Ideal of Science* seems to have played some kind of documentary role at the Institute for Social Research. It is a professionally typed complete monograph that even has an index of names. Grossmann however made numerous further additions and corrections to this typescript, often on added pages – largely in English but some remarks are in German and some quotations in French. He also cut up and used other copies of this typescript in various ways making further and also different additions and corrections, producing three more partial manuscripts (D, E, F – all wrongly dated 1938 in the Archives), which he renamed as "Descartes and the Social Origins of the Mechanistic Concept of the World." One of these manuscripts concentrates on the first part of the book; the other two elaborate later parts, but they all overlap. These indicate that Grossmann intended not just to edit and expand the manuscript but also to restructure it, add more material, and basically write a substantially restructured book – a book that was nowhere near completion when he died. Very many of these later additions were made in German, indicating perhaps that they were written at a time when Grossmann was already planning his return to Europe. Thus, while manuscript C does not represent Grossmann's last thoughts on the subject, it does represent a more or less completed work – one that Grossmann was not entirely satisfied with, but which, in contrast to his final thoughts in D–F, is coherent, intelligible and, with some serious editing, publishable.

The text presented here is basically C, the reworking of the translation of the German manuscript as edited, corrected and extended by Grossmann in the margins and on additional pages. Illegible additions and incomplete references were sometimes deciphered by consulting D–F. Grossmann's English, which was his fifth modern foreign language, is sometimes rough, often Germanic and seriously in need of editing. Corrections and additions written by him in German have been translated into English. On a few occasions, where the English text of C was obviously problematical, we have gone back to the early German-language draft (A) and have retranslated the garbled text. We have also edited the final manuscript as we would have done, had the author still been alive. We make no pretenses to a historical-critical edition or to a presentation of the source in print as it is in the archives; we have merely sought to present a readable text which contains as many of Grossmann's ideas as can be salvaged.

Grossmann: letter to Horkheimer and Pollock / Clark and Sarton review

The letter to Friedrich Pollock and Max Horkheimer of August 23, 1935 (from Valencia, Spain) was translated by Peter McLaughlin from the printed German version in: Horkheimer *Gesammelte Schriften* vol. 15 (Frankfurt/Main: Fischer, 1995) pp. 392–396.

The review of books by G.N. Clark and George Sarton was translated by Peter McLaughlin from the German version in *Zeitschrift für Sozialforschung 7* (1938) 233–237. A translation of part of this book review was published in *Science in Context, 1* (1987).

The Social and Economic Roots of Newton's *Principia*[1]

Boris Hessen

[151]

Introduction: Formulation of the Problem[2]

Both the work and the personality of Newton have attracted the attention of scientists of all nations and in all periods. The vast range of his scientific discoveries, the significance of his work to all subsequent developments in physics and technology, and the remarkable accuracy of his laws justifiably arouse special respect for his genius.

What placed Newton at the turning point of the development of science and enabled him to chart new paths forward?

Where is the source of Newton's creative genius? What determined the content and the direction of his work?

These are the questions that inevitably confront the researcher who aims not merely to gather materials relating to Newton but to penetrate into the very essence of his creative work. As Pope said in a well-known couplet:

> Nature and nature's laws lay hid in night;
> God said 'Let Newton be!' and all was light.

Our new culture, stated the famous British mathematician Professor Whitehead in a recent book *Science and Civilization,* owes its development to the fact that Newton was born in the same year that Galileo died. Just think what the course of human history would have been if these two men had not appeared in the world.[3]

The well-known English historian of science F. S. Marvin, a member of the presidium of this International Congress, concurred with this view in his article: "The Significance of the 17th Century," which appeared a couple of months ago in *Nature.*[4]

Thus the phenomenon of Newton is attributed to the benevolence of divine providence, and the mighty impetus that his work gave to the development of science and technology is attributed to his personal genius.

In this paper we shall present a radically different view of Newton and his work. [152]

We aim here to apply the method of dialectical materialism and Marx's conception of the historical process to an analysis of the genesis and development

G. Freudenthal, P. McLaughlin (eds.), *The Social and Economic Roots of the Scientific Revolution,* Boston Studies in the Philosophy of Science 278,
DOI 10.1007/978-1-4020-9604-4_2,

of Newton's work within the context of the period in which he lived and worked.

We shall give a brief exposition of Marx's basic assumptions that will be the guiding premises of our paper.

Marx expounded his theory of the historical process in the preface to the *Critique of Political Economy* and in the *German Ideology*. We shall attempt to convey the essence of Marx's views as far as possible in his own words.

Society exists and develops as an organic whole. In order to ensure its existence and development society must develop production.[5] In the social production of their life men enter into definite relations that are independent of their will. These relations correspond to a definite stage of development of their material productive forces.

The sum total of these relations of production constitutes the economic structure of society, the real foundation, on which rises a legal and political superstructure and to which correspond definite forms of social consciousness.

The mode of production of material life conditions the social, political and intellectual life process of society.

It is not the consciousness of men that determines their being, but, on the contrary their social being determines their consciousness. At a certain stage of their development the material productive forces of society come into conflict with the existing relations of production, or—what is but a legal expression for the same thing—with the property relations within which they have been at work hitherto.

From forms of development of the productive forces, these relations turn into their fetters. Then begins an age of social revolution. With the change of the economic foundation the entire immense superstructure is also transformed.

The prevailing consciousness during these periods must be explained by the contradictions of material life, by the existing conflict between the productive forces and the relations of production. [153]

Lenin remarked that this materialist conception of history removed the two chief defects in earlier historical theories.

Earlier historical theories examined only the *ideological* motives in the historical activities of human beings. Consequently, they were unable to reveal the true origins of those motives and regarded history as being driven by the ideological impulses of individual human beings, thereby blocking the way to recognition of the objective laws of the historical process. "Opinion governed the world." The course of history depended on the talents and the personal impulses of man. The individual created history.

Professor Whitehead's above-quoted view of Newton is a typical example of this limited understanding of the historical process.

The second defect that Marx's theory removes is the view that the subject of history is not the mass of the people, but individuals of genius. The most obvious representative of this view is Carlyle, for whom history was the story of great men.

According to Carlyle, the achievements of history are only the realisation of the thoughts of great men. The genius of the heroes is not the product of material conditions, but on the contrary the creative power of genius transforms those conditions, since it has no need for any external material factors.

In contradistinction to this view Marx examined *the movement of the masses* who make history and studied the social conditions of the life of the masses and the changes in those conditions.

Marxism, as Lenin emphasized, pointed the way to an all-embracing and comprehensive study of the process of the rise, development and decline of social systems. It explains this process by examining the totality of opposing tendencies, by reducing them to the precisely definable conditions of life and production of the various classes.[6]

Marxism eliminates subjectivism and arbitrariness in the selection of the various "dominant" ideas or in their interpretation, revealing that, without exception, all ideas stem from the condition of the material productive forces.

In class society the ruling class subjects the productive forces to itself and thus, becoming the dominant material force, subjects all other classes to its interests.[7]

The ideas of the ruling class are, in every historical age, the ruling ideas, and the ruling class distinguishes itself from all its predecessors by presenting its ideas as eternal truths. It wishes to reign eternally and bases the inviolability of its rule on [154] the eternal nature of its ideas.

In a class society the dominant ideas are separated from the relations of production, thus creating the notion that the material basis is determined by ideas.

Practice should not be explained by ideas, but on the contrary, ideological structures should be explained by material practice.

Only the proletariat, which aims to create a classless society, is free from a limited understanding of the historical process and produces a true, genuine history of nature and society.

The period during which Newton was at the peak of his activity coincided with the period of the English Civil War and Commonwealth.

A Marxist analysis of Newton's activity on the basis of the foregoing assumptions will consist first and foremost in understanding Newton, his work and his world outlook as a product of that period. [155]

The Economics, Technology and Physics of Newton's Era[8]

The segment of world history that has come to be known as medieval and modern history was first and foremost characterized by the rule of private property.

All the social and economic formations of this period feature this basic characteristic.

Consequently Marx regarded this period of human history as the history of the development of forms of private property, and distinguished three subsidiary periods within the larger era.

The first period is that of feudalism. The second period begins with the disintegration of the feudal system and is characterized by the emergence and development of merchant capital and manufacture.

The third period in the history of the development of private property is that of industrial capitalism. It gave birth to large-scale industry, the harnessing of the forces of nature to the goals of industry, mechanisation and the most detailed division of labour.

The dazzling flowering of natural science during the 16th and 17th centuries resulted from the disintegration of the feudal economy, the development of merchant capital, of international maritime relations and of heavy (mining and metallurgical) industry.

During the first centuries of the mediaeval economy, not only the feudal but also to a considerable extent the urban economy were based upon personal consumption.[9]

Production for the purpose of exchange was only beginning to emerge. Hence the limited nature of exchange and of the market, the insular and stagnant nature of the forms of production, the isolation of the various localities from the outside world, the purely local connections among producers: the feudal estates and the commune in the country, the guild in the towns. [156]

In the towns, capital was in kind, directly bound up with the labour of the owner and inseparable from him. This was corporation capital.[10]

In the mediaeval towns there was no strict division of labour among the various guilds, nor within those crafts among the individual workers.

The lack of intercourse, the sparse population and the limited extent of consumption hindered any further growth in the division of labour.

The next step in the division of labour was the separation of production from the form of exchange and the emergence of a special merchant class.

The boundaries of commerce were widened. Towns formed relations with each other. There arose a need for publicly safe roads, and a demand for good means of communication and transport.

The emerging links between towns led to the division of production among them. Each developed a special branch of production.

Thus the disintegration of the feudal economy led to the second period in the history of the development of private property, to the rule of merchant capital and manufacture.

The emergence of manufacture was the immediate consequence of the division of labour among the various towns.

Manufacture led to a change in relations between the worker and the employer. A monetary relation emerged between the capitalist and the worker.

Finally, the patriarchal relations between master and journeymen were destroyed.

Trade and manufacture created the haute bourgeoisie. The petty bourgeoisie was concentrated in guilds and, in the towns, was compelled to submit to the hegemony of the merchants and the manufacturers.

This period began in the mid-17th century and continued to the end of the 18th.

This is a schematic outline of the course of development from feudalism to merchant capital and manufacture.

Newton's activities fall within the second period in the history of the development of private property.

Consequently, we shall first investigate the historical demands presented by the emergence and development of merchant capital. [157]

Then we shall consider what technical problems were posed by the newly developing economy and what complex of physical problems and knowledge, essential for solving these technical problems, they generated.

We shall focus on three prominent spheres that were of decisive importance to the social and economic system we are investigating: communications, industry and war.

Communications

Trade had already reached a considerable level of development by the beginning of the Middle Ages. Nevertheless, land communications were in a very poor state. Roads were so narrow that not even two horses could pass. The ideal road was one on which three horses could travel side by side, where, in the expression of the time (the 14th century) "A bride could ride by without touching the funeral cart."

It is no wonder that goods were carried in packs. Road construction was almost non-existent. The insular nature of the feudal economy gave no impetus whatever to developing road construction. On the contrary, both the feudal barons and the inhabitants of places through which commercial transport passed were interested in maintaining the poor condition of the roads, because they had the *Grundrührrecht*,[11] the right to ownership of anything that fell on to their land from the cart or pack.

The speed of land transport in the 14th century did not exceed five to seven miles a day.

Naturally, maritime and water transport played a large role, both because of the greater load-capacity of ships and also because of their greater speed: the largest two-wheeled carts drawn by ten to twelve oxen hardly carried two tons of goods, whereas an average-sized ship could carry up to 600 tons. During the 14th century the journey from Constantinople to Venice took three times as long by land as by sea.

Nevertheless even the maritime transport of this period was very inadequate: as reliable methods of establishing a ship's position in the open sea had not yet been invented, they sailed close to the shores, which made their journey much slower.

Although the mariner's compass was first mentioned in the Arab book, *The Merchant's Treasury,* in 1242, it did not come into universal use until the second half of the 16th century. Geographical maritime maps made their appearance at about the same time.

But compass and charts can be rationally utilized only when the ship's position can be correctly established, i.e., when latitude and longitude can be determined. [158]

The development of merchant capital destroyed the isolation of the medieval town and the village commune, immensely extended the geographical horizon, and considerably accelerated the pace of life. It needed comfortable means of transport, improved means of communication, a more accurate measurement of time, especially in light of the ever accelerating pace of exchange, and precise tools of calculation and measurement.

Particular attention was directed to water transport: to maritime transport as a link between various countries, and to river transport as an internal link.

The development of river transport was also assisted by the fact that since antiquity waterways had been the most convenient and most investigated means of communications, and the natural growth of the towns was connected to the system of river communications. River transport was three times as cheap as haulage transport.

The construction of canals also developed as an additional means of internal transport and as a means of connecting maritime transport with the internal river system.

Thus the development of merchant capital confronted water transport with the following technical problems

In the Realm of Water Transport[12]

1. Increasing the tonnage capacity of vessels and their speed.
2. Improving the vessels' buoyancy: greater stability, sea-worthiness, a reduced tendency to rock, greater navigability and ease of manoeuvring, which was especially important for war-vessels.
3. Convenient and reliable means of determining position at sea: means of determining latitude and longitude, magnetic deviation, times of tides.
4. Improving internal waterways and linking them to the sea; building canals and locks.

Let us consider what physical prerequisites are necessary in order to solve these technical problems.

1. In order to increase the tonnage capacity of vessels it is necessary to know the fundamental laws governing bodies floating in fluids, since in order to estimate tonnage capacity it is necessary to know the method of estimating a vessel's water displacement. These are problems of hydrostatics.
2. In order to improve the buoyancy of a vessel it is necessary to know the laws governing the motion of bodies in fluids, which is an aspect of the laws governing the motion of bodies in a resistant medium—one of the basic problems of hydrodynamics. [159]

   The problem of a vessel's roll stability is one of the basic problems of the mechanics of mass points.
3. The problem of determining latitude consists in the observation of celestial bodies, and its solution depends on the existence of optical instruments and a knowledge of the chart of the celestial bodies and their motion—of celestial mechanics.

The problem of determining longitude can be most conveniently and simply solved with the aid of a chronometer. But as the chronometer was invented only in the 1730s, after the work of Huygens, longitude was determined by measuring the distance between the moon and the fixed stars.

This method, proposed in 1498 by Amerigo Vespucci, demands precise knowledge of the anomalies in the motion of the moon and constitutes one of the most complex problems of celestial mechanics. Determining the times of the tides according to locality and the position of the moon demands a knowledge of the theory of gravitation, which is also a problem of mechanics.

The importance of this problem is evident from the fact that in 1590, long before Newton gave the world his general theory of tides on the basis of the theory of gravity, Stevin drew up tables showing the time of the tides in any given place according to the position of the moon.

4. The construction of canals and locks demands a knowledge of the basic laws of hydrostatics, the laws governing the efflux of fluids, since it is necessary to calculate water pressure and the speed of its efflux. In 1598 Stevin, while studying the problem of water pressure, had already discovered that water could exert pressure on the bottom of a vessel greater than its weight; in 1642 Castelli published a special treatise on the flow of water in various sections of canals. In 1646 Torricelli was working on the theory of the efflux of fluids.

As we see, the problems of canal and lock construction also bring us to problems of mechanics (hydrostatics and hydrodynamics).

Industry

By the end of the Middle Ages (14th and 15th centuries) the mining industry was already developing into a large-scale industry. The mining of gold and silver in connection with the development of currency was stimulated by the growth of exchange. While the discovery of America was chiefly driven by the hunger for gold—since European industry, which had developed so vigorously during the 14th and 15th centuries, and the commerce it engendered, increased the demand for means of exchange—the demand for gold also drew particular attention to the exploitation of mines and other sources of gold and silver. [160]

The vigorous development of the war industry, which had made enormous advances since the invention of firearms and the introduction of heavy artillery, was a powerful stimulus to the mining of iron and copper. By 1350 firearms had become the customary weapon of the armies of eastern, southern and central Europe.

In the 15th century heavy artillery had reached a fairly high level of development. In the 16th and 17th centuries the war industry made enormous demands upon the metallurgical industry. In the months of March and April 1652 alone, Cromwell required 335 cannon, and in December a further 1,500 guns of a total weight of 2,230 tons, with 117,000 shells as well as 5,000 hand bombs.

It is therefore clear why the problem of how to exploit mines in the most effective way became a matter of prime importance.

The main problem is posed by the depth of the mines. The deeper the mines, the more difficult and dangerous it becomes to work in them.

A variety of devices are necessary for pumping out water, ventilating the mines, and raising the ore to the surface. It is also necessary to know how to construct mines correctly and find one's bearings in them.

By the beginning of the 16th century mining had already reached a considerable level of development. Agricola left a detailed encyclopedia of mining which shows how much technical equipment had come to be used in mining.

In order to extract the ore and water, pumps and hoists (windlasses and screws) were constructed; the energy of animals, the wind and falling water were all utilized. Ventilation pipes and blast-engines were constructed.[13] There was an entire system of pumps, since as the mines became deeper, water drainage became one of the most important technical problems.

In his book Agricola describes three kinds of water-drainage devices, seven kinds of pumps, and six kinds of installations for extracting water by means of a ladling or bucketing device, altogether around sixteen kinds of water-drainage machines.

The development of mining demanded vast equipment for processing the ore. [161] Here we encounter smelting furnaces, stamping mills, and machinery for separating metals.

By the 16th century the mining industry had become a complex organism whose organization and management required considerable knowledge. Consequently the mining industry immediately developed as a large-scale industry, free of the guild system, and hence not subject to the stagnation of the guilds. It was technically the most progressive industry and engendered the most revolutionary elements of the working class during the Middle Ages, i.e., the miners.

The construction of galleries demands considerable knowledge of geometry and trigonometry. By the 15th century scientific engineers were working in the mines.

Thus the development of exchange and of the war industry confronted the mining industry with the following technical problems:

1. The raising of ore from considerable depths.
2. Ventilation equipment in the mines.
3. Pumping water from the mines and drainage devices—the problem of the pump.
4. The transition from the crude, damp-blast method of production predominant until the 15th century to blast-furnace production, which, like ventilation, poses the problem of air-blast equipment.
5. Ventilation by means of air draught and special blast-engines.[14]
6. The processing of the ores with the aid of rolling and cutting machinery.

Let us consider the physical problems underlying these technical tasks.

1. The raising of ore and the problem of constructing hoists is a matter of designing windlasses and blocks, i.e., a variety of so-called simple mechanical machines.
2. Ventilation equipment demands a study of draughts, i.e., it is a matter of aerostatics, which in turn is part of the problem of statics.

3. The pumping of water from the mines and the construction of pumps, especially piston pumps, requires considerable research in the field of hydro- and aerostatics.
   Consequently Torricelli, Guericke and Pascal studied the problems of raising liquids in tubes and of atmospheric pressure.
4. The transition to blast-furnace production immediately gave rise to the phenomenon of large blast-furnaces with auxiliary buildings, water-wheels, bellows, rolling machines and heavy hammers. [162]
   These problems—the problem of hydrostatics and dynamics posed by the construction of water-wheels, the problem of air-bellows—like the problem of blast-engines for ventilation purposes, also require an investigation of the motion of air and air compression.
5. As in the case of other equipment, the construction of presses and heavy hammers driven by the power of falling water (or animals) requires a complex design of cogwheels and a transmission mechanism, which is also essentially a problem of mechanics. The science of friction and the mathematical arrangement of cogged transmission wheels developed in the mill.

Thus, if we disregard the great demands that the mining and metallurgical industries of this period made on chemistry, all these physical problems did not go beyond the bounds of mechanics.

War and War Industry

The history of war, Marx wrote to Engels in 1857,[15] demonstrates ever more graphically the correctness of our views on the connection between the productive forces and social relations.

Altogether the army is very important to economic development. It was in warfare that the guild system of corporations of artisans first originated. Here too was the first use of machinery on a large scale.

Even the special value of metals and their use as money at the beginning of the development of monetary circulation would seem to have been based on their significance in war.

Similarly, the division of labour within various branches of industry was first put into practice in the army. This, in condensed form, is the entire history of the bourgeois system.

From the time that gunpowder (which had been in use in China even before our era) became known in Europe, there was a rapid growth in firearms.

Heavy artillery first appeared in 1280, during the siege of Cordova by the Arabs. In the 14th century firearms passed from the Arabs to the Spaniards. In 1308 Ferdinand IV took Gibraltar with the aid of cannon.

Artillery spread from the Spaniards to other nations. By the mid-14th century firearms were in use in all countries of eastern, southern and central Europe.[16]

The first heavy guns were extremely unwieldy and could only be transported in sections. Even weapons of small calibre were very heavy, since no ratios had been established between the weight of the weapon and the projectile, or between the weight of the projectile and the charge.

Nevertheless firearms were used not only in sieges, but also on war-vessels. In 1386 the English captured two war-vessels armed with cannon.

[163] A considerable improvement in artillery took place during the 15th century. Stone balls were replaced by iron. Cannon were cast entirely from iron and bronze. Gun carriages and transportation were improved. The rate of fire was increased. The success of Charles VIII in Italy can be attributed precisely to this factor.

In the battle of Fornova the French fired more shots in one hour than the Italians fired in a day.

Machiavelli wrote his *Art of War* specially in order to demonstrate ways of resisting the effects of artillery by the skillful disposition of infantry and cavalry.

But of course the Italians were not satisfied with this alone, and they developed their own war industry. By Galileo's time the Arsenal at Venice[17] had attained a considerable level of development.

Francis I formed artillery into a separate unit and his artillery shattered the hitherto undefeated Swiss lancers.

The first theoretical works on ballistics and artillery date from the 16th century. In 1537 Tartaglia endeavoured to determine the trajectory of the flight of a projectile and established that the angle of 45 degrees allows the maximum flight distance. He also drew up firing tables for directing aim.

Vannoccio Biringuccio studied the process of casting and in 1540 he introduced considerable improvements in the production of weapons.

Hartmann invented a scale of calibres, by means of which each section of the gun could be measured in relation to the aperture, which set a specific standard in the manufacture of guns and paved the way to the introduction of firmly established theoretical principles and empirical rules of firing.

In 1690 the first artillery school was opened in France.

In 1697 Saint-Rémy published the first complete artillery primer.

By the end of the 17th century artillery in all countries had lost its mediaeval, guild character and was included as a component part of the army.

The variety of calibres and models, the unreliability of empirical rules of firing, and the almost total lack of firmly established ballistic principles had already become absolutely intolerable by the mid-17th century.[18]

Consequently, many experiments began to be carried out on the correlation between calibre and charge, the relation of calibre to weight and to the length of the barrel and on the phenomenon of recoil.

The science of ballistics advanced in tandem with the work of the most prominent physicists.

Galileo gave the world the theory of the parabolic trajectory of a projectile; Torricelli, Newton, Bernoulli and Euler studied the flight of a projectile through [164] the air, air resistance and the causes of deviation of the projectile.[19]

The development of artillery led in turn to a revolution in the construction of fortifications and fortresses, and this made enormous demands on the art of engineering.

The new form of fortifications (earthworks, fortresses) almost paralysed the effects of artillery in the mid-17th century, which in turn gave a powerful stimulus to its further development.

The development of the art of war posed the following technical problems.

Internal Ballistics

1. Study and improvement of the processes occurring in a firearm when fired.
2. The stability of the firearm combined with minimum weight.
3. A device for comfortable and accurate aim.

External Ballistics

4. The trajectory of a projectile through a vacuum.
5. The trajectory of a projectile through the air.
6. The dependence of air resistance upon the speed of the projectile.
7. The deviation of a projectile from its trajectory.

The physical bases of these problems are as follows:

1. In order to study the processes occurring in the firearm, it is necessary to study the compression and expansion of gases—which is basically a problem of mechanics—as well as the phenomena of recoil (the law of action and reaction).
2. The stability of a firearm poses the problem of studying the resistance of materials and testing their durability. This problem, which also has great importance for the art of construction, was resolved at this particular stage of development by purely mechanical means. Galileo devoted considerable attention to the problem in his *Mathematical Demonstrations*.
3. The problem of a projectile's trajectory through a vacuum consists in resolving the problem of the action of the force of gravity upon the free fall of a body and the superposition of its forward motion and its free fall. It is therefore not surprising that Galileo devoted much attention to the problem of the free fall of bodies. The extent to which his work was connected with the interests of artillery and ballistics can be judged if only from the fact that he began his *Mathematical Demonstrations* with an address to the Venetians praising the activity of the arsenal at Venice and pointing out that the work of that arsenal provided a wealth of material for scientific study.
4. The flight of a projectile through the air is one aspect of the problem of the motion of bodies through a resistant medium and of the dependence of that [165] resistance upon the speed of motion.
5. The deviation of the projectile from the estimated trajectory can be caused by a change in its initial speed, a change in the density of the atmosphere, or by the influence of the rotation of the earth. All these are purely mechanical problems.

6. Accurate tables for aiming at targets can be drawn up provided the problem of external ballistics is resolved and the general theory of a projectile's trajectory through a resistant medium is established.

Hence, if we exclude the actual process of producing the firearm and the projectile, which is a problem of metallurgy, the chief problems posed by the artillery of this period were problems of mechanics.

The Physical Themes of the Era and the Contents of the *Principia*[20]

Now let us systematically consider the physical problems presented by the development of transport, industry and mining.

First and foremost we should note that they are all *purely mechanical problems.*

We shall analyse, albeit in very general terms, the basic themes of research in physics during the period in which merchant capital was becoming the predominant economic force and manufacture was emerging, i.e., the period from the beginning of the 16th to the second half of the 17th century.

We do not include Newton's works on physics, since they will be analysed separately. By presenting the principal physical themes, we will be able to determine the problems that most interested physics in the period immediately preceding Newton and contemporary with him.

1. *The problem of simple machines, inclined planes and general problems of statics* were studied by: Leonardo da Vinci (end of 15th century); Cardano (mid-16th century); Guidobaldo (1577); Stevin (1587); Galileo (1589–1609).
2. *The free fall of bodies and the trajectory of a projectile* were studied by: Tartaglia (1530s); Benedetti (1587); Piccolomini (1597); Galileo (1589–1609); Riccioli (1651); Gassendi (1649); Accademia del Cimento.
3. *The laws of hydro- and aerostatics, and atmospheric pressure. The pump, the motion of bodies through a resistant medium*: Stevin, the engineer and inspector of the land and water installations of Holland (at the end of the 16th and beginning of the 17th centuries); Galileo, Torricelli (first quarter of the 17th century); Pascal (1647–1653); Guericke, military engineer in the army of Gustavus Adolphus, the constructor of bridges and canals (1650–1663); Robert [166] Boyle (1670s); Aaccademia del Cimento (1657–1667).
4. *Problems of celestial mechanics, the theory of tides.* Kepler (1609); Galileo (1609–1616); Gassendi (1647); Wren (1660s); Halley, Robert Hooke (1670s).

The problems enumerated above cover almost all the subjects of physics at that time.

If we compare these principal themes with the physical problems that emerged from our analysis of the technical demands presented by communications, industry and war, it becomes quite clear that these physical problems were mainly defined by those demands.

In fact the first group of problems constitutes the physical problems relating to lifting devices and transmission mechanisms that were important to the mining industry and the art of building.

The second group of problems is of great importance for artillery and constitutes the main physical problems relating to ballistics.

The third group of problems is of great importance to the problems of the drainage and ventilation of mines, the smelting of ore, canals and lock construction, internal ballistics and designing the shape of ships.

The fourth group is of enormous importance to navigation.

All these are fundamentally mechanical problems. This of course does not mean that in this period other aspects of the motion of matter were not studied. Optics also began to develop at this time, and the first observations on static electricity and magnetism were made.* Nevertheless both by their nature and by their relative weight these problems were of only secondary significance and lagged far behind mechanics in their level of study and mathematical development (with the exception of certain laws of geometrical optics, which were of considerable importance in the construction of optical instruments).

As for optics, it received its main impetus from the technical problems that were of importance, first and foremost, to marine navigation.†

We have compared the main technical and physical problems of the era with the topics studied by the leading physicists in the period we are investigating, and we [167] came to the conclusion that these topics were primarily determined by the economic and technical problems that the rising bourgeoisie placed on the agenda.

The development of the productive forces in the age of merchant capital presented science with a number of practical tasks and urgently demanded their solution.

Official science, based in the mediaeval universities, not only made no attempt to solve these problems, but actively opposed the development of the natural sciences.

In the 15th to the 17th centuries the universities were the scientific centres of feudalism. They were not only the bearers of feudal traditions, but the active defenders of those traditions.

In 1655, during the struggle of the master craftsmen with the journeymen's associations the Sorbonne actively defended the masters and the guild system, supporting them with "proofs from science and holy writ."

The entire system of pedagogy in the mediaeval universities constituted a closed system of scholasticism. There was no place for natural science in these universities. In Paris, in 1355, it was permitted to teach Euclidean geometry only on holidays.

The main "natural science" manuals were Aristotle's books, from which all the vital content had been removed. Even medicine was taught as a branch of logic. Nobody was allowed to study medicine unless he had studied logic for three years.

---

* Investigations into magnetism were directly influenced by the study of the deviation of the compass in the world's magnetic field, which had first been encountered during long-distance sea voyages. Gilbert had already given much attention to problems of the earth's magnetism. [Hessen's note]

† In this period optics developed from studying the problem of the telescope. [Hessen's note]

True, admittance to the medical examinations also involved a non-logical argument (evidence that the student was a legitimate child), but obviously this non-logical question alone was hardly sufficient for a knowledge of medicine, and the famous surgeon Arnold Villeneuve of Montpellier complained that even the professors in the medical faculty were unable, not only to cure the most ordinary illnesses, but even to apply a leech.

The feudal universities struggled against the new science just as fiercely as the obsolete feudal relations struggled against the new progressive modes of production.

For them, whatever could not to be found in Aristotle simply did not exist.

When Kircher (early 17th century) suggested to a certain provincial Jesuit professor that he should look at the newly discovered sunspots through a telescope, the [168] latter replied: "It is useless, my son. I have read Aristotle through twice and have not found anything there about spots on the sun. There are no spots on the sun. They are caused either by the imperfections of your telescope or by the defects of your own eyes."[21]

When Galileo invented the telescope and discovered the phases of Venus, the scholastic university philosophers did not even want to hear about these new facts, whereas the trading companies requested his telescope, which was superior to those made in Holland.

> "I think, my Kepler," Galileo wrote bitterly on August 19, 1610,[22] "we should laugh at the extraordinary stupidity of the multitude. What do you say to the foremost philosophers of the school here, who, though I have offered a thousand times of my own accord to show my studies, with the obstinacy of a sated viper have never wanted to look at the planets, nor the moon, nor the telescope? Truly as some close their ears, so do they close their eyes to the light of truth. These are great matters; yet they do not surprise me. People of this sort think that philosophy is some kind of book ... and that truths are to be sought, not in the world or in nature, but ... in the comparison of texts."

When Descartes resolutely came out against the Aristotelian physics of occult qualities and against the university scholasticism, he met with furious opposition from Rome and the Sorbonne.

In 1671 the theologians and physicians of the University of Paris sought a government resolution condemning Descartes' teaching.

In a biting satire Boileau ridiculed these demands of the learned scholastics. This remarkable document, which gives an excellent description of the state of affairs in the mediaeval universities, is appended in its entirety.[23]

Even in the second half of the 18th century the Jesuit professors in France were not prepared to accept Copernicus's theories. In 1760, in a Latin edition of Newton's *Principia*, Le Seur and Jacquier thought it necessary to add the following note: "In his third book Newton applies the hypothesis of the movement of the earth. The author's assumptions cannot be explained except on the basis of this hypothesis. Thus we are compelled to act in another's name. But we ourselves openly declare that we accept the decisions published by the heads of the church against the movement of the earth."[24]

The universities produced almost exclusively ecclesiastics and jurists.

The church was the international centre of feudalism and itself a large feudal proprietor, possessing no less than one third of the land in Catholic countries.

The mediaeval universities were a powerful weapon of church hegemony. [169]

Meanwhile, the technical problems that we have outlined above demanded enormous technical knowledge, and extensive mathematical and physical studies.

The end of the Middle Ages (mid-15th century) was marked by great advances in the development of the industry created by the mediaeval burghers.

Production, which was increasingly on a mass scale, was improved and diversified; commercial relations became more developed.

If after the dark night of the Middle Ages science again began to develop at a miraculous rate, we owe this to the development of industry (Engels).[25]

Since the time of the Crusades industry had developed enormously and had a mass of new achievements to its credit (metallurgy, mining, the war industry, dyeing), which supplied not only fresh material for observation but also new means of experimentation and enabled the construction of new instruments.

It can be said that from that time systematic experimental science became possible.

Furthermore, the great geographical discoveries, which in the last resort were also determined by the interests of production, supplied an enormous and previously inaccessible mass of material in the realm of physics (magnetic deviation), astronomy, meteorology and botany.

Finally, this period saw the appearance of a mighty instrument for distributing knowledge: the printing press.

The construction of canals, locks and ships, the construction of mines and galleries, their ventilation and drainage, the design and construction of firearms and fortresses, the problems of ballistics, the production and design of instruments for navigation, the development of methods for establishing the position of ships, all demanded people of a totally different type from those being produced by the universities at that time.

In the third quarter of the 16th century, Johann Mathesius already specified that the minimum of knowledge required by a mine-surveyor was proficiency in the method of triangulation and Euclidean geometry, the ability to use a compass, which was essential for constructing galleries, the ability to calculate the correct layout of the mine, and a knowledge of the construction of pumping and ventilation equipment.

He pointed out that engineers with a theoretical education were needed in order to construct galleries and work the mines, since this work was far beyond the powers of an ordinary, uneducated miner. [170]

Obviously, none of this could be learned in the universities of the time. The new science emerged in a struggle with the universities, as an extra-university science.

The struggle between the university and the extra-university science that served the needs of the rising bourgeoisie was a reflection in the ideological realm of the class struggle between the bourgeoisie and feudalism.

Step by step, science flourished along with the bourgeoisie. In order to develop its industry, the bourgeoisie required a science that would investigate the properties of material bodies and the manifestations of the forces of nature.

Hitherto science had been the humble servant of the church and had not been allowed to go beyond the limits set by the church.

The bourgeoisie had need of science, and science rebelled against the church together with the bourgeoisie (Engels).[26]

Thus the bourgeoisie came into conflict with the feudal church.

In addition to the professional schools (schools for mining engineers and for training artillery officers), the scientific societies outside the universities were the centers of the new science, the new natural sciences.

In the 1650s the famous Florentine Accademia del Cimento was founded, with the aim of studying nature by means of experiment. It included among its members scientists such as Borelli and Viviani.

The Academy was the intellectual heir of Galileo and Torricelli and continued their work. Its motto was *Provare e riprovare* (verify and verify again through experiment).[27]

In 1645 a circle of natural scientists was formed in London; they gathered weekly to discuss scientific problems and new discoveries.

It was from this gathering that the Royal Society developed in 1661. The Royal Society brought together the leading and most eminent scientists in England, and in opposition to the university scholasticism adopted as its motto: "Nullius in verba" (verify nothing on the basis of words).[28]

Robert Boyle, Brouncker, Brewster,[29] Wren, Halley, and Robert Hooke played an active part in the society.

One of its most outstanding members was Newton.

We see that the rising bourgeoisie brought natural science into its service, into the service of the developing productive forces.

Being at that time the most progressive class, it demanded the most progressive science. The English Revolution gave a mighty stimulus to the development of the productive forces. It became necessary not merely to resolve empirically particular [171] problems, but to establish a synthetic summary and solid theoretical basis for solving, by general methods, all the physical problems raised by the development of the new technology.

And since (as we have already demonstrated) the basic problems were mechanical ones,[‡] this encyclopedic survey of the physical problems amounted to creating a consistent structure of theoretical mechanics which would supply general methods for solving the problems of celestial and terrestrial mechanics.

It fell to Newton to carry out this work. The very title of his most important work—*Mathematical Principles of Natural Philosophy* (1687)—indicates that Newton set himself precisely this work of synthesis.

[‡] Optics also began to develop during this period, but the main research in optics was subordinated to the interests of maritime navigation and to astronomy. It is important to note that Newton came to study the spectrum by way of the phenomenon of the chromatic aberration in the telescope. [Hessen's note]

In his introduction to the *Principia* Newton pointed out that applied mechanics and teachings on simple machines had already been elaborated and that his task consisted not in considering "arts" and solving particular problems, but in providing a teaching about natural powers, the mathematical principles of philosophy.

Newton's *Principia* are expounded in abstract mathematical language and it would be futile to seek in them an exposition by Newton himself of the connection between the problems that he sets and solves and the technical demands from which they arose.

Just as the geometrical method of exposition was not the method Newton used to make his discoveries, but, in his opinion, was to serve as a worthy vestment for the solutions found by other means, so a work treating of "Natural Philosophy" should not contain references to the "low" source of its inspiration.

We shall attempt to show that the "terrestrial core" of the *Principia* consists precisely of the technical problems that we have analysed above and which fundamentally determined the themes of physical research in that period.

Despite the abstract mathematical character of exposition adopted in the *Principia*, not only was Newton by no means a learned scholastic divorced from life, but he firmly stood at the centre of the physical and technical problems and interests of his time.

Newton's well-known letter to Francis Aston gives a very clear notion of his broad technical interests. The letter was written in 1669 after he had received his professorship, just as he was finishing the first outline of his theory of gravity.[30] [172]

Newton's young friend, Aston, was about to tour various countries in Europe, and he asked Newton to instruct him how to utilize his journey most rationally and what was especially worthy of attention and study in the European countries.

We will cite a brief summary of Newton's instructions.

To thoroughly study the mechanism of steering and the methods of navigating ships.

To survey carefully all the fortresses he should happen upon, their method of construction, their power of resistance, their defense advantages, and in general to acquaint himself with military organisation.

To study the natural resources of the country, especially the metals and minerals, and also to acquaint himself with the methods of their production and refinement.

To study the methods of obtaining metals from ores.

To find out whether it was true that in Hungary, Slovakia and Bohemia, near the town of Eila or in the Bohemian mountains not far from Silesia, there were rivers whose waters contained gold.

To find out also whether the method of obtaining gold from gold-bearing rivers by amalgamation with mercury was still a secret, or whether it was now generally known.

In Holland a glass-polishing factory had recently been established; he must go to see it.

To find out how the Dutch protected their vessels from worm damage during their voyages to India.

To find out whether clocks were of any use in determining longitude during long-distance sea voyages.

The methods of transmuting one metal into another, iron into copper, for instance, or any metal into mercury, were especially worthy of attention and study.

It was said that in Chemnitz[31] and in Hungary, where there were gold and silver mines, it was known how to transmute iron into copper by dissolving the iron in vitriol, then boiling the solution, which on cooling yielded copper.

Twenty years previously the acid possessing this noble property had been imported into England. Now it was unobtainable. It was possible that they preferred to exploit it themselves in order to transmute iron into copper rather than to sell it.

These last instructions, dealing with the problem of transmuting metals, occupy almost half of this extensive letter.

That is not surprising. Alchemistic investigations still abounded in Newton's period. The alchemists are usually imagined to be a kind of magician seeking the philosopher's stone. In reality alchemy was closely bound up with the necessities of production, and the aura of mystery surrounding the alchemists should not conceal [173] from us the real nature of their research.

The transmutation of metals constituted an important technical problem, since there were very few copper mines at that time, and warfare and the casting of cannon demanded much copper.

The developing commerce made great demands on currency that the European gold mines were unable to satisfy. Together with the drive to the east in search of gold, the quest for means of transmuting common metals into copper and gold intensified.

Since his youth Newton had always been interested in metallurgical processes, and he later successfully applied his knowledge and skills in his work at the Mint.

He carefully studied the classics of alchemy and made copious extracts from these works, which show his great interest in all kinds of metallurgical processes.

During the period immediately preceding his work at the Mint, from 1683 to 1689, he carefully studied Agricola's work on metals, and the transmutation of metals was his chief interest.

Newton, Boyle and Locke conducted extensive correspondence on the question of transmuting metals and exchanged formulae for the transmutation of ore into gold.[32]

In 1692 Boyle, who had been one of the directors of the East India Company, communicated to Newton his formula for transmuting metal into gold (see Appendix 3)

When Montague invited Newton to work at the Mint he did so not merely out of friendship, but because he highly valued Newton's knowledge of metals and metallurgy.

It is interesting and important to note that whilst a wealth of material has been preserved relating to Newton's purely scientific activities, none at all has been preserved relating to his activities in the technical sphere.

Not even the materials that would indicate Newton's activities at the Mint have been preserved, although it is well known that he did much to improve the processes of casting and stamping coins.

In connection with Newton's bicentennial, Lyman Newell, who made a special study of the question of Newton's technical activities at the Mint, asked the director of the Mint, Colonel Johnson, for materials relating to Newton's activities in the sphere of the technical processes of casting and stamping.

In his reply Colonel Johnson said that no materials whatever on this aspect of Newton's work had been preserved.

All that is known is his long memorandum to the Chancellor of the Exchequer (1717) on the bi-metallic system and the relative value of gold and silver in various countries. This memoran- dum shows that Newton's circle of interests was [174] not restricted to technical questions of coin production, but extended to economic problems of currency circulation.

Newton took an active part in, and was an adviser to, the commission for the reform of the calendar, and among his papers is a work entitled "Observations on the Reform of the Julian Calendar," in which he proposes a radical reform of the calendar.

We cite all these facts as a counterweight to the traditional representation of Newton in the literature as an Olympian standing high above all the "terrestrial" technical and economic interests of his time, and soaring only in the lofty realm of abstract thought.

It should be noted, as I have already observed, that the *Principia* certainly afford justification for such a treatment of Newton, which, however, as we see, bears absolutely no relation to reality.

If we compare the range of interests briefly outlined above, we have no difficulty in noting that it embraces almost the entire complex of problems arising from the interests of transport, commerce, industry and war during his period, which we summarized above.

Now let us turn to an analysis of the contents of Newton's *Principia* and consider their relations to the topics of research in physics in that period.

The definitions and axioms or laws of motion expound the theoretical and methodological basic principles of mechanics.

The first book contains a detailed exposition of the general laws of motion under the influence of central forces. In this way Newton provides a preliminary conclusion to the work of establishing the general principles of mechanics begun by Galileo.

Newton's laws provide a general method for solving the great majority of mechanical problems.

The second book, devoted to the problem of the motion of bodies, addresses a number of problems closely connected with the complex of problems noted above.

The first three sections of the second book are devoted to the problem of the motion of bodies in a resistant medium and to the various instances of the dependence of resistance on speed: resistance proportional to the first and second powers of speed and resistance proportional partly to the first and partly to the second power.

In the scholium to the first section Newton notes that the linear cases are of more mathematical than physical interest and proceeds to a detailed examination of cases that were observed during the actual motion of bodies in air.[33] As we have shown above when analysing the physical problems of ballistics, whose development was [175] connected with the development of heavy artillery, the problems posed and solved by Newton are of fundamental significance to external ballistics.

The fifth section of the second book is devoted to the fundamentals of hydrostatics and the problems of floating bodies. The same section considers the pressure of gases and the compression of gases and liquids under pressure.

When analysing the technical problems posed by the construction of vessels, canals, water-drainage and ventilation equipment, we saw that all the physical aspects of these problems amount to the fundamentals of hydrostatics and aerostatics.

The sixth section deals with the problem of the motion and resistance of pendulums.

The laws governing the oscillation of mathematical and physical pendulums in a vacuum were discovered by Huygens in 1673 and applied by him to the construction of pendulum clocks.

We have seen from Newton's letter to Aston the importance of pendulum clocks for determining longitude.

The use of clocks for determining longitude led Huygens to the discovery of centrifugal force and of the change in acceleration by the force of gravity.

When the pendulum clocks brought by Richer from Paris to Cayenne in 1673 slowed down, Huygens was able at once to explain the phenomenon by a change in acceleration by the force of gravity. The importance Huygens himself attached to clocks is evident from the fact that his chief work is called *On Pendulum Clocks*.

Newton's works continue this course, and just as he progressed from the mathematical case of the motion of bodies in a resistant medium with resistance proportionate to speed to the study of an actual case of motion, so he progressed from the mathematical pendulum to an actual case of a pendulum's motion in a resistant medium.

The seventh section of the second book is devoted to the problem of the motion of fluids and the resistance met by a projected body.

It considers problems of hydrodynamics, including the problem of the efflux of fluids and the flow of water through tubes. As shown above, all these problems are of cardinal importance for the construction of canals and locks and in designing drainage equipment.

The same section investigates the laws governing the fall of bodies in a resistant medium (water and air). As we know, these problems are of considerable importance [176] in determining the trajectory of projectiles, either thrown or shot.

The third book of the *Principia* is devoted to the "System of the World." It is devoted to the problems of the motion of the planets, the motion of the moon and its anomalies, the acceleration by the force of gravity and its variations in connection with the problem of the irregular movement of chronometers during sea voyages and the problem of high and low tides.

As we have noted above, until the invention of the chronometer the motion of the moon was of cardinal importance for determining longitude. Newton returned to this problem more than once (in 1691). The study of the laws of the moon's motion was of cardinal importance for compiling accurate tables for determining longitude, and the English Council of Longitude instituted a high award for work on the moon's motion.

In 1713 Parliament passed a special bill to encourage research into the determination of longitude. Newton was one of the eminent members of the Parliamentary commission.

As we have pointed out in analysing the sixth section, the study of the motion of the pendulum, begun by Huygens, was of great importance to maritime navigation. In the third book Newton therefore studies the problem of the seconds pendulum, and analyses the motion of clocks during a number of ocean expeditions: that of Halley to St. Helena in 1677, Varin and Deshayes' voyage to Martinique and Guadeloupe in 1682, Couplet's voyage to Lisbon, etc., in 1697, and a voyage to America in 1700.

When discussing the origins of high and low tides, Newton analyses the height of tides in various ports and river mouths and discusses the problem of the height of tides in relation to the location of the port and the form of the high tide.

Even this cursory survey indicates the complete overlap between the topics that concerned physics in that era, which arose out of economic and technical needs, and the contents of the *Principia,* which constitute in the full sense of the word an outline and systematic solution of the entire range of the main physical problems. And since all these problems were of a mechanical nature, it is clear that Newton's chief work was precisely laying the foundation of terrestrial and celestial mechanics. [177]

The Class Struggle During the English Revolution and Newton's Worldview

It would, however, be a gross oversimplification to derive *every problem* studied by various physicists, and *every task* they solved, directly from economics and technology.

According to the materialistic conception of history, the final determining factor in the historical process is the production and reproduction of actual life.

But this does not mean that the economic factor is the *sole* determining factor. Marx and Engels severely criticized Barth precisely for such a primitive understanding of historical materialism.[34]

The economic situation is the basis. But the development of theories and the individual work of a scientist are also affected by various superstructures, such as political forms of the class struggle and its results, the reflection of these battles in the minds of the participants—in political, juridical, and philosophical theories, religious beliefs and their subsequent development into dogmatic systems.

Therefore, when analysing the subjects addressed by physics we took the central, cardinal problems that attracted the greatest attention of scientists in that period. But the foregoing general analysis of the economic problems of the period is inadequate for understanding how Newton's work proceeded and developed and for explaining all the features of his work in physics and philosophy. We must analyse more fully Newton's period, the class struggle during the English Revolution, and the political, philosophical and religious theories as reflections of that struggle in the minds of the contemporaries.

When Europe emerged from the Middle Ages, the rising urban bourgeoisie was its revolutionary class. The position that it occupied in feudal society had become too narrow for it, and its further free development had become incompatible with the feudal system.[35]

The great struggle of the European bourgeoisie against feudalism reached its peak in three important and decisive battles:

The Reformation in Germany, with the subsequent political uprisings of Franz von Sickingen and the Great Peasant War.
The Revolution of 1649–1688 in England.
The Great French Revolution.

[178]

There is, however, a great difference between the French Revolution of 1789 and the English Revolution.

In England, feudal relations had been undermined since the Wars of the Roses. The English aristocracy at the beginning of the 17th century was of very recent origin. Out of 90 peers, sitting in Parliament in 1621, 42 had received their peerages from James I, whilst the titles of the others dated back no earlier than the 16th century.

This explains the close relationship between the upper aristocracy and the first Stuarts. This feature of the new aristocracy enabled it to compromise more easily with the bourgeoisie.

It was the urban bourgeoisie that began the English Revolution and the middle peasantry (yeomanry) brought it to a victorious end.

1688 was a compromise between the rising bourgeoisie and the former great feudal landlords. Since the times of Henry VII, the aristocracy, far from opposing the development of industry, had, on the contrary, tried to benefit from it.

The bourgeoisie was becoming an acknowledged, though modest, part of the ruling classes of England.[36]

In 1648 the bourgeoisie, together with the new aristocracy, fought against the monarchy, the feudal nobility and the dominant church.

In the Great French Revolution of 1789, the bourgeoisie, in alliance with the people, fought against the monarchy, the nobility and the dominant church.

In both revolutions the bourgeoisie was the class that actually headed the movement.

The proletariat and the non-bourgeois strata of the urban population either did not yet have different interests from those of the bourgeoisie or did not yet constitute an independently developed class or part of a class.

Therefore, wherever they opposed the bourgeoisie, as, for instance, in 1793–1794 in France, they fought only for the attainment of the interests of the bourgeoisie, even if not in the manner of the bourgeoisie.

All French terrorism was nothing but a plebeian way of dealing with the enemies of the Revolution: absolutism and feudalism. The same may be said of the Levellers movement during the English Revolution.

The revolutions of 1648 and 1789 were not English or French revolutions. They were revolutions on a European scale. They represented not merely the victory of one particular class over the old political order, but they heralded the political order of the new European society. [179]

> The bourgeoisie was victorious in these revolutions, but the *victory of the bourgeoisie* was at that time the *victory of a new social order*, the victory of bourgeois ownership over feudal ownership, of nationality over provincialism, of competition over the guild, of the division of land over primogeniture, of the rule of the landowner over the domination of the owner by the land, of enlightenment over superstition, of the family over the family name, of industry over heroic idleness, of bourgeois law over medieval privileges (Marx)[37]

The English Revolution of 1649–1688 was a bourgeois revolution.

It brought into power the "capitalist and landlord profiteers."[38] The Restoration did not mean the reestablishment of the feudal system. On the contrary, in the Restoration the owners of land destroyed the feudal system of land relations. In essence, Cromwell was already doing the work of the rising bourgeoisie. The pauperization of the population, as the precondition for the emergence of a free proletariat, intensified after the revolution. It was in this change of the ruling class that the true meaning of the revolution is to be found. The emerging new socio-economic system produced a new governing class. Herein lies the main difference between Marx's interpretation and that of traditional English historians, particularly Hume and Macaulay.

Like a true Tory, Hume viewed the importance of the 1649 revolution and the Restoration, and then the revolution of 1688, only in relation to the destruction and reestablishment of order.

He severely condemned the upheaval caused by the first revolution and welcomed the Restoration as the reestablishment of order. He sympathized with the 1688 revolution as a constitutional act, although he did not consider that it had simple restored the old freedom. It had begun a new constitutional era, giving "an ascendant to popular principles."[39]

To Macaulay the revolution of 1688 was closely connected with the first revolution. But for him, the revolution of 1688 was "the glorious revolution" precisely because it was a constitutional one.

He wrote his history of 1688 immediately after the events of 1848, and his fear of the proletariat and its possible victory is evident throughout. He proudly and joyfully relates that, when depriving James II of his throne, Parliament observed all the detailed precedents and even sat in the ancient halls in robes prescribed by ritual.

Law and constitution are regarded as extra-historical essences with no connection [180] to the dominant class, a view that prevents an understanding of the true essence of the revolution.

Such was the distribution of class forces after the English Revolution. The fundamental philosophical trends in the period immediately before and after the English Revolution were:

Materialism, whose beginning can be traced to Bacon, was represented in Newton's period by Hobbes, Toland, Overton, and partly by Locke.

Idealistic sensualism, represented by Berkeley (H. More was closely associated with this view).

In addition, a fairly strong trend of moral philosophy and Deism, represented by Shaftesbury and Bolingbroke.

All these philosophical trends existed and developed in the complex conditions of the class struggle whose main features have been outlined above.

From the time of the Reformation the church became one of the chief bulwarks of the King's power. The church organisation was a component part of the state system, and the King was the head of the State Church. James I was fond of saying: "No Bishop, no King."

Every subject of the English King had to belong to the State Church. Anyone who did not belong to it was regarded as committing an offence against the state.

The struggle against the absolute power of the King was at the same time a struggle against the centralism and absolutism of the dominant State Church, and therefore the political struggle of the rising bourgeoisie against absolutism and feudalism was waged under the slogans of a struggle for religious democracy and tolerance.

The collective name "Puritans" applied to all supporters of the purification and democratization of the ruling church. However, among the Puritans a distinction should be made between the more radical Independents and the more conservative Presbyterians. These two trends formed the basis of political parties.

The supporters of the Presbyterians came mainly from among the well-to-do merchants and the urban bourgeoisie. The Independents drew their supporters from the ranks of the rural and urban democracy.

Thus both the class struggle of the bourgeoisie against absolutism and the struggle between the different tendencies within the ranks of the bourgeoisie and peasantry were waged under religious slogans.

The religious tendencies of the bourgeoisie were yet further strengthened by the development of materialistic teachings in England.

Let us briefly review the main stages of the development of materialism in this [181] period and its most important representatives.

Bacon was the father of materialism. His materialism arose out of a struggle with medieval scholasticism. He wanted to release humanity from the old traditional prejudices and to create a method for controlling the forces of nature. His teachings contain the germs of the many-sided development of this doctrine. "Matter smiles with its sensuous, poetic glamour at Man in his entirety" (Marx).[40]

In the hands of Hobbes, materialism became abstract and one-sided. Hobbes did not develop Bacon's materialism, but only systematized it.

Sensuality lost its bright colours and was transformed into the abstract sensuality of a geometrician. All the diverse forms of motion were sacrificed to mechanical motion. Geometry was proclaimed as the dominant science (Marx).[41]

The living spirit was excised from materialism, and it became misanthropic. This abstract, calculating, formally mathematical materialism could not stimulate revolutionary action.

That is why the materialistic theory of Hobbes accorded with his monarchical views and defense of absolutism. After the victory of the Revolution of 1649 Hobbes went into exile.[42]

But alongside the materialism of Hobbes there existed another materialistic movement, indissolubly bound up with the true revolutionary movement of the Levellers and headed by Richard Overton.

Richard Overton was the loyal companion-in-arms of the Levellers' leader, John Lilburne, the fiery exponent of revolutionary ideas and brilliant political pamphleteer. Unlike Hobbes, he was a practical materialist and revolutionary.

The fate of this warrior-philosopher is curious. Whilst the name of Hobbes is widely known and to be found in all the philosophy textbooks, not a single word can be found about Overton, not only in the most detailed bourgeois primer of philosophy, but even in the most complete biographical encyclopedias.[43] Thus does the bourgeoisie takes revenge on its political opponents.[44]

Richard Overton did not write much. He exchanged too often the pen for the sword and philosophy for politics. His treatise *Man Wholly Mortal*[45] was first published in 1643, and the second edition appeared in 1655. It is a blatantly materialistic and atheistic essay. Immediately after its appearance it was condemned and banned by the Presbyterian Church.

The manifesto of the Presbyterian Assembly "against unbelief and heresy" called down all curses on Richard Overton's head. "The chief representative of the terrible teaching of materialism," declares the manifesto, "that denies the immortality of the soul, is Richard Overton, the author of the book on the mortality of man."[46] [182]

We will not go into the details of Overton's teaching and its fate—a most interesting page in the history of English materialism—but will only mention one point from the publication mentioned, in which Overton formulated very clearly the basic principles of his materialistic worldview.

In criticizing the opposition between the body as inert matter and the soul as the active, creative principle, Overton writes:

> The *Form* is the *Form* of the *Matter,* and the *Matter* the *Matter* of the *Form*; neither of themselves, but each by other, and both together make *one Being.*[47]

> "All that is created, is elemental." (Overton uses the term "elements" in the sense of the ancient Greeks: water, air, earth) "But all that is created is material: for that which is not material, is nothing."[48]

Unlike in England, materialism on French soil was the theoretical banner of French republicans and terrorists, and formed the basis of the "Declaration of the Rights of Man."

In England the revolutionary materialism of Overton was the teaching of only one extreme group, while the main struggle was waged under religious slogans.

English materialism as preached by Hobbes proclaimed itself to be a philosophy fit for scientists and educated people, in contrast to religion, which was good enough for the uneducated masses, including the bourgeoisie.

Together with Hobbes, materialism, shorn of its actual revolutionary nature, came to the defense of royal power and absolutism and encouraged the repression of the people.

Even with Bolingbroke and Shaftesbury the new deistic form of materialism remained an esoteric, aristocratic doctrine.

Therefore the "misanthropic" materialism of Hobbes was hateful to the bourgeoisie both for its religious heresy and for its aristocratic connections.

Accordingly, in opposition to the materialism and deism of the aristocracy, it was those Protestant sects who had provided the cause and the fighters against the Stuarts who also provided the main fighting forces of the progressive middle class (Engels).[49]

But still more hateful to the bourgeoisie than Hobbes's esoteric materialism was Overton's materialism, under whose banner the political struggle against the bourgeoisie was waged, a materialism that turned into militant atheism and fearlessly opposed the very bases of religion. It was in these circumstances that Newton's worldview was formed.[50]

Newton was a typical representative of the rising bourgeoisie, and his worldview [183] reflected the characteristic features of his class. We may quite rightly apply to him the description that Engels applied to Locke. He too was a typical child of the class compromise of 1688.[51]

Newton was the son of a small farmer. Until his appointment as Warden of the Mint (1699), he had a very modest position in the university and in society. He also belonged to the middle class through his connections, but philosophically he was closest to Locke, Samuel Clarke and Bentley.

In his religious beliefs Newton was a Protestant and there are many grounds for assuming that he belonged to the Socinian sect.[52] He was an ardent supporter of religious democracy and tolerance. We shall see below that Newton's religious beliefs were a component part of his worldview.

In his political views Newton belonged to the Whig Party. During the second revolution Newton was a Member of Parliament for Cambridge from 1689 to 1690. When the conflict arose over the possibility of swearing allegiance to "the illegitimate ruler"—William of Orange—which even led to riots in Cambridge, Newton, who as Member of Parliament for Cambridge University had to bring the University to swear allegiance, insisted on the necessity of swearing allegiance to William of Orange and recognizing him as King.

In his letter to Dr. Covel[53] Newton adduced three arguments in favour of swearing allegiance to William of Orange, which were to remove any doubts in this regard on the part of members of the University who had previously sworn fidelity to the deposed King.

Newton's reasoning and arguments are strongly reminiscent of Macaulay's and Hume's opinions cited above.

This ideological cast of mind of Newton, who was a child of his class, explains why the latent materialistic germs of the *Principia* did not grow to become a consistent system of mechanical materialism, like the physics of Descartes, but were interwoven with his idealistic and theological views, to which, on philosophical questions, even the materialistic elements of Newton's physics were subordinated.

The significance of the *Principia* is not limited to technical matters alone. Its very name indicates that it forms a system, a worldview. Therefore it would be incorrect to confine an analysis of the contents of the *Principia* merely to determining its intrinsic connection with the economics and technology of that period, which served the needs of the rising bourgeoisie. [184]

Modern natural science owes its independence to its freedom from teleology. It recognizes only the causal study of nature.

One of the battle slogans of the Renaissance was: "True knowledge is knowledge by causes" (*vere scire per causas scire*).[54]

Bacon emphasized that the teleological view is the most dangerous of the *idola*. The true relations of things are found in mechanical causation. "Nature knows only mechanical causation, to the investigation of which all our efforts should be directed."[55]

A mechanistic conception of the universe necessarily leads to a mechanistic conception of causation. Descartes laid down the principle of causation as "an eternal truth."

Mechanistic determinism came to be generally accepted on English soil, although it was often interwoven with religious dogma (for instance of the "Christian necessarian" sect, to which Priestley belonged). This peculiar combination—so characteristic of thinkers of the English type—is also found in Newton.

The universal acceptance of the principle of mechanical causation as the sole and basic principle for the scientific investigation of nature was brought about by the mighty development of mechanics. Newton's *Principia* is a grandiose application of this principle to our planetary system. "The old teleology has gone to the Devil,"[56] but so far only in the realm of inorganic nature, of terrestrial and celestial mechanics.

The basic idea of the *Principia* consists in the conception of the motion of the planets as a result of the compounding of two forces: one directed towards the sun, and the other that of the original impulse. Newton left this original impulse to God but "forbade Him further interference in His solar system" (Engels).[57]

This unique "division of labour" in the government of the universe between God and causation was characteristic of the way in which the English philosophers interwove religious dogma with the materialistic principles of mechanical causation.

The acceptance of the modality of motion, and the rejection of moving matter as *causa sui* was inevitably bound to bring Newton to the conception of the original impulse. From this perspective, the conception of divinity in Newton's system is by no means incidental but is organically connected with his views on matter and motion, as well as with his views on space, in the development of which he was greatly influenced by Henry More.

It is at this point that the entire weakness of Newton's general philosophical conception of the universe becomes apparent. The principle of pure mechanical causation leads to the notion of the divine impulse. "The bad infinity" of the universal chain of mechanical determinism ends in the original impulse, thus opening the door to teleology.

[185]

Thus, the importance of the *Principia* is not confined to purely physical problems, but is also of great methodological interest.

In the third book of the *Principia* Newton expounds a "conception of the universe." The general scholium to the third book (third edition) proves the necessity of a divine power as the organizing, moving and directing element of the universe.

We shall not go into the question of the authorship of this scholium nor of the role of Cotes and Bentley in the publication of the *Principia*. There is extensive literature on this question, but Newton's letters quoted below undeniably prove that Newton's theological views were by no means a mere appendage to his system and were not forced upon him by Cotes or Bentley.

When Robert Boyle died in 1692[58] he left a sum yielding £50 per annum in order that every year eight lectures would be delivered in one of the churches in England proving the irrefutability of Christianity and repudiating unbelief.

Bentley, Chaplain of the Bishop of Worcester, had to deliver the first series of these lectures. He decided to devote the seventh and eighth to proving the necessity of the existence of divine providence, basing the proof on a consideration of the physical principles of the creation of the world as stated in Newton's *Principia*.

While preparing these lectures, he encountered a number of physical and philosophical difficulties, which he requested the author of the *Principia* to explain.

Newton replied in detail to Bentley's questions in four letters which provide a valuable source of information on Newton's views on the cosmological problem.

The chief difficulty Bentley asked Newton about was how to repudiate the materialistic argument, already propounded by Lucretius, that the creation of the world could be explained by purely mechanical principles, if it is assumed that matter possesses an innate property of gravity and is evenly distributed in space.

In his letters Newton pointed out in detail to Bentley how this materialistic argumentation can be overcome.

It is not difficult to see that this discussion was essentially about the theory of the evolution of the universe, and on this question Newton was resolutely opposed to the materialistic conception of evolution.

[186]

"When I wrote my Treatise about our Systeme," wrote Newton to Bentley, "I had an Eye upon such Principles as might work with considering Men, for the beliefe of a Deity."[59]

If matter were uniformly distributed in finite space, then, owing to its force of gravity, it would accumulate into one large spherical mass. But if matter were distributed in infinite space, then it could, in obedience to the force of gravity, form masses of varying magnitude.

However, in no case can it be explained by natural causes how the luminous mass—the sun—is in the centre of the system and precisely in the position in which it is placed.

> Therefore the only possible explanation lies in the acknowledgment of a divine creator of the universe, who wisely distributed the planets in such a manner that they receive the light and warmth necessary to them.

Going further into the question of whether planets could be set in motion as a consequence of natural causes, Newton pointed out to Bentley that planets could be set in motion as a consequence of the force of gravity, which was a natural cause, but could never achieve periodical rotation along closed orbits, since this requires also a tangential component. Therefore, Newton concludes, the actual paths of the planets and their formation can in no way be explained by natural causes, and hence, an enquiry into the structure of the universe leads to the presence of an intelligent divine principle.

Furthermore, when discussing the question of the stability of the solar system, Newton pointed out that such a marvellously organized system, in which the speed and mass of bodies are selected in such a manner as to maintain stable equilibrium, could only be created by divine reason.

This conception and Newton's appeal to divine reason as the supreme cause, organizer and prime moving force of the universe is by no means incidental but is the inevitable consequence of his conception of the principles of mechanics.

Newton's first law of motion attributed to matter the faculty of maintaining that state in which it exists.

As Newton considered only the mechanical form of motion, his conception of the state of matter is synonymous with the state of rest or mechanical translation.

Matter that is not acted upon by external forces can exist either in a state of rest or in a state of rectilinear, uniform motion. If a material body is at rest, then only an external force can remove it from that state.

If, however, a body is in motion, then only an external force can change that motion. [187]

Thus, motion is not an immanently inherent attribute of a body, but is a mode which matter may or may not possess.

In this sense Newton's matter is inert in the full meaning of the word. An external impulse is always necessary to set it in motion or to alter or stop this motion.

Moreover, since Newton accepts the existence of an absolute, motionless space, inertia is possible for him as is also absolute rest, and thus the existence of absolutely motionless matter, not merely motionless within the given frame of reference, is physically possible.

It is clear that such a conception of the modality of motion must inevitably lead to the introduction of an external motive force, and in Newton this role is performed by God.

It is very important to note that, in principle, not only is Newton not opposed to the idea of endowing matter with specific attributes, but, contrary to Descartes, declares density and inertia to be "innate properties of matter."

Thus, by depriving motion of the character of being an attribute of matter, and recognizing it only as a mode, Newton deliberately deprives matter precisely of that inalienable property without which the structure and origin of the world cannot be explained by natural causes.

If we contrast Newton's point of view with that of Descartes, the difference in their beliefs is immediately apparent.

> "I freely acknowledge," the latter declares in his *Principia*, "that I recognize no matter in corporeal things apart from that which the geometers call quantity, and take as the object of their demonstrations, i.e. that to which every kind of division, shape and motion is applicable. Moreover, my consideration of such matter involves absolutely nothing apart from these divisions, shapes and motions; and even with regard to these, I will admit as true only what has been deduced from indubitable common notion so evidently that it is fit to be considered as a mathematical demonstration. And since all natural phenomena can be explained in this way, as will become clear in what follows, I do not think that any other principles are either admissible or desirable in physics."[60]

In his physics, Descartes does not recognize any supernatural causes. Therefore Marx points out that the mechanistic French materialism was close to Descartes' physics, in opposition to his metaphysics.

Descartes' physics could play that role only because "within his physics, matter is the sole substance, the sole basis of being and of knowledge" (Marx).[61]

In the third part of his *Principia* Descartes also gives a picture of the development [188] of the universe. The difference in Descartes' position consists in his detailed consideration of the historical genesis of the universe and the solar system in accordance with the principles mentioned above.

It is true that Descartes also considers motion only as a mode of matter, but, in contrast to Newton, for him the supreme law is the law of conservation of quantity of motion.

Individual material bodies can acquire and lose motion, but the general quantity of motion in the universe is constant.

Descartes' law of the conservation of quantity of motion includes the assumption that motion is indestructible.

It is true that Descartes understood indestructibility in a purely quantitative sense, and this mechanical formulation of the law of conservation of motion is not accidental but arises from the fact that Descartes, like Newton, considers that all varieties of motion consist of mechanical displacement. They do not consider the problem of the transformation of one form of motion into another, and, as we shall see in the second part of this paper, there are profound reasons for this.

Engels' great merit lies in the fact that he considered the process of the motion of matter as the eternal passing of one form of material motion into another. This enables him not only to establish one of the basic theses of dialectic materialism, i.e., the inseparability of motion from matter, but also to raise the conception of the law of conservation of energy and quantity of motion to a higher level.

We shall return to this problem in the second part of this paper.[62]

Descartes, like Newton, also introduced God, but he needed God only to prove that the quantity of motion in the universe remains constant.

He not only refused to admit the conception of an external impetus imparted by God to matter, but, on the contrary, considered that constancy is one of the principal properties of the deity; hence, we cannot assume any inconstancy in his creations, since by assuming inconstancy in his creations we also assume inconstancy in him.

Thus Descartes' reason for introducing a deity is different from Newton's, but his conception also requires a deity since Descartes, too, does not maintain an entirely consistent view of the self-movement of matter.

During the period when Descartes and Newton were elaborating their conceptions of matter and motion, although somewhat later (the 1690s), we find in John Toland a far more consistent materialistic conception of the relation between matter and motion.

Criticizing the views of Spinoza, Descartes and Newton, Toland directed his chief attack against the conception of the modality of motion. [189]

> "*Motion*," contended Toland in his fourth letter to Serena, "*is essential to Matter*, that is to say, as inseparable from its Nature as Impenetrability or Extension, and that it ought to make a part of its Definition."

> "This Notion alone," Toland quite justly avers, "accounts for the same Quantity of Motion in the Universe . . . it solves all the Difficultys about the moving Force. . ."[63]

The doctrine of the self-movement of matter was fully developed in the dialectical materialism of Marx, Engels and Lenin.

The entire progress of modern physics demonstrates the truth of this doctrine. Modern physics is increasingly confirming the view that motion and matter are inseparable.

Modern physics rejects absolute rest.

The universal significance of the law of the conservation and transformation of energy increasingly corroborates Engels' conception of the correlation of forms of motion of matter. This is the only conception that provides a true understanding of the law of the transformation of energy, as it synthesises the quantitative aspect of this law with its qualitative aspect, uniting it organically with the self-movement of matter.

The way in which the law of inertia and the conception of inert matter are connected with Newton's absolute space has been indicated above.

However, Newton did not confine himself to a physical conception of space, but also provided a philosophical-theological conception.

Dialectical materialism considers space as a form of existence of matter. Space and time are the fundamental conditions for the existence of all being, and therefore space is inseparable from matter. All matter exists in space, but space exists only in matter. Empty space separated from matter is only a logical or mathematical abstraction, the fruit of our thought, to which no real thing corresponds.

According to Newton's thesis, space can be separated from matter, and absolute space preserves its absolute properties precisely because it exists independently of matter.

Material bodies exist in space, as in a kind of container. Newton's space is not a form of the existence of matter, but only a container that is independent of these bodies and exists independently.

Such is the conception of space as laid down in the *Principia*. Unfortunately, we cannot enter here into a detailed analysis of this conception. We will only note that such a conception is closely connected with the first law of motion. [190]

Having thus defined space as a container, separated from matter, Newton, naturally, asks himself what is the essence of this container.

In solving this question Newton concurs with H. More, who held the view that space is "the sensorium of God" (*sensorium dei*).

In this matter Newton also differs fundamentally from Descartes, who developed the conception of space as a physical body.

The unsatisfactory nature of Descartes' conception lies in the fact that he identified matter with a geometric object.

Whilst Newton separated space from matter, Descartes, by materialising geometrical forms, deprived matter of all properties except extension. This, of course, is also incorrect, but this conception did not lead Descartes in his physics to the same conclusions as Newton.

What is there in space devoid of matter? asks Newton in Query 28 of his *Optics*. How can it be that in Nature everything is ordered and whence arises the harmony of the world? Does it not follow from the phenomena of Nature itself that there is an incorporeal, intelligent, omnipresent being for whom space is his sensorium, through which he perceives things and comprehends their very essence?[64]

Thus we see that in this question too Newton firmly adopts the viewpoint of theological idealism.

Thus the idealistic views of Newton are not incidental, but organically bound up with his conception of the universe.

Whilst there is a distinct dualism in Descartes' physics and metaphysics, Newton, particularly in his later period, not only demonstrates no desire to separate his physical conception from his philosophical one, but, on the contrary, even attempts in his *Principia* to justify his religious-theological views.

In so far as the *Principia* for the most part arises from the demands of the economy and technology of the era and investigates the laws of the motion of material bodies, it undoubtedly contains elements of healthy materialism.

But the general defects of Newton's philosophical conception outlined above, and his narrow mechanical determinism, not only do not permit him to develop these elements, but on the contrary thrust them into the background of his general religious-theological conception of the universe. [191]

Hence, in his philosophical views, as in his religious and political views, Newton was a child of his class. He ardently opposed materialism and unbelief.

In 1692, after the death of his mother and the fire that destroyed his manuscripts, Newton was in a state of depression. At that time he wrote to Locke, with whom he corresponded on various theological matters, a caustic letter on his philosophical system.

In his letter of 16 September 1693 he asked Locke to forgive him for that letter and for having thought that Locke's system offended moral principles. Newton particularly asked forgiveness for having considered Locke a follower of Hobbes.[65] Here is confirmation of Engels' statement that Hobbes's materialism was hateful to the bourgeoisie.

Overton's materialism could not even be mentioned—after all, he was almost a Bolshevik.

When Leibniz, in his letters to the Princess of Wales, accused Newton of materialism because he considered space as the sensory of a deity, by which it perceives things, which consequently do not wholly depend on it and are not created by it, Newton fiercely protested against such accusations. Clarke's polemics with Leibniz were aimed at rehabilitating Newton from this accusation (see Appendix 5).

If in the realm of physics Newton's research remained mainly within the bounds of one form of motion, that is, mechanical displacement, and therefore contained no conception of development and transition from one form of motion to another, then the conception of development is also entirely absent from his views on nature as a whole.

Newton concludes the first period of the new natural science in the field of the inorganic world. It is a period when the available material was mastered. He achieved great results in the realm of mathematics, astronomy and mechanics, particularly thanks to the work of Kepler and Galileo, which Newton completed.

But a historical view of nature is absent. It does not exist as a system in Newton. Natural science, which is basically revolutionary, comes to a halt in face of a conservative nature that remains throughout the ages in the state in which it was created.

Not only is there no historical view of nature in Newton, but his system of mechanics does not even contain a law of the conservation of energy. At first sight, this is even harder to understand since the law of conservation of energy is a simple mathematical consequence of the central forces that Newton considered. [192]

Furthermore, Newton considers, for instance, cases of oscillation, for which Huygens, when studying the question of the centre of oscillations, had implicitly formulated the law of the conservation of energy.

It is quite obvious that it was not any lack of mathematical genius or limitation in his physical horizon that prevented Newton from enunciating this law, even in the form of an integral of living forces.[66]

In order to explain this we must consider the question from the viewpoint of our Marxist conception of the historical process. Such an analysis will enable us to link this question to the problem of the transformation of one form of motion into another, the solution to which was provided by Engels. [193]

Engels' Conception of Energy and the Lack of the Law of Conservation of Energy in Newton

In analysing the problems of the interrelations between matter and motion in Newton, we saw that Toland took the view that motion was inseparable from matter. Nevertheless, the simple recognition of the inseparability of matter from motion is still far from resolving the problem of studying the forms of motion of matter.

In nature we observe an endless variety of forms of motion of matter. If we pause to consider the forms of motion of matter studied by physics we see that here too are a number of different forms of motion (mechanical, thermal, electromagnetic).

Mechanics studies the form of motion that consists in the simple displacement of bodies in space.

Nevertheless, in addition to this form of motion there are a number of other forms of motion of matter, in which mechanical displacement recedes to the background by comparison with new specific forms of motion.

Although the laws of the motion of electrons are connected with their mechanical displacement, they do not amount to their simple change of place in space.

Consequently, in distinction to the mechanical worldview, which regards the main task of natural science as the reduction of all forms of motion of matter to the one form of mechanical displacement, dialectical materialism regards the principal task of natural science as the study of the forms of motion of matter in their interconnections, interactions and development.

Dialectical materialism understands motion as change in general. Mechanical displacement is only one, partial form of motion.

In nature, in real matter, absolutely isolated, pure forms of motion do not exist. Every real form of motion, including, of course, mechanical displacement, is always bound up with the transformation of one form of motion into another.

Hitherto physics has remained within the bounds of studying one form of motion, the mechanical form, and, as we have seen, this is what constitutes the distinctive nature of physics in Newton's period; the problem of the interrelations between this and other forms of motion could not really be posed. And when such a problem was posed there was always a tendency to hypostatise precisely this most simple and most fully studied form of motion and to present it as the sole and universal aspect [194] of motion.

Descartes and Huygens adopted this position, and Newton essentially associated himself with it.

In the introduction to the *Principia* Newton notes, "If only we could derive the rest of the phenomena of nature from mechanical principles by the same kind of reasoning." (Newton deduced the motion of the planets from these laws in the third book.) "For many things lead me to have suspicion," he continues, "that all phenomena may depend on certain forces by which the particles of bodies, by causes not yet known, either are impelled towards one another and cohere in regular figures, or are repelled from one another and recede."[67]

The development of large-scale industry made it necessary to study new forms of motion of matter and exploit them for the needs of production.

The steam engine gave enormous impetus to the development of the study of the new, thermal form of motion. The history of the development of the steam engine is of importance to us in two regards.

First we shall investigate why the problem of the steam engine emerged during the development of industrial capitalism and not during the development of merchant capital. This will explain why the steam engine became the central object of investigation only in the period immediately after Newton, even though the invention of the first steam engine dates from Newton's period (Ramsey's patent in 1630).

Thus we see that the connection between the development of thermodynamics and the steam engine is the same as that between the technical problems of Newton's period and his mechanics.

But the development of the steam engine is also of interest for another reason.

In distinction from mechanical machines (the pulley, the windlass, the lever) in which one kind of mechanical motion is converted into another kind of the same mechanical displacement, by its very essence the steam engine is based on the conversion of one form of motion (thermal) into another (mechanical).

Thus, the development of the steam engine also inevitably raises the problem of the conversion of one form of motion into another, which we do not find in Newton and which is closely bound up with the problem of energy and its conversion.

We shall first investigate the main stages in the development of the steam engine in connection with the development of the productive forces.

Marx noted that the mediaeval trade of the first merchant towns was of an intermediary character. It was founded on the barbarism of the producing nations, for whom those towns and the merchants played the role of middlemen. [195]

So long as merchant capital played the role of middleman in the exchange of products between undeveloped countries, commercial profit not only appeared as out-bargaining and cheating, but directly originated from them.

Later merchant capital exploited the difference between the prices of production of various countries. In addition, as Adam Smith emphasizes, during the first stage of its development merchant capital is chiefly a contractor and supplies the needs of the feudal landlord or the oriental despot, concentrating the main mass of surplus-product in its own hands and being relatively less interested in the prices of commodities.[68]

This explains the enormous profits of mediaeval trade. The Portuguese expedition of 1521 purchased cloves for two or three ducats and sold them in Europe at 336 ducats. The total cost of the expedition amounted to 22,000 ducats, the receipts were 150,000 ducats, the profits 130,000, i.e., about 600 per cent.

At the beginning of the 17th century the Dutch purchased cloves at 180 guldens for 625 pounds, and sold them in the Netherlands for 1,200 guldens.

The greatest percentage of profit came from those countries that were completely subject to Europeans. But even in the trade with China, which had not lost its independence, the profits reached 75 to 100 per cent.

When merchant capital possesses overwhelming hegemony everywhere, it constitutes a system of despoliation.

The high rates of profit were maintained in the 17th and the beginning of the 18th centuries.

This was because the extensive trade of the late Middle Ages and the beginning of the modern era was mainly monopolistic commerce. The British East India Company was closely connected with state power. Cromwell's navigation act strengthened the monopoly of British trade. It was from that time that the gradual decline of Holland as a naval power began and a solid basis was laid to England's maritime hegemony.

Thus, so long as the dominant form of capital was merchant capital, attention was mainly directed, not so much to improving the actual process of exchange, but to consolidating the monopolistic position and dominating the colonies.

Developing industrial capitalism immediately turned its attention to the process of production. The free competition within the country, which the British bourgeoisie achieved in 1688, immediately made it necessary to consider the question of costs of production.

As Marx observed, large-scale industry universalized competition and made protective tariffs a mere palliative.

It was necessary not only to produce sufficient quantities of high-quality commodities, but to produce them as cheaply as possible.

The process of reducing the cost of the production of commodities was directed [196] along two lines: the ever increasing exploitation of labour power (the production of absolute surplus value) and the improvement of the production process itself (relative surplus value). The invention of machines not only failed to reduce the working day but, on the contrary, as a powerful means of increasing the productivity of labour, as an instrument of capital, it became a means for excessively extending the working day.

We shall trace this process in the steam engine. But before turning to an analysis of the history of the development of the steam engine, we must elucidate what we mean by an machine since on this question the Marxist point of view differs radically from that of other researchers.

At the same time, in order to elucidate the essence of the industrial revolution, which made the steam engine so prominent, it is necessary to have a clear understanding of the role played by the steam engine in the industrial revolution.

It is widely believed that the steam engine created the industrial revolution. Such an opinion is erroneous. Manufacture developed out of handicrafts in two ways. On the one hand it arose from the combination of heterogeneous independent handicrafts, which lost their independence, and on the other hand it arose from the co-operation between craftsmen in the same craft, which broke down the particular process into its component parts and led to a division of labour within manufacture.

The starting point in manufacture is labour power.

The starting point in large-scale industry is the tool. Of course, the problem of the motive power is also important for manufacture, but the revolutionization of the entire process of production, which had been prepared by the detailed division of labour within manufacture, was brought about not by the motive power but by the machine driven tool.

Every machine consists of three basic parts: the motor, the transmission mechanism and the tool.[69]

The essence of a historical view of the definition of a machine is precisely the fact that in different periods a machine has different purposes.

Vitrivius' definition of a machine remained valid until the industrial revolution. For him a machine was "a coherent combination of joinery most capable of moving loads."[70]

Consequently the basic instruments serving these ends: the inclined plane, the windlass, the pulley, the lever, were called simple machines.

In his introduction to the *Principia,* Newton attributes the teachings about five simple machines—the lever, the wheel, the pulley, the windlass, the wedge—to the applied mechanics developed by the ancients.

This is the source of the widespread opinion in English literature that an instrument is a simple machine and a machine, a complex instrument. [197]

However, it is not entirely a question of simplicity and complexity. The essence of the matter is that the introduction of a machine-driven tool designed to grip and expediently change the object of labour brought about a revolution in the very process of production.

The other two parts of the machine exist in order to set the tool in motion.

Thus, it is clear that a great gulf divides the machines known to Vitruvius, which accomplish only the mechanical displacement of the finished products, from the machines of large-scale industry, whose function consists in a complete transformation in the original material of the product.

The fruitful nature of Marx's definition is especially clear if we compare it with the definitions of a machine found in the literature.

In his *Theoretical Kinematics* Reuleaux[71] defines a machine as "a combination of bodies capable of resistance, which is so arranged that by its means mechanical natural forces can be compelled to act under certain motions."

This definition is equally applicable to Vitruvius' machine and to the steam engine. However, there are difficulties arise when this definition is applied to the steam engine.

Sombart's definition of a machine suffers from the same defect. Sombart calls the machine a means or a complex of means of labour, served by man, the purpose of which is the mechanical rationalization of labour. What distinguishes the machine as a means of labour from a tool is precisely the fact that the former, is served by man, whereas the latter serves man.[72]

This definition is inadequate precisely because it bases the distinction between a tool and a machine on the fact that the one serves man and the other is served by man. This definition, which at first sight is based on a socio-economic character, not only fails to distinguish between the period in which the simple tool predominates and the period in which the machine method of production predominates, but creates the quite absurd notion that the essence of the machine consists in its being served by man.[73]

Thus an imperfect steam engine demanding the continual service of a man (in Newcomen's first engines a boy had continually to open and close a tap) will be a machine, while a complex automaton producing bottles or electric bulbs will be a tool, since it hardly requires any servicing.

Marx's definition of a machine draws attention to the fact that it caused a revolution in the very process of production.

The motor is a necessary and very important component part of the machinery of industrial capitalism, but it does not determine its fundamental character. [198] When John Wyatt invented his first spinning machine he did not even mention how it was driven. “A machine in order to spin without the aid of fingers” was his programme.[74]

It was not the development of the motor and the invention of the steam engine that created the industrial revolution of the 18th century, but on the contrary the steam engine gained such enormous importance precisely because the division of labour that was emerging in manufacture and its increasing productivity made the invention of a machine-driven tool both possible and necessary, and the steam engine, which had been born in the mining industry, found a field awaiting its application as a motive power.

Arkwright’s spinning jenny was at first driven by means of water. However the use of water power as the predominant form of motive power involved great difficulties.

It could not be increased arbitrarily; if there was a shortage of it, it could not be replenished; sometimes it dried up; and it remained of purely local character.

Only with the invention of Watt’s machine did the machine textile industry, which was already fairly well developed, receive the motor that was essential for it at that particular stage of development.

Thus the machine textile industry is by no means a consequence of the invention of the steam engine.

The steam engine was born in the mining industry. As early as 1630 Ramsay was granted a patent in England “to raise water from low pits by fire.”

In 1711 the Proprietors of the Invention for Raising Water by Fire was formed for exploiting Newcomen’s engine in England.

The greatest service rendered by England’s thermal (steam) engine, Carnot wrote in his work *On the Motive Power of Heat*, was undoubtedly the revival of the working of the coal mines, which threatened to cease entirely in consequence of the continually increasing difficulty of drainage and of raising the coal.[75]

The steam engine gradually became an important factor in production. It was then noticed that it could be made more economical by reducing the consumption of steam, and consequently the consumption of water and fuel.

Even before Watt’s work Smeaton was investigating the consumption of steam in different steam engines, setting up a special laboratory for this purpose in 1769. He found that steam consumption in different engines varies from 176 to 76 kg per horsepower hour. Savery succeeded in building an engine of the Newcomen type [199] with a steam consumption of 60 kg per horsepower hour.

By 1767 fifty-seven steam engines with a total power of 1,200 horsepower were already at work around Newcastle alone.

It is no wonder that the problem of economy was one of the main problems confronting Watt.

Watt’s patent, taken out in 1769, begins thus: “My method of lessening the consumption of steam, and consequently fuel, in fire engines, consists in the following principles.”[76]

Watt and Boulton concluded an agreement with an owner of coal mines, according to which they would be paid one-third of the sum saved by the reduced expenditure on fuel.

According to this agreement, they received over two thousand pounds[77] a year from this mine alone.

The chief inventions of the textile industry were made in the period 1735–1780, thus creating an immediate demand for a motor.

In his patent taken out in 1784 Watt described the steam engine as a universal motor of large industry.

The main problem was the technical rationalization of the steam engine. In order to realize this task in practice it was necessary to make a detailed study of the physical processes that occurred in the engine.

Unlike Newcomen, Watt, in the laboratory of Glasgow University, made a detailed study of the thermo-dynamic properties of steam, thus laying the basis for thermodynamics as a branch of physics.

He carried out a number of experiments on the boiling temperature of water under various pressures in relation to change in the expansion of steam. Then he investigated the latent heat in steam formation and developed and tested Black's theory.

Thus the main problems of thermo-dynamics, the teaching about the latent heat in steam formation, the dependence of boiling point on pressure and the magnitude of the latent heat in steam formation, began to be scientifically elaborated by Watt.

It was this detailed study of the physical processes in the steam engine that enabled Watt to go further than Smeaton, who, despite his goal of investigating the steam engine in the laboratory, was limited to making purely empirical, superficial improvements to Newcomen's engine, since he had no knowledge of the physical properties of water vapours.

Thermodynamics not only received an impetus to its development from the steam engine, but in fact developed from the study of that engine.

It became necessary to study not only the particular physical processes in the steam engine, but the general theory of steam engines, the general theory of the maximum efficiency of steam engines. This work was carried out by Sadi Carnot. [200]

The general theory of the steam engine and the theory of the maximum efficiency led Carnot to the necessity of investigating general thermal processes, to the discovery of the second principle of thermodynamics.

The study of steam engines, said Carnot in his work *On the Motive Power of Heat* (1824), is of the greatest interest, as their importance is enormous and their use is continually increasing. Clearly they are destined to produce a great revolution in the civilized world.[78]

Carnot remarked that, despite various kinds of improvements, the theory of the steam engine had made but little progress.

Carnot formulated his task of elaborating a general theory of the steam engine in such a way that the practical problems he set in order to discover a general theory of maximum efficiency are quite clear.

The question has often been raised, he wrote, whether the motive power of heat is limited or infinite; by motive power we mean the work a motor can perform.

Is there any limit to the possible improvements, a limit that the nature of things will not allow to be surpassed by any means whatever? Or, on the contrary, can these improvements be carried on indefinitely?[79]

Machines which do not receive their motion from heat, but have for a motor the force of men, animals, a waterfall, an air current, can be studied, Carnot observed, by means of theoretical mechanics.

Here, all cases are foreseen, all imaginable motions are referred to their general principles (which was made possible by Newton's work on mechanics), firmly established and applicable in all circumstances.

No such theory exists in the case of heat engines.

We cannot have such a theory, Carnot stated, until the laws of physics are extended enough, generalized enough, to make known beforehand all the effects of heat acting in a determinate manner on any body.[80]

Here the connection between technology and science, between the investigation of the general laws of physics and the technical problems raised by economic development is established with extraordinary clarity.

But the history of the steam engine is important to us in another connection as well.

Historically, the investigation of various forms of physical motion of matter took place in the following sequence: mechanics, heat, electricity.

We have seen that the development of industrial capitalism presented technology with the demand to create a universal motor. [201]

This demand was preliminarily supplied by the steam engine, which had no competitors until the invention of the electric motor.

The problem of the theory of the efficiency maximum of steam engines led to the development of thermodynamics, i.e., to the study of the thermal form of motion.

This, therefore, is the explanation for the historical sequence in the study of forms of motion: the study of the thermal form of motion—thermodynamics—developed in the wake of mechanics.

We shall now proceed to a consideration of the importance of the steam engine from the perspective of the transformation of one form of motion into another.

Whilst Newton never even posed the problem of the law of the conservation and conversion of energy, Carnot was compelled to pose it, although still in an unclear form.

The reason for this was that Carnot's study of the steam engine focused precisely on the conversion of thermal into mechanical energy.

The category of energy as one of the basic categories of physics appeared at the time when the problem of the correlations between various forms of motion emerged. And as the forms of motion investigated by physics became more varied, so the category of energy acquired ever more significance.

Thus the historical development of the study of physical forms of motion of matter should provide the key to understanding the origin, significance and interconnection of the categories of physics.

A historical study of forms of motion should be conducted from two perspectives. We must study the historical sequence of the forms of motion as they appear in the development of the science of physics in human society. We have already shown the connection between the mechanical and the thermal form of motion from the perspective of their historical genesis in human society. The study of these forms follows the sequence in which they were brought to the forefront by human practice.

The second perspective is to study the "natural science of the development of matter." The process of studying the development of inorganic matter in the microcosm and the macrocosm should provide the key to understanding the connection between the various forms of motion of inorganic matter and reciprocal conversions of one to another, and should lay a sound basis for a natural classification of forms of motion of matter. This principle should lie at the basis of a Marxist classification of sciences.

Every science analyses a single form of motion or a series of forms of motion that are interconnected and transformed into one another. [202]

The classification of sciences is none other than a hierarchy of the forms of motion of matter in accordance with their essential order, in other words, in accordance with their natural development and the passing of one form of motion into another, as they occur in nature.

Hence, this principle of a Marxist classification of science bases classification on the great idea of the development and the transformation of one form of motion of matter into another form. (Engels.)[81]

Herein consists Engel's remarkable notion of the interconnection and hierarchy of forms of motion of matter.

The conception of energy is indissolubly bound up with the conversion of one form into another, with the problem of measuring this conversion. Modern physics emphasizes precisely the quantitative aspect of this conversion and postulates the conservation of energy in those conversions.

We recall, as was shown in the previous chapter, that the constancy and invariability of quantity of motion were already stated by Descartes. The new element introduced into physics by the work of Mayer and Helmholtz lay in the discovery of the transformation of forms of motion along with the conservation of energy in these conversions.

It was this, and not the simple postulation of constancy, that was the new element.

As a result of this discovery, the different isolated forces of physics (heat, electricity, mechanical energy), which until then had been seen as comparable to the invariable species of biology, were transformed into interconnected forms of motion that convert one into another according to definite laws.

Like astronomy, physics came to the inevitable conclusion that the end result was the eternal circulation of moving matter. That is why Newton's period, which was acquainted with only one form of motion—the mechanical—and was primarily interested, not in the conversion of one form into another, but only in the conversion and modification of one and the same form of motion—mechanical displacement—(let us recall Vitrivius' definition of a machine and Carnot's observations) did not, and could not, consider the problems of energy.

As soon as the thermal form of motion appeared on the scene, and precisely because it appeared on the scene when it was indissolubly bound up with the problem of its conversion into mechanical motion, the problem of energy came to the forefront. The very way in which the problem of the steam engine was formulated ("to raise water by fire") clearly points to its connection with the problem of the conversion of one form of motion into another. It is not by chance that Carnot's classic work is entitled: *On the Motive Power of Heat.* [203]

Engels' treatment of the law of the conservation and conversion of energy emphasizes the qualitative aspect of the law of conservation of energy, in contradistinction to the predominant treatment in contemporary physics that reduces it to a purely quantitative law—the quantitative constancy of energy during its transformations. The law of the conservation of energy, of the indestructibility of motion, should be understood not only in a quantitative but also in a qualitative sense. It contains not only the postulation that energy cannot be destroyed or created, which is one of the basic prerequisites of the materialistic conception of nature, but a dialectical treatment of the problem of the motion of matter. From the perspective of dialectical materialism, the indestructibility of motion consists not only in the fact that matter moves within the limits of one form of motion, but also in the fact that matter itself is capable of producing from itself all the endless variety of forms of motion in their spontaneous transformations into one another, in their self-movement and development.

We see that only the conception of Marx, Engels and Lenin provides the key to understanding the historical sequence of the development and investigation of forms of motion of matter.

If Newton did not consider or solve the problem of the conservation of energy, this, of course, was not because he lacked genius.

Great men in all spheres, no matter how remarkable their genius, formulate and resolve those problems that have been placed on the agenda by the historical development of the forces and relations of production in their time. [204]

The Machine-Wreckers in Newton's Age and the Present-Day Wreckers of the Productive Forces

We have come to the end of our analysis of the *Principia*. We have shown how its physical content arose out of the tasks of that era, which were placed on the agenda by the class that was coming to power.

The historically inevitable transition from feudalism to merchant capital and manufacture, and from manufacture to industrial capitalism, stimulated an unprecedented development of the productive forces, and this in turn gave a powerful impetus to the development of scientific research in all spheres of human knowledge.

Newton happened to live in the very age when new forms of social relations, new forms of production, were being created.

In his mechanics he was able to solve the complex of physical and technical problems placed on the agenda by the rising bourgeoisie.

But he came to a halt, helpless, before nature as a whole. Newton was familiar with the mechanical displacement of bodies, but he rejected the view that nature is in a process of unceasing development. Still less can we hope to find in him any view of society as a developing whole, even though it was an age of transition that gave rise to his main work.

Has the movement of the historical process ceased since Newton's time? Of course not, for nothing can check the forward movement of history.

After Newton, Kant and Laplace were the first to make a breach in the view of nature as eternal and unchanging throughout the ages. They showed, albeit in a far from complete form, that the solar system is the product of historical development.

It was through their works that the notion of development, which was subsequently to become the basic and guiding principle of all teaching on nature, entered into natural science for the first time.

The solar system was not created by God, the movement of the planets is not the result of a divine impulse. It not only preserves its state solely as a consequence of natural causes, but also came into existence through their influence alone. Not only does God have no place in a system whose existence is based on the laws of mechanics, but he is unnecessary even as an explanation of its origin. [205]

"I have had no need to include any hypothesis of a deity in my system, Your Highness," so Laplace is said to have replied when Napoleon asked him why he had omitted all reference to the role of God in his *System of the World*.

The progressive development of the productive forces gave rise to progressive science.

The transition from domestic handicraft industry to manufacture and from manufacture to large-scale machine industry, which was only beginning in Newton's age, was greatly accelerated during the following century. It was completed by the monopolistic imperialist phase of capitalism, which is the threshold to new, socialist forms of development.

As one phase of the capitalist method of production was replaced by another, so the very views on technology and science held by the ruling class in capitalist society changed.

On coming to power the bourgeoisie struggled mercilessly against the old guild and handicraft modes of production. With an iron hand it introduced large-scale machine industry, shattering in its course the resistance of the obsolete feudal class and the still unorganized protest of the newborn proletariat.

Science and technology are powerful weapons of struggle for the bourgeoisie, and it is interested in developing and perfecting these weapons.

The bard of industrial capitalism of this period (Ure)[82] portrayed the struggle of the bourgeoisie for new methods of production in the following terms:

> Then the combined malcontents, who fancied themselves impregnably entrenched behind the old lines of division of labour, found their flanks turned and their defenses rendered useless by the new mechanical tactics, and were obliged to surrender at discretion.

> Examining further the significance of the invention of the spinning machine, he said: "A creation destined to restore order among the industrious classes . . . This invention confirms the great doctrine already propounded, that when capital enlists science into her service, the refractory hand of labour will always be taught docility."[83]

Ure spoke for the bourgeoisie that was coming to power as it built new methods of production on the blood and bones of the "refractory hand of labour."

On coming to power the bourgeoisie revolutionized all modes of production. It tore the old feudal bonds to shreds and shattered the archaic forms of social relations that fettered the further development of the productive forces. In that period it was revolutionary because it brought with it new and more advanced methods of production.

Over a period of a century it changed the face of the earth and brought into existence new, powerful productive forces. New, hitherto unexplored forms of motion of matter were discovered. [206]

The immense development of technology was a powerful stimulus to the development of science, and the rapidly developing science in turn fertilized the new technology.

And this unprecedented flourishing of the productive forces, the tremendous growth of material culture, brought about the unprecedented impoverishment of the masses of the people and a terrible growth in unemployment.

It is not surprising that these contradictions in the predominant capitalist methods of production should have attracted the attention, not only of the state officials in the capitalist countries, but also of their scientists.

In Newton's period the bourgeoisie called for new methods of production. In his memorandum on the reform of the Royal Society, Newton urged the state authorities to support science, which contributed so much to the study of nature and the creation of new productive forces.

Today the situation is very different.

In 1930/31 *Nature* published a number of leading articles dealing with the questions we are considering. These articles consider problems that are now agitating the whole world. Of these articles, we will consider two that express most clearly the point of view of English natural scientists. One is entitled "Unemployment and Hope," the other "Science and Society."

This is how these articles depict the tasks of industry, its aims and course of development.

Discussing the question of unemployment, which is rending capitalist society, *Nature* defines the role of machines as follows:

> There is, indeed, in the present situation much to excuse a passing reflection that perhaps, after all, the people of Erewhon were wiser than ourselves in destroying their machines, lest, as Marx predicted, the machines reversed the original relation and the workmen became the tool and appendage of a lifeless mechanism.[84]

Modern science and technology create machines of remarkable precision and productivity, with an extraordinarily complex and delicate structure. And it now appears that the machine wreckers of Newton's period were wiser than we, who create machines of unprecedented complexity and power.

The above quotation not only distorts the ideas of Marx, but also misinterprets the movement of the machine-wreckers.

Let us first re-establish the true historic circumstances and actual reasons that drove the workers to wreck the machines. [207]

The workers' struggle against the machine merely reflected the struggle between wage labourers and capitalists. The working class of that period did not struggle against the machines as such but against the position to which the developing capitalist order was relegating it in the new society.

During the 17th century the whole of Europe experienced the workers' anger against carding machines. The first wind-power saw-mill was destroyed in London at the end of the 1670s.

The first decade of the 19th century was marked by the mass movement of the Luddites against the power loom. As industrial capitalism developed, it transformed labour power into a commodity. Forced out of industry by machinery, the worker could not find a purchaser for his labour, and was comparable to paper money that had gone out of currency. The growing working class, which had not yet developed a class consciousness, directed its hatred against the external forms of capitalist relations—the machines.

But this reactionary form of protest in fact expressed a revolutionary protest against the system of wage labour and private ownership of the means of production.

The worker was indeed becoming an appendage to the machine, not because machines had been invented, but because these machines served the interest of the class that owned the means of production.

The call to machine-wrecking will always be a reactionary slogan, and the wisdom of the inhabitants of Erewhon consisted not in their destroying the machines, but in their protest against the slavery of wage labour.

> "The comfort and the welfare of the few," continues the leading article, "on this view, may, however, be too dearly purchased when we consider the lot of the displaced workers, and, perhaps, still more the repression of individuality and the retarded development which, as Marx predicted, have often accompanied mass production."

Thus, in the opinion of *Nature*, improvement in the means of production inevitably leads to the repression of individuality and the suffering of the masses of the people.

Here it is permissible to ask: Why was it that during Newton's time, when there was an enormous development in the means of production, scientific circles not only did not call for a curb on this development, but, on the contrary, made every effort to encourage every new discovery and invention; and the organ of the leading natural scientists in Newton's period, *Philosophical Transactions*, was full of descriptions of these new inventions? [208]

Before answering this question, we will see what methods this journal of British naturalists proposes for solving the crisis of production and unemployment, which, so it believes, are the results of the overdevelopment of the productive forces.

These methods are outlined in the leading article "Unemployment and Hope." We quote the corresponding section *in extenso*:[85]

> The aims of industry are, or should be, ... chiefly two (1) to furnish a field for ... growth of character; and (2) to produce commodities to satisfy man's varied wants, mostly of a material kind, though of course there are large exceptions outside the material category, and the term 'material' is here used in no derogatory sense. Attention has hitherto been directed mainly to (2) and the primary aim of industry has been ignored. Such one-sided view of industry coupled with a too narrow use of the much abused word 'evolution' ...

> has, led to over-concentration on quantity and mass production and a ridiculous neglect of the human element and there can be no doubt that had a little thought been given to the first aim then the second would have been much more completely and satisfactorily attained; also unemployment would not have been heard of . . .
>
> The prevailing idea . . . appears to be that industry is evolving and must evolve towards one fixed type, for example, that of large-scale production. . .. The best form or type of industry . . . may consist of many different and constantly changing forms, distinguished above all things by adaptability and elasticity—a living organism.
>
> Elasticity further means the possibility of reviving, under new and improved forms to meet modern conditions, two at least of the older types of industry which are supposed to have been superseded or rendered obsolete by modern large-scale production, namely: (1) small cottage industries or handicrafts. . .; (2) a combination of manufacturing with agricultural or garden industry. . . Industry still has its roots firmly and deeply fixed in the past, and foolishly to tear up a great part of those roots as old and useless is the surest way to weaken the industrial tree. Perchance the source of the unemployment curse is to be found here.
>
> The restitution of these two principles of an older industrial order, so essentially and characteristically English, under improved forms made possible by modern scientific achievement, including notably electrical power distribution, would furnish, in the first place, a new and almost infinite field for human employment of all kinds, absorbing all or most of the present [209] unemployed. . . By unemployed we mean chiefly the unemployed in Great Britain only, but it would be vastly better to extend our consideration to cover unemployment throughout the whole world . . .
>
> The application of these two principles to unemployment is, of course, only one part of their scope, for they have a far wider range even than this, especially in counteracting one of the greatest evils of modern industry, namely, extreme specialism, monotonous work, and lack of scope for developing skill, with all that that implies . . .
>
> It is probable that, under the more bracing atmosphere of varied work and interest and skill thus envisaged, the inventive faculties of mankind would be greatly stimulated, and a much needed spur be given to originality.

Thus, according to *Nature*, the remedy for healing the wounds of capitalist society, the means of eliminating all the contradictions of a system based on wage labour and individual ownership of the means of production, is a return to those forms of industry that directly preceded the age of industrial capitalism.

We have demonstrated above that it was these very forms that engendered the advances in Newton's period; and although they were a step forward by comparison with feudal methods of production, manufacture and small handicraft industry, at the present time the slogan "Back to small handicraft industry" is profoundly reactionary.

The fetishism of the commodity system, which Marx so brilliantly exposed, lies in the fact that the relations of material things created by human society are isolated from human relations and are considered as inherent to the things themselves.

This fetishism can be deciphered and exposed by understanding that it is not things as such that create relations, but that the relations between things created in the process of social production simply express a particular social relation between people, which they conceive of in the fantastical form of relations between things.

The views cited above are also a certain form of fetishism. Machinery, the means of production, the organization of production into large-scale machine production are considered in isolation, outside the social relations of the particular economic system in which the given mode of production exists and by which it is created.

Improving the instruments of labour brings misfortune to the great mass of the population, we are told. The machine transforms the worker into its mere appendage. It kills individuality. Let us return to the good old days. [210]

No, we reply. It is not the improvement in the means of production that causes the impoverishment and unprecedented sufferings of the masses. It is not the machines that transform the worker into a blind appendage of a mechanism, but those social relations that exploit machinery in such a way as to turn the worker into a mere appendage to it.

The solution lies not in returning to old, long since obsolete modes of production, but in changing the entire system of social relations, a change that is just as radical as the transition from feudal and handicraft methods of production to industrial capitalism was in its time.

Private property passes through three stages of development: feudalism, merchant capital and manufacture, industrial capitalism.

At every stage of development in the process of production of their lives, people involuntarily enter into specific relations of production that correspond to the stage of development of the productive forces. At a certain stage of their development, the productive forces come into antagonism with the existing relations of production or, in juridical terms, with the property relations within which they developed. Having previously been their forms of development, the latter become their fetters.[86]

The further development of productive forces is only possible through a radical reconstruction of all relations of production.

The transition from one form of production to another is characterized first and foremost by such a reconstruction.

At every new stage the change in social relations brings about a further rapid growth in the productive forces.

And, conversely, a crisis in the growth of the productive forces indicates that they are unable to continue to develop within the framework of the given social system.

The remedy that we cited above, which amounts to curbing the productive forces by a return to the old forms of production, is merely an expression of the contradiction between the productive forces in capitalist society and the relations of production based on private ownership of the means of production.

Science develops out of production, and those social forms that become fetters upon the productive forces likewise become fetters upon science.

Genuine methods for transforming society cannot be found through brilliant inspiration or guesswork, nor through a return to "the good old days" which in distant historical perspective appear to be a peaceful idyll, but which in reality were a bitter class struggle and the repression of one class by another.

Thus it has always been, and so it was in the age when Newton lived and worked, and to whose forms of production it is proposed that we return.

We have seen that the obsolete system of social relations of that period, speaking through their universities, also recommended restricting science, which was shattering the stagnant forms of feudal ideology and was entering into the service of a new mode of production.

What we are now witnessing is the repetition, on a new basis, of the fundamental antagonism between the forces and relations of production that Marx so brilliantly and lucidly revealed and explained.

Whilst the newly emerging proletariat spontaneously protested by wrecking machines and resisting inventions and science, today, armed with Marx's, Engels' and Lenin's method of dialectical materialism, the proletariat clearly sees the path towards the liberation of the world from exploitation of man by man.

The proletariat knows that genuine scientific knowledge of the laws of the historical process leads with iron necessity to the conclusion that the change from one social system to another is inevitable—to the change from capitalism to socialism.[87]

The proletariat exposes all the fetishes of class society and sees, behind the relations between things, the relations between the human beings who create these things.

Having learnt the real nature of the historical process, the proletariat does not remain a mere spectator. It is not only the object, but the subject of the process.

The great historical significance of the method created by Marx lies in the fact that knowledge is not regarded as the passive, contemplative perception of reality, but as the means for actively reconstructing it.

For the proletariat science is a means and instrument for this reconstruction. That is why we are not afraid to expose the "worldly origin" of science, its close connection to the mode of production of material existence.

Only such a conception of science can truly liberate it from those fetters in which it is inevitably trapped in bourgeois class society.

Not only does the proletariat have no fear of the development of the productive forces, but it alone is capable of creating all the conditions for their unprecedented flourishing, and also for the flourishing of science.

The teachings of Marx and Lenin have come to life. The socialist reconstruction of society is not a distant prospect, not an abstract theory, but a definite [212] plan for the great works being accomplished by the population of one-sixth of the globe.

And as in all eras, by reconstructing social relations we reconstruct science.

The new method of research, which in the persons of Bacon, Descartes and Newton gained victory over scholasticism and led to the creation of a new science, was the result of the victory of the new mode of production over feudalism.

The building of socialism not only absorbs into itself all the achievements of human thought, but, by setting science new and hitherto unknown tasks, charts new paths for its development and enriches the storehouse of human knowledge with new treasures.

Only in socialist society will science genuinely belong to all mankind. New paths of development are opening up before it, and its victorious march has no bounds neither in infinite space or in eternal time.

Appendix 1

Galileo Gallilei: Discorses and mathematical demonstrations concerning two new sciences

Salviati: Frequent experience of your famous arsenal, my Venetian friends, seems to me to open a large field to speculative minds for philosophizing, and particularly in that area which is called mechanics, inasmuch as every sort of instrument and machine is continually put in operation there. And among its great number of artisans there must be some who, through observations handed down by their predecessors as well as those which they attentively and continually make for themselves, are truly expert and whose reasoning is of the finest.
Sagredo: You are quite right. And since I am by nature curious, I frequent the place for my own diversion and to watch the activity of those whom we call "key men" (*Proti*) by reason of a certain preeminence that they have over the rest of the workmen. Talking with them has helped me many times in the investigation of the reason for effects that are not only remarkable but also abstruse, and almost unthinkable.
(Galileo Galilei, *Two New Sciences* (transl. Stillman Drake), Madison, WI: Univ. of Wisconsin Press, 1974, p. 11)

Appendix 2

Nicolas Boileau: Arrest

Donné en la grand'chambre du Parnasse, en faveur des maîtres-es-arts, medecins et professeurs de l'Universite de Stagyre au pays des chimeres: pour le maintien de la doctrine d'Aristote.
Veu par la Cour la Requeste présentée par les Regens, Maîtres-és-Arts, Docteurs et Professeurs de l'Université, tant en leurs noms que comme Tuteurs, et deffenseurs de la Doctrine de Maître *en blanc* Aristote, ancien Professeur Royal en Grec dans le College du Licée, et Precepteur du feu Roy de querelleuse memoire Alexandre dit le Grand, acquereur de l'Asie, Europe, Afrique et autres lieux; contenant que depuis quelques années, une inconnuë nommée la Raison, auroit entrepis d'entrer par force dans les Ecôles de ladite Université, et pour cet effet à l'aide de certains Quidams factieux prenans les surnoms de Gassendistes, Cartesiens, Malebranchistes et Pourchotistes, gens sans aveu, se seroit mise en estat d'en expulser ledit Aristote ancien et paisible possesseur desdites Ecôles, contre lequel, Elle et ses Consorts auroient déja publié plusieurs livres, traités, dissertations et raisonnemens diffammatoires, voulant assujettir ledit Aristote à subir devant Elle l'examen de sa Doctrine; ce qui seroit directement opposé aux loix, us et coûtumes de ladite Université, où ledit Aristote auroit toûjours esté reconnu pour Juge sans appel et non comptable de

ses opinions. Que même sans l'aveu d'icelui elle auroit changé et innové plusieurs choses en et au dedans de la Nature, ayant osté au Cœur la prerogative d'estre le principe des nerfs, que ce Philosophe lui avoit accordée liberalement et de son bon gré, et laquelle Elle auroit cedée et transportée au Cerveau. Et ensuite, par une procedure nulle, de toute nullité, auroit attribué audit Coeur la charge de recevoir le chile appartenante cydevant au Foye; comme aussi de faire voiturer le Sang par tout le corps, avec plein pouvoir audit Sang d'y vaguer, errer et circuler impunément par les veines et arteres, n'ayant autre droit ni titre pour faire lesdites vexations que la seule Experience, dont le témoignage n'a jamais esté reçu dans lesdites Ecôles. Auroit aussi attenté ladite Raison, par une entreprise inouïe, de déloger le Feu de la plus haute region du ciel, et pretendu qu'il n'avoit là aucun domicile, nonobstant les certificats dudit Philosophe et les visites et descentes faites par luy sur les lieux. Plus, par un attentat et voye de fait énorme contre la Faculté de Medecine, se seroit ingerée de guerir, et auroit réellement et de fait guery quantité de fièvres intermitentes, comme tierces, double-tierces, quartes, triple-quartes, et même continuës, avec vin pur, poudres, ecorce de Quinquina, et autres drogues inconnuës audit Aristote et à Hippocrate son devancier, et ce sans saignée, purgation ny evacuation precedentes; ce qui est non seulement irregulier, mais tortionnaire et abusif; ladite Raison n'ayant jamais esté admise ny agregée au Corps de ladite Faculté, et ne pouvant par consequent consulter avec les Docteurs d'icelle, ni estre consultée par eux, comme Elle ne l'a en effet jamais esté. Nonobstant quoy, et malgré les plaintes et oppositions réïterées des sieurs Blondel, Courtois, Denyau, et autres deffenseurs de la bonne Doctrine, elle n'auroit pas laissé de se servir toûjours desdites drogues, ayant eu la hardiesse de les employer sur les Medicins mêmes de ladite Faculté, dont plusieurs, au grand scandale des regles, ont esté gueris par lesdits remedes. Ce qui est d'un exemple tres-dangereux, et ne peut avoir esté fait que par mauvaises voyes sortilege et pacte avec le diable. Et non contente de ce, auroit entrepris de diffammer et de bannir des Ecôles de Philosophie les formalités, materialités, entités, virtualités, ecceités, Petreités, Policarpeités, et autres estres imaginaires, tous enfans et ayans cause de deffunt Maistre Jean Scot leur pere. Ce qui porteroit un préjudice notable, et causeroit la totale subversion de la Philosophie Scolastique dont elles font tout le Mystere, et qui tire d'elles toute sa subsistance, s'il n'y estoit par la Cour pourvû.

Veu les libelles intitulés Physique de Rohault, Logique de Port-Royal, Traités du Quinquina, même l'*Adversus Aristoteleos* de Gassendi, et autres pieces attachées à ladite Requeste, signée CHICANEAU, Procureur de la dite Université. Oüy le Rapport du Conseiller Commis. Tout consideré.

La Cour ayant égard à ladite Requeste, a maintenu et gardé, maintient et garde ledit Aristote en la pleine et paisible possession et joüissance desdites Ecôles. Ordonne qu'il sera toûjours suivi et enseigné par les Regens, Docteurs, Maîtres-és-Arts et Professeurs de ladite Université. Sans que pour ce ils soient obligés de le lire ni de sçavoir sa langue et ses sentimens. Et sur le fond de sa doctrine, les renvoye à leurs cahiers. Enjoint au Cœur de continuer d'estre le principe des Nerfs, et à toutes personnes de quelque condition et profession qu'elles soient de le croire tel, nonobstant toute experience à ce contraire. Ordonne pareillement au Chile d'aller droit au Foye sans plus passer par le Coeur, et au Foye de le recevoir. Fait deffense au Sang d'estre plus vagabond, errer ni circuler dans le corps, sous peine d'estre entierement livré et abandonné à la Faculté de Médecine. Deffend à la Raison et à ses adherans de plus s'ingerer à l'avenir de guerir les fiévres tierces, doubletierces, quartes, triple-quartes ni continuës, par mauvais moyens et voyes de sortileges, comme vin pur, poudres, ecorce de Qinquina, et autres drogues non approuvées ny connuës des Anciens. Et en cas de guérison irreguliere par icelles drogues, permet aux Medecins de ladite Faculté de rendre, suivant leur methode ordinaire, la fiévre aux Malades, avec casse, séné, sirops, juleps, et autres remedes propres à ce; et de remettre lesdits Malades en tel et semblable état qu'ils estoient auparavant; pour estre ensuite traités selon les regles, et s'ils n'en rechappent, conduits du moins en l'autre monde suffisamment purgés et êvacués.

Remet les entités, identités, virtualités, ecceités, et autres pareilles formules Scotistes, en leur bonne fâme etrenommée. A donné acte aux sieurs Blondel, Courtois et Denyau de leur opposition au bon sens. A reintegré le feu dans la plus haute region du ciel, suivant et conformément aux descentes faites sur les lieux. Enjoint à tous Regens, Maîtres-és-Arts et Professeurs d'enseigner comme ils ont accoûtumé, et de servir pour raison de ce, de tels raisonnemens qu'ils aviseront bon estre; et aux Repetiteurs Hibernois et autres leurs Supposts, de leur prêter main-forte, et de courir sus aux Contre-venans peine d'estre privés du droit de disputer sur les Prolegomenes de la Logique. Et afin qu'à l'avenir il n'y soit contrevenu, a banni à perpetuité la Raison des Ecoles de ladite Université; luy fait deffenses d'y entrer, troubler, ni inquieter ledit Aristote en la possession et joüissance d'icelles, à peine d'estre declarée Janseniste, et amie des nouveautez. Et à cet effet sera le present Arrest lû et publié aux Mathurins de Stagyre à la premiere Assemblée qui sera faite pour la Procession du Recteur, et affiché aux portes de tous les Colleges du Parnasse, et par tout où besoin sera. Fait ce trente-huitième jour d'Aoust onze mil six cens soixante et quinze.
(Nicolas Boileau, *Oeuvres Complètes*, Paris: Gallimard, 1966, pp. 327–330)

Appendix 3

From David Brewster, Memoirs of Sir Isaac Newton

While Newton was corresponding with Locke in 1692, the process of Boyle for "multiplying gold," by combining a certain red earth with mercury, became the subject of discussion. Mr. Boyle having "left the inspection of his papers" to Locke, Dr. Dickison, and Dr. Cox, Mr. Locke became acquainted with the particulars of the process we have referred to. Boyle had, before his death, communicated this process both to Locke and Newton, and procured some of the red earth for his friends. Having received some of this earth from Locke, Newton tells him, that though he has "no inclination to prosecute the process," yet, as he had "a mind to prosecute it," he would "be glad to assist him," though "he feared he had lost the first and third of the process out of his pocket." He goes on to thank Locke for "what he communicated to him out of his own notes about it," and adds in a postscript, that "when the hot

weather is over, he intends to try the beginning, (that is the first of the three parts of the recipe) though the success seems improbable."[1] In Locke's answer of the 26th July,[2] he sends to Newton a transcript of two of Boyle's papers, as he knew he wished it; and, it is obvious from their letters, that both of them were desirous of "multiplying gold." In Newton's very interesting reply[3] to this communication, he "dissuades Locke against incurring any expense by a too hasty trial of the recipe." He says, that several chemists were engaged in trying the process, and that Mr. Boyle, in communicating it to himself, "had reserved a part of it from my knowledge, though I knew more of it than he has told me." This mystery on the part of Boyle is very remarkable. In "offering his secret" to Newton and Locke, he imposed conditions upon them, while in the case of Newton at least, he did not perform his own part in the arrangement. On another occasion, when he communicated two experiments in return for one, "he cumbered them," says Newton, "with such circumstances as startled me, and made me afraid of any more." It is a curious fact, as appears from this letter, that there was then a Company established in London to multiply gold by this recipe, which Newton "takes to be the thing for the sake of which Mr. Boyle procured the repeal of the Act of Parliament against multipliers." The pretended truths in alchemy were received by men like Boyle on the same kind of evidence as that by which the phrenology and clairvoyance of modern times have been supported. Although Boyle possessed the golden recipe for twenty years, yet Newton could not find that he had "either tried it himself, or got it tried successfully by any body else; for," he says, "when I spoke doubtingly about it, he confessed that he had not seen it tried, but added, *that a certain gentleman was now about it, and it succeeded very well so far as he had gone, and that all the signs appeared, so that I needed not doubt of it.*"
(David Brewster, *Memoirs of the Life, Writings, and Discoveries of Sir Isaac Newton*, Edinburgh: Constable, 1855, vol. 2, pp. 120–122)

Appendix 4

Newton: Letter to Locke (Sept. 16, 1693)

Sir,—Being of opinion that you endeavoured to embroil me with women, and by other means, I was so much affected with it, as that when one told me you were sickly and would not live, I answered, 'twere better if you were dead. I desire you to forgive me this uncharitableness; for I beg your pardon for my having hard thoughts of you for it, and for representing that you struck at the root of morality, in a principle you laid in your book of ideas, and designed to pursue in another book, and that I took you for a Hobbist.[1] I beg your pardon also for saying or thinking that there

[1] Newton to Locke 7 July, 1692, *Correspondence* III, 215.
[2] Locke to Newton 26 July, 1692, *Correspondence* III, 216.
[3] Newton to Locke 2 August 1692, *Correspondence* III, 217.

was a design to sell me an office, or to embroil me.—I am your most humble and unfortunate servant.
Is. Newton.
At the Bull, in Shoreditch, London, Sept. 16, 1693.

[1]The System of Hobbes was at this time very prevalent. According to Dr. Bentley, "the taverns and coffee-houses, nay, Westminster-Hall, and the very churches, were full of it", and he was convinced, from personal observation, that "not one English infidel in a hundred was other than a Hobbist."—Monk's Life of Bentley: 31.
(David Brewster, *Memoirs of the Life, Writings, and Discoveries of Sir Isaac Newton*, Edinburgh: Constable, 1855, vol. 2, pp. 148–149)

Appendix 5

G.W. Leibniz: First Letter to Clarke

1. Natural religion itself, seems to decay (in England) very much. Many will have human souls to be material: others make God himself a corporeal being.
2. Mr. Locke, and his followers, are uncertain at least, whether the soul be not material, and naturally perishable.
3. Sir Isaac Newton says, that space is an organ, which God makes use of to perceive things by. But if God stands in need of any organ to perceive things by, it will follow, that they do not depend altogether upon him, nor were produced by him.
4. Sir Isaac Newton, and his followers, have also a very odd opinion concerning the work of God. According to their doctrine, God Almighty wants to wind up his watch from time to time: otherwise it would cease to move. He had not, it seems, sufficient foresight to make it a perpetual motion. Nay, the machine of God's making, is so imperfect, according to these gentlemen; that he is obliged to clean it now and then by an extraordinary concourse, and even to mend it, as a clockmaker mends his work; who must be so much more unskilful a workman, as he is oftener obliged to mend his work and set it right. According to my opinion, the same force and vigour remains always in the world, and only passes from one part of matter to another, agreeably to the laws of nature, and the beautiful pre-established order. And I hold, that when God works miracles, he does not do it in order to supply the wants of nature, but those of grace. Whoever thinks otherwise must have a very mean notion of the wisdom and power of God.
   (*The Leibniz-Clarke Correspondence* (ed. H.G. Alexander), Manchester: Manchester University Press, 1956, pp. 11–12)

Bibliography

D.K. Agafonov, *Современная механика.* [*Modern Technology*] vol. 3 (1912) of: *Итоги науки в теории и практике* [Results of Science in Theorie and Practice] vols. 1–12. Edited by M.M. Kovalevskij, N.N. Lange, et al. Moscow: Publishing House "Mir", 1911–1914

Agricola (Georg Bauer), *De re metallica.* [1556] (English translation by Herbert Hoover, New York, NY: Dover, 1950)

Archiv für die Geschichte der Naturwissenschaften und der Technik. Leipzig: Vogel 1–9 (1908/09–1920/22). [continued as *Archiv für Geschichte der Mathematik, der Naturwissenschaften und der Technik.* Leipzig, N.S. 1=10–4=13 (1927/28–1930/31)]

Eduard Bernstein, *Demokratie und Sozialismus in der Großen Englischen Revolution.* 4th ed. Stuttgart: Dietz, 1922

David Brewster, *Memoirs of the life, writings, and discoveries of Sir Isaac Newton.* 2 vols., Edinburgh: Constable, 1855

A.A. Bogdanov and I.I. Stepanov, *Курс политической экономии.* [Course of political economy] Leningrad, 1924

Sadi Carnot, *Réflexions sur la puissance motrice du feu et sur les machines propres à développer cette puissance.* Paris: Bachelier, 1824 [Russian translation 1923]

Ludwig Darmstaedter, *Handbuch zur Geschichte der Naturwissenschaften und der Technik.* Berlin, 1908

Hans Delbrück. *Geschichte der Kriegskunst im Rahmen der politischen Geschichte.* Berlin: Stilke, 1900ff [Russian]

Encyklopädie der mathematischen Wissenschaften: mit Einschluss ihrer Anwendungen. vol. 4. *Mechanik.* Leipzig: Teubner, 1908

Albert Einstein, *Newtons Mechanik und ihr Einfluß auf die Gestaltung der theoretischen Physik,* in: *Die Naturwissenschaften* 15, No. 12, 273–276 [Russian]

Friedrich Engels, *Der deutsche Bauernkrieg.* [Russian]

Friedrich Engels, *Army, Navy, Artillery.* (Articles from the *New American Cyclopedia* 1858–1860) [Russian]

Friedrich Engels, *Socialism: Utopian and scientific.* (translated by Edward Aveling) [Russian]

Galileo Galilei, *Unterredungen und mathematische Demonstrationen über zwei neue Wissenszweige, die Mechanik und die Fallgesetze betreffend* translated by A. von Oettingen, Leipzig: Engelmann, 1890–1904 (Ostwald's Klassiker der exakten Wissenschaften, vol. 11)

Georg Kaufmann, *Die Geschichte der deutschen Universitäten.* Bd. 1: *Vorgeschichte.* Stuttgart: Cotta 1888. Bd. 2: *Entstehung und Entwicklung der deutschen Universitäten bis zum Ausgang des Mittelalters.* Stuttgart: Cotta, 1896

Karl Kautsky, *Vorläufer des neueren Sozialismus* (vol. 1: *Kommunistische Bewegungen im Mittelalter.* 2nd ed. Berlin, Stuttgart: Dietz, 1909 [Russian translation, Moscow, 1919]

Karl Kautsky, *Thomas More und seine Utopie*: *mit einer historischen Einleitung.* 5th ed. Stuttgart: Dietz, 1922 [Russian translation, Moscow 1924]

Josef Kulischer, *Allgemeine Wirtschaftsgeschichte des Mittelalters und der Neuzeit.* (vol. 1: *Das Mittelalter*, vol. 2: *Die Neuzeit*) Munich, Berlin: Oldenbourg, 1928–1929 [Russian translation]

V.I. Lenin, *Карл Маркс* [Karl Marx (1918)] Lenin Institute [*Collected Works 21*, Moscow: Progress, 1960, 43–49]

N.A. Ljubimov, *История физики. Опыт изучения логики открытий в их истории* T. 1–3, Saint-Peterburg, 1892–1896], [History of Physics]

N. Lukin-Antonov, *Из истории революционных армий.* [History of Revolutionary Armies] Moscow, 1923

Paul Mantoux, *La Révolution industrielle au 18e siècle.* Paris: Société Nouvelle de Librairie et d'Édition, 1905 [Russian translation 1925]

F.S. Marvin, The significance of the seventeenth century. *Nature* 127, 7 February, 1931, 191–192

Karl Marx, *Das Kapital. Kritik der politischen Ökonomie,* vol. 1, chapt. 13: "Maschinerie und große Industrie" and chapt. 24: "Die sogenannte ursprüngliche Akkumulation" [Russian]

Karl Marx, *Das Kapital*, vol. 3, chapt. 20: "Geschichtliches über das Kaufmannskapital" [Russian translation]

Karl Marx, Friedrich Engels, *Die heilige Familie* [Russian translation 1907]

Karl Marx, Friedrich Engels, *Marx und Engels über Feuerbach* [first part of German Ideology]. Edited by D. Rjazanov, in: *Marx-Engels-Archiv. Zeitschrift des Marx-Engels-Instituts in Moskau 1,* 1926, 205–306 [Russian]

Karl Marx, Friedrich Engels, *Письма, под ред. АдораТского* [Selected Letters]. Edited by V.V. Adoratskij. Moscow, 3rd ed. 1928, 4th ed. 1931

James Clerk Maxwell, *Matter and motion.* London: Society for Promoting Christian Knowledge, 1876 [Russian]

Franz Mehring, *Eine Geschichte der Kriegskunst*. Stuttgart: Singer, 1908 [Russian translation, Moscow 1924]

Nature Nr. 2995, Supplement to issue of March 26, 1927

Isaac Newton, *Philosophiae naturalis principia mathematica.* [Russian translation by A.N. Krylov, Saint Petersburg, 1916]

Isaac Newton, *Opticks: or, a treatise of the reflections, refractions, inflections and colours of light.* [Russian translation by S.I. Vavilov, Moscow, 1927]

William Hyde Price, *The english patents of monopoly*. London: Constable 1906

Hastings Rashdall, *The Universities of Europe in the middle ages*, 3 vols. Oxford: Clarendon, 1895

Franz Reuleaux, *Theoretische Kinematik. Grundzüge einer Theorie des Maschinenwesens*. Braunschweig: Vieweg, 1875

Ferdinand Rosenberger. *Die Geschichte der Physik in Grundzügen mit synchronistischen Tabellen der Mathematik, der Chemie und beschreibenden Naturwissenschaften sowie der allgemeinen Geschichte*. Braunschweig: Viehweg, 1887–1890 [Russian translation]

Ferdinand Rosenberger, *Isaac Newton und seine physikalischen Principien: ein Hauptstück aus der Entwicklungsgeschichte der modernen Physik*. Leipzig, 1895

Русский астрономический календарь за 1927, Nishni-Novgorod, 1927 [*Russian Astronomical Calendar*, 1927]

Werner Sombart, *Der moderne Kapitalismus. Historisch-systematische Darstellung des gesamteuropäischen Wirtschaftslebens von seinen Anfängen bis zur Gegenwart.* 2nd ed. Munich and Leipzig: Duncker & Humblot, 1916–1927

Werner Sombart, *Die vorkapitalistische Wirtschaft*. Berlin: Duncker & Humblot, 1928 [Russian translation]

A.N. Savin, *История Англии в XVIII. в.* [History of England in the 18th Century] Moscow, 1912

A.N. Savin, *История Англии в новое время.* [History of modern England] Moscow, 1924

A.A. Svechin, *История военного искусства.* Moscow, 1920 [History of Warcraft]

Z. Tseitlin, *Наука и гипотеза.* [Science and Hypothesis] Moscow, 1926. (Change transliteration according to English convention)

Notes

1. The Russian title reads: "The Socio-Economic Roots of Newton's Mechanics." The original English version had the footnote on the first page: "The quotations cited in this essay have been translated from Russian. The chief exceptions are the quotations from *Nature* in Chap. 5."
2. The original English version had the section title "Marx's Theory of the Historical Process."
3. Whitehead's original reads: "Our modern civilization is due to the fact that in the year when Galileo died, Newton was born. Think for a moment of the possible course of history supposing that the life's work of these two men were absent." A.N. Whitehead, "The First Physical Synthesis" in *Science and Civilization*, ed. F.S. Marvin, Oxford University Press, 1923, pp. 161–178. This Whitehead passage is quoted by F.S. Marvin in the essay cited below in footnote 4.
4. F.S. Marvin, "The Significance of the 17th Century," *Nature 127,* Feb. 7, 1931, which is a (laudatory) book review of G.N. Clark's *The Seventeenth Century,* Oxford: Oxford University Press, 1929. Marvin was the senior figure among the organizers of the London Congress; Clark was the opening speaker of the first session of the Congress.
5. The following paragraphs contain a close paraphrase of Lenin's remarks on the materialist conception of history in "Karl Marx" (1918) in Lenin's *Collected Works 21*, pp. 43–91 (LW 21, 43–91, here LW 21, pp. 55–57), which in turn extensively quotes Marx from the Preface to the *Contribution to the Critique of Political Economy, Karl Marx/Friedrich Engels Collected Works*, vol. 29 (London: Lawrence & Wishart, 1975): CW 29, pp. 262–263; *Marx-Engels Werke*, Berlin: Dietz, 1964 (MEW 13, p. 9).
6. This and the next paragraph: LW 21, 57.
7. See *German Ideology,* CW 5, pp. 59–60 (MEW 3, pp. 46–48).
8. The original English version has a different order in the title: "Economics, Physics and Technology."
9. This page summarizes part of the section on Feuerbach in the *German Ideology,* CW 4, pp. 66–69 (MEW 3, pp. 51–56).
10. Marx's term is 'ständisches Kapital'.
11. German in the Russian original; literally, 'ground-touching-right'.
12. Heading not in the Russian.
13. Sentence added in the Russian.
14. Sentence added in the Russian.
15. These four paragraphs paraphrase the third paragraph of a letter of Marx to Engels (Sept. 25, 1857), CW 40, p. 186 (MEW 29, p. 192).
16. Two sentences added in the Russian.
17. Corrected according to the Russian. The original English version contains the translator's slip "Florence". This mistake took on a life of its own, since R.K. Merton (*Science, Technology and Society in Seventeenth-Century England* [1938] 2nd ed. New York: Harper, 1970, pp. 148, 187, 275) cited Hessen on the "Florentine Arsenal." H.F. Cohen (*The Scientific Revolution. A Historiographical Inquiry*, University of Chicago Press, 1994, p. 331) makes this slip the centerpiece of his diatribe against Hessen.
18. Sentence added in Russian.
19. With the exception of the reference to the Arsenal at Venice, the last nineteen paragraphs report facts taken from Friedrich Engels' article "Artillery" in the *New American Cyclopedia*, vol. 2, 1858, in: *Karl Marx/Friedrich Engels, Collected Works*, vol. 18. London: Lawrence & Wishart, 1982, pp. 188–210, here pp. 189–196. According to this edition, a collection of such articles was published in volume 11, part 2 of the Russian edition of Marx and Engels, *Works,* in 1933.
20. The Russian adds a section heading here. There are thus six sections in the Russian, five in the original English.
21. Athanasius Kircher (1602–1680); source not traced.
22. Galileo to Kepler, Aug. 19. 1610. Galileo Galilei, *Opere,* vol. 10, ed. by A. Favaro, Florence: Barbèra, 1890–1909, pp. 421–423; translated from the Latin.
23. See Appendix 2. Reference to appendix added in the Russian.

24. *Philosophiae Naturalis Principia Mathematica,* ed. by Thomas Le Seur and François Jacquier, 2nd ed., Cologne, 1760 (1st ed., Geneva, 1739–1742): "Declaratio. Newtonus in hoc tertio Libro telluris motae hypothesim assumit. Autoris propositiones aliter explicari non poterant, nisi eadem quoque facta hypothesi. Hinc alienam coacti sumus gerere personam. Caeterum latis a summis Pontificibus contra telluris motum Decretis nos obsequi profitemur." (vol. 3, front matter).
25. *Dialectics of Nature*, CW 24, p. 465 (MEW 20, pp. 456–457): "If after the dark night of the Middle Ages was over, the sciences suddenly arose anew with undreamt-of force, developing at a miraculous rate, once again we owe this miracle to production. [Marginal note:] Hitherto, what has been boasted of is what production owes to science, but science owes infinitely more to production."
26. The last three paragraphs paraphrase the Introduction to the English edition of *Socialism, Utopian and Scientific,* CW 24, p. 290 (MEW 19, p. 533): "Moreover, parallel with the rise of the middle class went on the great revival of science; astronomy, mechanics, physics, anatomy, physiology were again cultivated. And the bourgeoisie, for the development of its industrial production, required a science which ascertained the physical properties of natural objects and the modes of action of the forces of Nature. Now up to then science had but been the humble handmaid of the Church, had not been allowed to overlap the limits set by faith, and for that reason had been no science at all. Science rebelled against the Church; the bourgeoisie could not do without science, and, therefore, had to join in the rebellion."
27. Translation added in the Russian.
28. Horace (Quintus Horatius Flaccus), *Epistles* I, i, line 14. *Nullius addictus iurare in verba magistri*, "committed to affirming the doctrines of no master."
29. Presumably a slip of the pen referring to David Brewster, the nineteenth-century biographer of Newton and historian of the Royal Society cited in Appendices 3 and 4.
30. May 18, 1669, *The Correspondence of Isaac Newton*, vol. 1, ed. by H.W. Turnbull, Cambridge University Press, 1959, pp. 9–11. As G.N. Clark points out (1937, p. 365), the letter was in fact written a few months *before* Newton received his professorship. The letter was also apparently never sent. See R. Westfall, *Never at Rest. A Biography of Isaac Newton,* Cambridge: Cambridge University Press, 1980, p. 193.
31. Newton writes "Schemnitium" which the editors (1959) report refers not to Chemnitz but to Schemnitz (Selmeczbánya) in Hungary.
32. Appendix 3.
33. Sentence added in the Russian.
34. Engels' letter to Konrad Schmidt, Aug. 5, 1890, CW 49, p. 6 (MEW 37, pp. 435f): "And if this man has not yet discovered that while the material mode of existence is the *primum agens* this does not preclude the ideological spheres from reacting upon it in their turn, though with a secondary effect, he cannot possibly have understood the subject he is writing about."
35. This paragraph is a quote from the Introduction to the English edition of Engels' *Socialism Utopian and Scientific*, CW 24, p. 289 (MEW 19, pp. 532–533). "When Europe emerged from the Middle Ages the rising middle class of the towns constituted its revolutionary element. It had conquered a recognised position within medieval feudal organisation, but this position, also had become too narrow for its expansive power. The development of the middle class, the *bourgeoisie*, became incompatible with the maintenance of the feudal system; the feudal system had to fall." The next paragraphs are based on this work; see CW 24, pp. 290–294 (MEW 19, pp. 533–538).
36. Introduction to the English edition of *Socialism Utopian and Scientific:* "From that time, the bourgeoisie was a humble, but still a recognized, component of the ruling classes of England" CW 27, p. 293 (MEW 19, p. 536).
37. Except for one sentence on the Levellers, the last 5 paragraphs paraphrase Marx "The Bourgeoisie and the Counter-Revolution," *Neue Rheinische Zeitung, No. 169,* Dec. 15, 1848, CW 8, p. 161 (MEW 6, p. 107): "In both revolutions the bourgeoisie was the class that really headed the movement. The *proletariat* and the *non-bourgeois strata of the middle class* had either not yet evolved interests which were different from those of the bourgeoisie or they did not yet constitute independent classes or class sub-divisions. Therefore, where they opposed the bourgeoisie, as they did in France in 1793 and 1794, they fought only for the attainment of the aims of the bourgeoisie, even

if not *in the manner* of the bourgeoisie. *All French terrorism* was nothing but a *plebeian way* of dealing with the *enemies of the bourgeoisie*, absolutism, feudalism and philistinism.
"The revolutions of 1648 and 1789 were not *English* and *French* revolutions, they were revolutions of a European type. They did not represent the victory of a *particular* class of society over the *old political order*; they *proclaimed the political order of the new European society.* The bourgeoisie was victorious in these revolutions, but the *victory of the bourgeoisie* was at that time the *victory of a new social order*, the victory of bourgeois ownership over feudal ownership, of nationality over provincialism, of competition over the guild, of the division of land over primogeniture, of the rule of the landowner over the domination of the owner by the land, of enlightenment over superstition, of the family over the family name, of industry over heroic idleness, of bourgeois law over medieval privileges."

38. "Die 'glorious Revolution'... brachte mit dem Oranier Wilhelm III. die grundherrlichen und kapitalistischen Plusmacher zur Herrschaft." (MEW 23, p. 751; CW 35, pp. 713–714): "The 'glorious Revolution' brought into power, along with William of Orange, the landlord and capitalist appropriators of surplus value."
39. Hume: *History of England*, Chapt. LXXI: "By deciding many important questions in favour of liberty, and still more, by that great precedent of deposing one king, and establishing a new family, it gave such an ascendant to popular principles, as has put the nature of the English constitution beyond all controversy. And it may justly be affirmed, without any danger of exaggeration, that we, in this island, have ever since enjoyed, if not the best system of government, at least the most entire system of liberty, that ever was known amongst mankind."
40. *Holy Family,* MEW 2, p. 135, retranslated from the German; cf. CW 4, p. 128.
41. *Holy Family,* CW 4, p. 128 (MEW 2, p. 135): "In its further evolution, materialism becomes *one-sided. Hobbes* is the man who *systematises Baconian* materialism. Knowledge based upon the senses loses its poetic blossom, it passes into the abstract experience of the *geometrician. Physical* motion is sacrificed to *mechanical* or *mathematical* motion; *geometry* is proclaimed as the queen of sciences. Materialism takes to *misanthropy.* If it is to overcome its opponent, *misanthropic, fleshless* spiritualism, arid that on the latter's own ground, materialism has to chastise its own flesh and turn *ascetic.* Thus it passes into an *intellectual entity*; but thus, too, it evolves all the consistency, regardless of consequences, characteristic of the intellect."
42. Hobbes went into exile in the Fall of 1640 and returned to England in 1651.
43. G.N. Clark ("Social and Economic Aspects of Science," *Economic History 3*, 1937, 362, fn) points out that Overton is in fact mentioned in the *Dictionary of National Biography*; but, as he does not point out, the *Encyclopedia Britannica* (11th ed.) has no entry—nor does the Edinburgh *Enyclopedia.*
44. Sentence added in the Russian.
45. Richard Overton, *Man Wholly Mortal*, London, 1655; Hessen's quotes from and about Overton are taken from Eduard Bernstein, *Sozialismus und Demokratie in der großen englischen Revolution.* Chapt. 8., 4th ed. Stuttgart: Dietz, 1922, pp. 115–119.
46. Cited according to Bernstein 1922, p. 117: "Der Hauptvertreter der füchterlichen Lehre des Materialismus oder der Leugnung der Unsterblichkeit der Seele ist R.O., der Verfasser des Traktats über des Menschen Sterblichkeit."
47. Overton 1655, p. 10.
48. Overton 1655, pp. 20–21.
49. The last 5 paragraphs paraphrase the Introduction to the English edition of *Socialism Utopian and Scientific,* CW 27, pp. 293–294 (MEW 19, p. 536): "There was another factor that contributed to strengthen the religious leanings of the bourgeoisie. That was the rise of materialism in England. This new doctrine not only shocked the pious feelings of the middle class; it announced itself as a philosophy only fit for scholars and cultivated men of the world, in contrast to religion, which was good enough for the uneducated masses, including the bourgeoisie. With Hobbes, it stepped on the stage as a defender of royal prerogative and omnipotence; it called upon absolute monarchy to keep down that *puer robustus sed malitiosus* ["robust but malicious boy"]—to wit, the people. Similarly,

with the successors of Hobbes, with Bolingbroke, Shaftesbury, etc., the new deistic form of materialism remained an aristocratic, esoteric doctrine, and, therefore, hateful to the middle class both for its religious heresy and for its anti-bourgeois political connections. Accordingly, in opposition to the materialism and deism of the aristocracy, those Protestant sects which had furnished the flag and the fighting contingent against the Stuarts continued to furnish the main strength of the progressive middle class, and form even today the backbone of 'the Great Liberal Party'."

50. Sentence added in the Russian.
51. Engels to Conrad Schmidt, Oct. 27, 1890, CW 49, p. 62 (MEW 37, pp. 492–493): "In philosophy, for example, this is most easily demonstrated in respect of the bourgeois period. Hobbes was the first modern materialist (in the eighteenth-century sense), but an absolutist at a time when, throughout Europe, absolute monarchy was at its heyday and, in England, was embarking on a struggle with the populace. In religion as in politics, Locke was the product of the class compromise of 1688. The English deists and their more logical successors, the French materialists, were the true philosophers of the bourgeoisie—and, in the case of the French, even of the bourgeois revolution. German philosophy, from Kant to Hegel, is permeated by the German philistine—now in a positive, now in a negative, sense. But in every epoch philosophy, as a definite sphere of the division of labour, presupposes a definite fund of ideas inherited from its predecessors and from which it takes its departure. And that is why economically backward countries can nevertheless play first fiddle where philosophy is concerned—France in the eighteenth century as compared with England, upon whose philosophy the French based themselves and, later on, Germany as compared with both."
52. Sentence added in the Russian.
53. Feb. 21, 1689 (*The Correspondence of Isaac Newton,* vol. III, ed. by H.W. Turnbull, Cambridge University Press, 1961, p. 12): "1. Fidelity & Allegiance sworn to the King, is only such a Fidelity & Obediance as is due to him by the law of the Land. For were that Faith and Allegiance more then the law requires, we should swear ourselves slaves & the King absolute: whereas by the Law we are Free men notwithstanding those oaths.
    "2. When therefore the obligation by the law to Fidelity and allegiance ceases, that by oath also ceases ...
    "3. Fidelity & Allegiance are due by the law to King William & not to King James."
54. Francis Bacon, *Novum Organum* bk. II §2, *The Works of Francis Bacon,* vol. 1, ed. by J. Spedding, R. Ellis, and D.D. Heath, London, 1861.
55. Quote not located.
56. Engels, *Dialectics of Nature*, CW 25, p. 475 (MEW 20, p. 466).
57. Engels, *Dialectics of Nature*, CW 25, p. 480 (MEW 20, p. 471): "Newton allowed Him the 'first impulse' but forbade Him further interference in his solar system." The second clause of the sentence—which is false—is quoted only in the Russian.
58. Boyle died on Dec. 30, 1691 (English style) which was already 1692 on the continent.
59. Letter to Bentley, Dec. 10, 1692, *Correspondence* III, 233.
60. Descartes, René, *Principia Philosophiae*, part II, §64, *Oeuvres de Descartes,* ed. by Ch. Adam and P. Tannery, Paris: Vrin, 1964–1974. *The Philosophical Writings of Descartes*, 3 vols. (transl. by J. Cottingham, R. Stoothoff, D. Murdoch, A. Kenny) Cambridge University Press, vol. 1, p. 247.
61. *Holy Family,* CW 4, p. 125 (MEW 2, p. 133).
62. Sentence added in the Russian.
63. John Toland, *Letters to Serena,* London, 1704, pp. 158–160. The passage is not set in quotes in the Russian.
64. *Isaac Newton, Opticks* (New York: Dover, 1952, pp. 369–370): "What is there in places almost empty of Matter ... Whence is it that Nature doth nothing in vain; and whence arises all that Order and Beauty which we see in the World? ... And these things being rightly dispatch'd, does it not appear from Phaenomena that there is a Being incorporeal, living, intelligent, omnipresent, who in infinite Space, as it were in his Sensory, sees the things themselves intimately, and throughly perceives them, and comprehends them wholly by their immediate presence to himself ..."
65. See Appendix 4.
66. "Living force" (*vis viva*) is Leibniz's term for the conserved magnitude mv^2.

67. Newton *Principia* Preface (not Introduction): *The Principia. Mathematical Principles of Natural Philosophy,* (transl. by I.B. Cohen and A. Whitman) Berkeley, Los Angeles: University of California Press, 1999, pp. 382–383.
68. *Capital,* vol. 3, Chapt. 20, CW 37, pp. 328–329 (MEW 25, pp. 343).
69. Karl Marx, *Capital*, CW 35, p. 376 (MEW 23, p. 393): "All fully developed machinery consists of three essentially different parts, the motor mechanism, the transmitting mechanism, and finally the tool or working machine." See also Marx's remarks in a letter to Engels, Jan. 28, 1863 (MEW 30, pp. 320–322; CW 41, pp. 449–451).
70. Marcus Vitruvius Pollio: *De Achitectura* 10.1.1: "Machina est continens e materia coniunctio maximas ad onerum motus habens virtutes."
71. Franz Reuleaux, *Theoretische Kinematik. Grundzüge einer Theorie des Maschinenwesens,* Braunschweig: Vieweg, 1875, p. 38 (not marked as quote). Retranslated from the German. In the original English version the sense of the passage was lost in the translation from German to Russian to English: "Eine Maschine ist eine Verbindung widerstandsfähiger Körper, welche so eingerichtet ist, dass mittelst ihrer mechanische Naturkräfte genöthigt werden können, unter bestimmten Bewegungen zu wirken."
72. Werner Sombart's well-known distinction reads: "A tool is a means of labor that serves to support human labor (sewing needle). A machine is a means of labor that is supposed to replace human labor, that thus does what a human would do without it (sewing machine)" (*Der moderne Kapitalismus*, vol. I, part 1, p. 6).
73. The last sentence of the previous paragraph seems to allude to a passage in Marx's *Capital*: "In handicrafts and manufacture, the workman makes use of a tool, in the factory, the machine makes use of him" (CW 35, p. 425; MEW 23, p. 445). In this paragraph Hessen interprets the passage not as a definition of machine and tool but as a characterization of two different relations of the workman to these means of production.
74. *Capital,* CW 35, p. 375 (MEW 23, p. 392): "When in 1735, John Wyatt brought out his spinning machine, and began the industrial revolution of the 18th century, not a word did he say about an ass driving it instead of a man, and yet this part fell to the ass. He described it as a machine 'to spin without fingers.' "
    Marx's discussion of the definition of a machine ocurrs at the beginning of the chapter on machinery and modern industry. (CW 35, pp. 374–379; MEW 23, pp. 391–396) Marx's criterion for the distinction between tool and machine is that the latter is "emancipated from the organic limits that hedge in the tools of a handicraftsman" (CW 35, p. 377): "The machine, which is the starting-point of the industrial revolution, supersedes the workman, who handles a single tool, by a mechanism operating with a number of similar tools, and set in motion by a single motive power, whatever the form of that power may be" (CW 35, p. 379).
75. Sadi Carnot, *Réflexions sur la puissance motrice du feu: et sur les machines propres à développer cette puissance,* Paris: Bachelier, 1824, p. 3: "Le service le plus signalé que la machine à feu ait rendu à l'Angleterre est sans contredit d'avoir ranimé l'exploitation des ses mines de houille, devenue languissante et qui menaçait de s'eteindre entièrement à cause de la difficulté toujours croissante des épuisemens et de l'extraction du combustible."
76. *The Origin and Progress of the Mechanical Inventions of James Watt*, London: Murray, 1854, p. 18.
77. In Russian: 45 thousand marks.
78. Carnot 1824, p. 2. "L'étude de ces machines est du plus haut intérêt, leur importance est immense, leur emploi s'accroît tous les jours. Elles paraissent destinées à produire une grande révolution dans le monde civilisé."
79. Carnot 1824, pp. 6–7: "L'on a souvent agité la question de savoir si la puissance motrice de la chaleur est limitée, ou si elle est sans bornes; si les perfectionnemens possibles des machines à feu ont un terme assignable, terme que la nature des choses empêche de dépasser par quelque moyen que ce soit, ou si au contraire ces perfectionnemens sont susceptibles d'une extension indéfinie."
80. Carnot 1824, pp. 9–10: "On ne la possédera que lorsque les lois de la physique seront assez étendues, assez généralisées, pour faire connaître à l'avance tous les effets de la chaleur agissant d'une manière déterminée sur un corps quelconque."

81. Engels, *Dialectics of Nature*, CW 25, p. 528 (MEW 20, p. 514): "*Classification of the sciences*, each of which analyses a single form of motion, or a series of forms of motion that belong together and pass into one another, is therefore the classification, the arrangement, of these forms of motion themselves according to their inherent sequence, and herein lies its importance." Reference to Engels not in the Russian.
82. Andrew Ure (1778–1857) English chemist and economist: "Dr. Ure, the Pindar of the automatic factory" (Marx, *Capital,* CW 35, p. 421; MEW 23, p. 441).
83. Andrew Ure, *The Philosophy of Manufactures or an Exposition of the Scientific, Moral and Commercial Economy of the Factory System of Great Britain*, London, 1835, pp. 368–370, quoted in *Capital,* CW 35, p. 439 (MEW 23, p. 460).
84. Following quotes corrected according to "Science and Society," *Nature 126,* No. 3179, Oct. 4, 1930, p. 497.
85. Minor errors in the following quotes have been corrected according to W.G. Linn Cass, "Unemployment and Hope," *Nature 125,* No. 3146, Feb. 15, 1930, pp. 226–227. The passages quoted in the Russian and English versions are for the most part the same but do not entirely coincide.
86. Hessen paraphrases the preface to Marx, *Contribution to the Critique of Political Economy* (CW 29, pp. 263–264; MEW 13, p. 9; and LW 21, pp. 55–57).
87. Last phrase added in the Russian.

The Social Foundations of the Mechanistic Philosophy and Manufacture*

Henryk Grossmann

Borkenau's Theory

[161]

The following is not meant to be a résumé of the contents of Borkenau's book *The Transition from the Feudal to the Bourgeois World-Picture*. Rather, these are but a few economic-historical and sociological reflections on some problems connected with the book's main subject, leaving aside problems involving philosophy and the history of ideas

Borkenau wants to show that the metamorphosis of the image of nature in the course of historical development "can only be understood from the changes in the image of the world in general" (p. 15). These again do not only depend on the experiences derived from the process of production, but also on the "general categories" which, by virtue of their being organizing concepts, hold the world-picture together. All experience is as such subject to change through categories which are themselves [162] changing in the course of history: "which experience is being sought and accepted,

---

* The concept 'manufacture' is taken from Marx, *Capital*, Vol. I, Chap. 14 (chap. 12 in the fourth German edition). Since this concept is not familiar in current English usage, a brief quotation from that chapter may prove helpful in understanding Grossmann's views. The term refers to production characterized by division of labor in which the labor process is not yet simplified enough that a mechanism can replace the skill of the craftsman.

"That co-operation which is based on division of labour, assumes its typical form in manufacture, and is the prevalent characteristic form of the capitalist process of production throughout the manufacturing period properly so called. That period, roughly speaking, extends from the middle of the sixteenth to the last third of the eighteenth century... For a proper understanding of the division of labour in manufacture, it is essential that the following points be firmly grasped. First, the decomposition of a process of production into its various successive steps coincides, here, strictly with the resolution of a handicraft into its successive manual operations. Whether complex or simple, each operation has to be done by hand, retains the character of a handicraft, and is therefore dependent on the strength, skill, quickness, and sureness, of the individual workman in handling his tools. The handicraft continues to be the basis. This narrow technical basis excludes a really scientific analysis of any definite process of industrial production, since it is still a condition that each detail process gone through by the product must be capable of being done by hand and of forming, in its way, a separate handicraft" (Karl Marx, *Capital*, Vol. 1, transl. S. Moore and E. Aveling, edited by F. Engels (1887), Moscow: Progress Publishers, 1965, pp. 338–339; see also pp. 342–348) [G.F./P.M.].

G. Freudenthal, P. McLaughlin (eds.), *The Social and Economic Roots of the Scientific Revolution,* Boston Studies in the Philosophy of Science 278,
DOI 10.1007/978-1-4020-9604-4_3,

whatever appears to be evident, empirical or nonsensical – all this depends on the dominant categories." The book undertakes the task of demonstrating this historical change in the basic categories in relation to the natural sciences and "to render comprehensible their connection with social life" (p. 15). Borkenau wants to present the process of the reification of consciousness, as described in the introductory part of the book (pp. 15–96). He leads us from the flowering of scholasticism via the late Renaissance and Francis Bacon to the threshold of Cartesian philosophy – the principal topic of the book. Dealing with the beginning of this development, which started out with Thomas Aquinas, Borkenau explicitly states the priority of the social sphere. The natural law is guided by the "natural," i.e. class-determined, order of society with its hierarchical structure, and the world is understood analogously as a harmonious ordering of its parts in its ultimate relation to God. Since all nature exists for society and the latter is and should be a harmoniously arranged cosmos of fixed structures, "the" scholasticism's conception of natural law is static: "The Thomist system excludes modern dynamics and all of modern natural science which is based thereon" (p. 34).

With the erosion of feudalism due to the advent of the monetary system and of capitalism, the optimistic-harmonious view of the universe in Thomist doctrine is replaced by the pessimistic doctrine of the separation and antagonism between [non-]rational affects and the natural law. There follows a gradual transformation of the concept of natural law and the reversal of the order of precedence between natural and human law. During the Renaissance, human fate was regarded as accidental, at the mercy of unpredictable external forces. Yet, even here in this wicked world, God's influence becomes visible in the contemplation of harmoniously ordered nature. Nature, which in classic scholasticism ranked lowest in the divine plan of the world, attains a higher rank, and human society should be understood – and legitimized – only by the cognition of nature. The reversal of the hierarchy is complete.

Contrary to scholasticism, the Renaissance took upon itself the task of concretely exploring nature. But the Renaissance was not interested in the cognition of nature as such – the cognition of the causal connection between the parts of nature by means of quantitative measuring methods – but in the "interpretation of the concrete world as a whole, as a system of harmonic measures" (p. 65). The mathematical proportion of the whole universe was to be shown in the seemingly chaotic flow of nature; only from this viewpoint are all parts of nature important and is the concrete exploration of nature significant. [163]

This attitude of the Renaissance to the exploration of nature is understandable. Within the monetary and commercial capitalism of the Renaissance period, there still was lacking any attempt at elaborating quantitative methods for the exploration of nature. Therefore the philosophy of that period remained qualitative. Only with the development of industrial capitalism and its first expression – *manufacture* – did the quantitative methods evolve; "only the application of capitalist methods in the production process enabled an observation of nature by quantitative methods" (p. 54). Since manufacture, which already arose in the sixteenth century, only developed in the seventeenth century, it is understandable that the formation of the modern world-picture, based upon exact quantitative methods, only became possible at the beginning of the seventeenth century. The culmination of the "process of reification" of consciousness was reached in Descartes; for him it is an established fact that everything that happens in human life (apart from thinking) is of a purely external contingency which, however, is governed by laws which conform to reason. With this "the bell is ringing in the birth of the modern concept of natural law" (p. 358) and simultaneously of the mechanistic world view.

The mechanistic world view prevailed because of the thorough revolution of social relations which took place at the turn of the sixteenth to the seventeenth century. Italy had been affected by it only temporarily; therefore the mechanistic exploration "was soon suffocated there by the blows of the Counter-Reformation" (p. 14). In France, Holland and England the development was different. "In all these three countries it is at that great turning point that the industrial bourgeoisie and its related class of the gentry first appeared on the stage as an independent power and soon occupied the center of the stage. . . . This historic change immediately preceded the emergence of the mechanistic world-picture; it brought it about" (p. 14). Yet Borkenau did not describe this "revolution of social relations," which was decisive for the emergence of the mechanistic world-picture, for any of the countries with which he was dealing. Neither did he demonstrate why Italy had been "only temporarily affected" by this revolution. Rather, we have to deduce from incidental remarks, scattered throughout his book, how Borkenau perceives this situation which is so decisive for his research.

The specific carrier of the Renaissance world view is the monetary and commercial capital, viz. in the first instance the "booty-capitalism," (p. 215) the "adventuresome capitalism" (pp. 155, 157) which – in contrast to the "solid" manufacturing capital (p. 155) of the later period – remained exclusively in the sphere of circulation (p. 89) and stood apart from the capitalistic labor process and thus from its rational shaping (p. 155); this view of a class remote from the labor process could only be a harmoniously balanced ideology, an estheticism which despised the life of the masses. Only when the monetary capital entered the sphere of production, which – despite the repeated endeavors in this direction during the sixteenth century – did [164] not have "the first decisive success" before the beginning of the seventeenth century, did the "first period of capitalistic industry, the period of manufacture" arise (pp. 89, 90). This also implied an important revolution in the history of science as well as in the history of philosophy. For the monetary capital, remote from the labor process, could not create rational techniques; the latter was "adequate to capitalism alone and was realized . . . for the first time during the period of manufacture" (p. 90).

The representatives of these new manufacturing techniques are not the "religiously indifferent capitalists," but the "upward-struggling Calvinist little people." The rational techniques of manufacture "emerged from the efforts toward the rationalization of handicraft" (p. 90), whereas the monetary capitalists are lacking "any motivation to rationalize the techniques systematically" (p. 90). Borkenau declares, though, that "during the Renaissance innumerable inventions had been made by practitioners, some of them of the highest significance; but incidentally and without the possibility of systematical perfection" (pp. 88, 89). But things had been different in the case of manufacturing techniques: "The simple basic forms of modern techniques," which "became the foundations of the mechanistic world-picture ... developed quite apart from the inventions of the Renaissance" (p. 90).

The essence of these new techniques, which were so important for the new world-picture was, "as is well known," nothing but an extreme division of labor, retaining at the same time all the craftsmanlike essentials of the production process (p. 2). Through division of labor, the skilled worker in manufacture was replaced by unskilled laborers whose work consisted of "the execution of a quite simple movement" (p. 7). Thus there was no longer any need for specialized training, work lost its quality and "became mere quantity." This meant that qualified work was replaced by "general human" or "abstract work," which constitutes the basic concept of modern mechanics. Thus it is evident that manufacture constitutes a necessary condition for the development of the basic concepts of modern Galilean mechanics, "in that for the first time it created abstract work and abstract matter" (p. 13).

Galilean mechanics or "one related to it" was, however, a condition of the mechanistic world picture, since this new philosophy was nothing but the demonstration "that all processes in nature can be explained in a mathematical-mechanistic way" (p. 10), that all natural phenomena can be reduced to meaningless changes of matter, i.e. to impact and motion (p. 12). This interlocking chain of deductions provides evidence that the mechanistic world-picture is only "an extrapolation from the manufacturing processes to the cosmos as a whole" (p. 12). The mechanistic world-picture prevailed simultaneously with modern mechanics and modern philosophy (p. 10): "The rejection of qualitative philosophy, the creation of the mechanistic world-picture is a radical change that started around 1615 and had its culmination in Descartes' *Discours* (1637), Galileo's *Discorsi* (1638), and Hobbes's *Elements* (1640)" (p. 13). [165]

The Reality of Historical Development

The historian has methodological doubts from the very beginning: Does history really take so rectilinear a course as Borkenau would have it? Do the single stages of the process really follow each other in such a sequence that one can speak of *the* world-picture of Scholasticism, of *the* Renaissance and of modern times as clearly distinct concepts? And are there never any regressions – often lasting for centuries – which also should be taken into account and explained? Yet doubts arise not only with regard to the succession in time but also to the proximity in space: do not different world-pictures coexist in every period, e.g. in the Scholastic, hence rendering the scholar's work even more complicated; does he not also have to explain this particular coexistence? Are these world-pictures not equally differentiated as the social circumstances of the times? And furthermore: Should it not be assumed that the various disciplines develop at a quite uneven pace; that in northern Italy, e.g., astronomy, mathematics, and mechanics attained a higher stage of development than anatomy and other disciplines? It appears to us that the real task is the tracing of the concrete connections between the various areas of material social life and the individual disciplines.

We might have expected that Borkenau, adducing characteristic examples from the history of the sciences and their particular disciplines, would demonstrate and directly explain their prevailing basic categories and their metamorphoses by studying the historical material to be analyzed. In order to fulfill the task outlined in the title of the book, to demonstrate the "transition from the feudal to the bourgeois world-picture," it would have been necessary to describe, in the separate spheres of positive knowledge of nature, those social and intellectual processes through which the feudal world-picture was upset and the germ of the modern world picture was developed simultaneously. But Borkenau got stuck in generalities: the empty formula of the eroding influence of the upcoming monetary and commercial capitalism upon the harmonious stratified hierarchical feudal order is supposed to explain phenomena which can only be elucidated by a closer study of the networks of facts of material life! Borkenau feels correctly that, with such a formula as the methodological instrument of analysis, the task cannot be fulfilled, and he actually narrows [166] this task with the aid of a syllogism: the concept of natural law is *the* fundamental category of our image of nature. Instead of presenting the change of categories in the history of sciences, he gives the history of the development of the *concept* of natural law, i.e. the "history of the *word*" (p. 19).

The examination of Borkenau's construction as to its content of reality appears to us even more necessary since in itself it constitutes a revolution in the previously prevailing conceptions. A few of the most important points will be enumerated here:

1. The assumption seems to suggest itself that mechanistic philosophy and scientific mechanics derived their basic mechanical concepts from the observation of mechanisms, of *machines*. Borkenau however deduces the rise of mechanical conceptions not from the machines but from the division of human labor in the crafts.

2. The beginning of modern natural sciences, viz. of a complex of knowledge having at its disposal both exact methods of investigation and the formulation of fundamental laws governing a certain sector of nature, has been placed, usually, in the second half of the fifteenth century, but the *beginnings* of exact research date even farther back. Borkenau negates more than 150 years of the history of science with its "increasingly speedy" progress, and postpones the birth of modern science to the turn of the sixteenth to the seventeenth century.

3. According to Borkenau the elaboration of exact scientific methods, which some scholars already ascribe to the Arabs of the Middle Ages, viz. at least to the twelfth and thirteenth centuries, arises between the sixteenth and seventeenth centuries together with the dissemination of the division of labor in manufacture. Here he even negates three to four centuries of development. Before considering Borkenau's further deviations from the previous state of knowledge, we have to delve more deeply into this question of the beginnings of scientific mechanics.

Presenting here the beginnings and development of scientific mechanics since the end of the fifteenth century would lead us too far. Suffice it to mention the name of Leonardo da Vinci. Today, after the publication of the most important heritage of Leonardo's manuscripts from the library of the Institut de France (1881–1901), of the *Codice Trivulziano* (1891) and *Codice Atlantico* from the Abrosiana of Milan [167] (1894), of the *Windsor manuscripts* (1901), of the *Codice sul volo degli uccelli* (1893) – so important to theoretical mechanics – and especially of the mechanical manuscripts of the Kensington Museum in London (1901),[1] and after the pioneering research by P. Duhem into Leonardo,[2] determining the time of Galileo and Descartes as the beginnings of scientific mechanics would mean ignoring at least fifty years of scientific research. "Nowhere," says Borkenau, "does the Renaissance seek knowledge for mere knowledge's sake" (p. 73). It was only interested in the symbolism of the circle with God as its center of attraction, whereas scientific research was but a by-product of this attitude. Only where the circular form was applicable, as in astronomy, did science progress as far as the formulation of precise laws; beyond astronomy, therefore, the attempt at framing the phenomena in mathematically exact laws proved unavailing. The contribution of the Renaissance to our contemporary knowledge of nature was "pure natural history; an accumulation of an immense, often valuable, mass of material, an empirical acceptance" (p. 72), and "an entirely unmathematical method of observation" (p. 80). There followed in the second half of the sixteenth century an upward turn in the sciences which described nature; the use of experiments was demanded but not systematically carried out (p. 80).

One need only lay one's hand on Leonardo da Vinci's manuscripts, only consult any general presentation of Leonardo's scientific achievements (e.g. G. Séailles' book)[3] in order to become convinced that every sentence in the above description of "the" Renaissance is quite incredible. It is an established fact that Leonardo in his research used exact quantitative methods, that he stressed the general applicability of mathematics. Libri, the learned historian of the mathematical sciences in Italy, reports: "Léonard étudiait la mécanique et la physique avec le secours de l'algèbre et de la géometrie ... et appliqua cette science à la mécanique, à la perspective et à la theorie des ombres."[4] It is also certain that Leonardo not only always demanded [168] the systematic use of experiments but also actually carried them out in various fields of science – see the book on the flight of birds.[5] There is no doubt nowadays that

---

[1] Leonardo da Vinci, *Problèmes de géométrie et d'hydraulique. Machines hydrauliques. Application du principe de la vis d'Archimède. Pompes, machines d'épuisement et de dragage*, Paris, 1901, Vols. I–III.

[2] P. Duhem, *Les origines de la statique*, Paris, 1905/6, Vols. I/II; *Études sur Léonard de Vinci*, Paris, 1906, Vols. I/II.

[3] *Léonard de Vinci, l'artiste et le savant*, Paris, 1906.

[4] *Histoire des sciences mathématiques en Italie*, Paris, 1840, Vol. III, p. 46. Leonardo writes: "Qu'il ne me lise pas celui qui n'est pas mathématicien, car je le suis toujours dans mes principes." (Peladan, *Léonard de Vinci, Textes choisis*, Paris, 1907, p. 34), and "La mechanica e il paradiso delle science matematiche perche con quella si viene al frutto matematico" (Duhem, *Les origines de la statique*, Vol. I, p. 15).

[5] Leonardo writes: "When dealing with scientific problems I first make some experiments, because I intend to pose the question according to experience, and then to prove why the bodies are compelled to act in the manner demonstrated. This is the method according to which one should

Leonardo not only knew the contents of the most important basic laws of mechanics, hydrostatics, hydrodynamics, optics, aerodynamics, and several other sciences, and precisely formulated these laws, but that he also already laid the foundations of a comprehensive mechanistic world-picture.[6]

Leonardo knew the basic principle of mechanics, the law of inertia, the impossibility of the perpetuum mobile, and he fought opposing views, even though until now the discovery of the principle of the impossibility of the perpetuum mobile was ascribed to Simon Stevin (1605).[7] Leonardo recognized "la loi d'équilibre de la balance ou du levier."[8] With regard to the parallelogram of forces, he provides an equivalent, mathematically precisely formulated solution: "le moment d'une résultante de deux forces est égal à la somme des moments des composants."[9]

According to Borkenau (who refers to Duhem) the "beginning of the calculations of the center of gravity" was in the mid-seventeenth century (p. 35). Duhem really shows that Leonardo had already made these calculations.[10] And before him M. Cantor, in his "Vorlesungen über die Geschichte der Mathematik," had stated that Leonardo correctly determined the center of gravity of a pyramid with a triangular base.[11] [169]

"Cent ans avant Stevin et avant Galilée Léonard établit . . . la chute d'un corps qui suit la pente . . . d'un plan incliné."[12] Leonardo provides an exact calculation of the speed of the fall on an inclined plane. "There is no doubt," says Hermann Grothe as early as 1874, "that by the end of the fifteenth century Leonardo had already clearly and distinctly formulated many laws of mechanics, and these bestow on

---

proceed in all explorations of the phenomena of nature." (Cf. August Heller, *Geschichte der Physik von Aristoteles bis auf die neueste Zeit*, Stuttgart, 1882, Vol. I, p. 237. Kurt Lasswitz, *Geschichte der Atomistik*, Hamburg 1890, Vol. II, p. 12.) "This clear insight," adds Lasswitz, "into the essence of the experimental-mathematical method of natural sciences determines da Vinci's procedure and success."

[6] "Cent ans avant Galilée," says G. Séailles (op. cit. p. 220), "Léonard a posé les vrais principes de la mécanique; cent cinquante ans avant Descartes il a pressenti en elle l'idéal de la science. Il semble qu'il lui ait dû l'idée même de sa méthode. Observer les phénomènes, les reproduire artificiellement, découvrir leurs rapports, appliquer à ces rapports la mesure, enfermer ainsi la loi dans une formule mathématique qui lui donne la certitude déductive d'un principe que confirment ses conséquences, c'est la méthode même de Léonard et celle de la mécanique."

[7] E. Mach, *Populär-wissenschaftliche Vorlesungen*, Leipzig, 1903, p. 169. [E. Mach, *Popular Scientific Lectures*, transl. by Th. J. McCormack, La Salle: The Open Court, 1943, p. 140f.]

[8] P. Duhem, *Les origines de la statique*, Vol. I, p. 19.

[9] Op. cit., Vol. I, p. 32. Cf. Vol. II. pp. 347f.

[10] Op. cit., Vol. II, p. 111.

[11] Leipzig, 1899, Vol. II, pp. 302, 570. Séailles says: "Commandin (1565) et Maurolycus (1685) se disputaient jusqu'ici l'honneur de ces découvertes" (op. cit., p. 225). The calculations of Maurolycus, though executed in 1548, were only published in 1685. (Cf. Libri, *Histoire des sciences mathématiques*, Vol. III, p. 115.)

[12] Séailles, op. cit., p. 229. Similarly Eug. Dühring, *Kritische Geschichte der Principien der allgemeinen Mechanik*, 3rd ed., Leipzig, 1887, pp. 12–17.

Leonardo ... at least equal importance for mechanics as was ascribed to Stevinus – and moreover the precedence."[13]

Leonardo's pioneering work in the field of comparative anatomy is based on the realization that the functions of the animal body and the motion of its limbs are governed by the laws of mechanics. "The whole world, including living things, is subject to the laws of mechanics; the earth is a machine, and so is man. He regards the eye as a camera obscura, ... he determines the crossing point of the reflected rays."[14]

In aerodynamics he provides a mechanical theory of air pressure. "Plus étonnantes sont les expériences sur le frottement et les lois qu'il sut en déduire ... Ainsi, deux siècles avant Amonton (1699), trois siècles avant Coulomb (1781), de Vinci avait imaginé leurs expériences et en avait fixé à peu près les mêmes conclusions."[15] In hydrodynamics and hydrostatics, Leonardo discovers the basic mechanic laws of liquids. "Il faut rectifier sur ce point l'histoire de la science positive." Leonardo "a l'idée nette de la composition moléculaire de l'eau ...; un siècle et demi avant Pascal, il observe les conditions d'équilibre de liquides placés dans des vases communicants."[16] In hydrodynamics: "plus de cent ans avant le traité de Castelli (*Della misura dell'aqua corrente*, 1638), Leonardo cherche la quantité d'eau qui [170] peut s'écouler par une ouverture pratiquée à la paroi d'un canal" ... "Il calcule la vitesse de l'écoulement de l'eau ... Il donne la théorie des tourbillons, il en produit d'artificiels pour les mieux observer."[17] "La mise au jour des manuscrits de Léonard de Vinci recule les origines de la science moderne de plus d'un siècle ... Conscience de la vraie méthode ... union féconde de l'expérience et des mathématiques, voilà ce que nous montrent les carnets du grand artiste. Pratiquée avec génie, la nouvelle logique le conduit à plusieurs des grandes découvertes attribuées a Maurolycus, Commandin, Cardan, Porta, Stevin, Galilée, Castelli."[18]

And now the essential point: according to Borkenau, the concept of mechanical work has its origins at the beginning of the seventeenth century only, in connection with the division of industrial labor and with highly skilled work being replaced by "general human" work. In fact, the concept of mechanical work was already well known to Leonardo by the end of the fifteenth century and he developed it from observing the effect of machines which replace human performance. In comparing the work of machines with the human work it replaces, both types of

---

[13] *Leonardo da Vinci als Ingenieur und Philosoph*, Berlin, 1874, p. 21; cf. p. 92. Similarly M. Herzfeld, *Leonardo da Vinci*, Leipzig, 1904, CXIII. The influence of Italian science is demonstrable in other areas of Stevin's work as well: he introduced into Holland the Italian bookkeeping system whose beginnings in Florence go back to the book by Luca Paccioli (1494), Leonardo's friend. (Cf. E. L. Jäger, *Luca Paccioli und Simon Stevin*, Stuttgart, 1876.)

[14] M. Herzfeld, op. cit., CXXII, CXV.

[15] Séailles, op. cit., p. 231.

[16] Op. cit., pp. 232–34. Leonardo writes: "Le superficie di tutti i liquidi immobili, li quali infra loro sieno congiunti, sempre sieno d'equale altezza," independent of the *width* and *shape* of the vessels, and he shows that the height of the columns of liquid is inversely proportional to their *weight* (density). (Cf. A. Heller, *Geschichte der Physik*, Vol. I, p. 242.)

[17] Séailles, op. cit., pp. 235/236.

[18] Op. cit., pp. 369/370.

work are reduced to a common denominator, to the concept of mechanical work. Thus Leonardo calculates the work of a water wheel which activates a machine.[19] From the knowledge of the basic laws of mechanics he already calculates the amount of work to be performed by machines while building them, and in the case of the rolling mill for iron rods which he constructs he calculates both the load and the force (work) to be applied to pull the iron along under the rollers. As theoretical basis for this calculation he uses his study "Elementi machinali" (which has evidently been lost), to which he often refers.[20] A similar case is that of a spinning jenny which he invented around 1490.[21] And there is more. Leonardo does not confine himself to such calculations, he even constructs an apparatus for this purpose: "Pour calculer l'effet des machines il inventa un dynamomètre; il détermina le maximum [171] de l'action des animaux en combinant leur poids avec la force musculaire."[22]

It would be superfluous to add further examples. Is Borkenau perhaps of the opinion that Leonardo da Vinci's achievements need not be taken into account because his writings were not published, and therefore had no influence on the emergence of scientific mechanics? Did Venturi not believe, when he rediscovered Leonardo's manuscripts in 1797, that, because they were unkown, mankind had been compelled to make his discoveries a second time? But would not such an argument be irrelevant, since the primary problem lies in the questions of why and how could Leonardo da Vinci already lay the foundations of scientific mechanics in the end of the fifteenth century?

Besides, P. Duhem has established – and therein in particular lies the importance and originality of his research – that Leonardo's mechanics did not remain as unknown as hitherto presumed; that, e.g., Galileo frequently quotes Girolamo Cardano, who himself was undoubtedly influenced by Leonardo;[23] that a long list of writers, conscious or unconscious plagiarists, knew Leonardo's mechanics and made use of its results throughout the sixteenth century; and that, through their very intermediary, his influence had a mighty effect on the works of Stevin, Kepler, Descartes, Roberval, Galileo, Mersenne, Pascal, Fabri, Christian Huygens and others. These intermediaries, such as Cardano, Tartaglia, Benedetti, or bold plagiarists like Bernardino Baldi, rendered an important service to mankind in rescuing Leonardo's ideas and inventions from oblivion and introducing them into the wide mainstream of science.[24]

---

[19] Duhem, *Les origines de la statique*, Vol. I, p. 21.

[20] H. Grothe, *Leonardo da Vinci als Ingenieur*, op. cit., p. 77. Cf. furthermore August Heller, op. cit., p. 242.

[21] H. Grothe, op. cit., p. 82. Leonardo calculates, e.g., the force (work) required for hammering in nails and bolts, regarding them as wedges. A. Heller, op. cit., p. 242.

[22] G. Libri, *Histoire des sciences mathématiques*, Vol. III, p. 42. Cf. there also appendix VII, p. 214: "Della forza dell'uomo."

[23] Duhem, *Les origines de la statique*, Vol. I, pp. 40, 44.

[24] Op. cit., Vol. 1, pp. 35, 147; *Études sur Léonardo da Vinci*, Vol. I, pp. 108, 127. Olschki (*Galilei und seine Zeit*, Halle, 1928) however says that pre-Galilean mechanics had quite a different character (he did not show wherein this difference lies) and that, therefore, the predating of the origins of scientific mechanics is due to the "malice" of Galileo's "detractors." But this is not a matter of

The fact that modern mechanics already originated with Leonardo da Vinci at the end of the fifteenth century has been recognized for these past fifty years by a [172] majority of leading scholars; for example, K. Lasswitz wrote in 1890: "Thus the new mechanics, and modern scientific thinking generally, begins with the admirable genius of Leonardo da Vinci who was so incredibly far ahead of his time."[25] The task lies, though, in rendering the "incredible" "credible," i.e. not conceiving of the phenomenon of Leonardo da Vinci as an "individual phenomenon beyond the context of historical development" (Lasswitz), but rendering it comprehensible from the social development of his epoch.

However – and here we come to our decisive conclusion – if Borkenau nevertheless does not want to recognize the significance of Leonardo's mechanics, if he rejects the views of a Venturi, Libri, Grothe, Duhem, G. Séailles, and many others who see in Leonardo the originator of modern mechanics already at the end of the fifteenth century, then such rejection must be substantiated. By failing to do so, he conceals the whole problem! In his book, wherein he deals with so many secondary figures of the Renaissance, the name of Leonardo da Vinci is not even mentioned.

4. Just as revolutionary as his view of the chronological beginning and substantial origin of modern science is Borkenau's opinion on the processes in social and economic history, which were conditional for the development of modern science and of the mechanistic world-picture. Even if the capitalist methods of production only became general in the sixteenth century, and one can only speak of the "capitalist era" at that time, the beginnings of the capitalist method of production (and these are of prime importance in elucidating the rise of the bourgeois world-picture) date much farther back. In contrast to Marx's view, that in Italy "we meet the first beginnings of capitalist production as early as the fourteenth or fifteenth century, sporadically, in certain towns of the Mediterranean,"[26] Borkenau says that not before the turn of the seventeenth century did the introduction of monetary capital into the sphere of production "have its first decisive success." Only at that time, therefore, did the "first period of capitalist industry, the period of manufacture" begin. Here, Borkenau skips three hundred years of capitalist development in Western Europe.

5. Wherever capitalist production is taking place, the abolition of serfdom and of the stratified-feudal order has been long since achieved through monetary capital. [173] Since capitalist production exists in Italy in the fourteenth century, the dissolution of the stratified-feudal structure through the mercantile and monetary capital must have taken place much earlier, viz. during the twelfth and thirteenth centuries – as could be read (until now) in every history book of Italian economics. Let us only mention the development of monetary and commercial capital in the proud Italian republics of the twelfth and thirteenth centuries, and the protracted trade

---

"detracting" from Galileo, Descartes, Pascal, or Stevin, but of understanding an historical epoch as a whole.

[25] K. Lasswitz, *Geschichte der Atomistik*, Vol. II, p. 12.

[26] *Das Kapital*, 3rd ed., Vol. I, pp. 739, 740. [*Marx-Engels-Werke*, Vol. 23, pp. 743, 744; *Capital*, Vol. I, pp. 715–716.]

wars between Amalfi and Pisa, between Pisa and Genoa, and between Genoa and Venice.[27] Due to the greatly intensified circulation of money and goods in thirteenth-century Italy, the quantity of silver specie in circulation no longer sufficed, so that in 1252 Florence was compelled to start minting gold coins (hence the name "Florin"). J. Burckhardt describes how, as early as the twelfth century, the Italian nobility lived in the towns together with the burghers and, having become quite bourgeois, turned to commerce.[28] There are licensed banks in Genoa, since the thirteenth century, with a highly developed system of deposits and clearing concentration.[29] When industrial capitalism began to develop in northern Italy, feudalism had long since completely disintegrated due to the invasion of monetary and banking capital. These historical research findings are also disregarded by Borkenau. According to him, capitalist production methods were nonexistent in Italy prior to the beginning of the seventeenth century; the dissolution of the stratified-feudal order through the incursions of monetary and merchant capital into Italy had not taken place in the twelfth and thirteenth, but only in the sixteenth century, and then the mental attitude of the Renaissance, the character of its scientific research, and its philosophy, are explained by the destructive influence of the inflow of monetary capital.

6. Past research into the history of economics presented the view that the industrial-capitalist development of Italy, which started in the fourteenth century and which developed in a steeply ascending line until the middle of the fifteenth century, suffered a heavy setback after the discovery of America and the blockade of the East European trade routes by the Turks: As a consequence of the shift in the axis of international trade from the Mediterranean to the Atlantic Ocean, Italy [174] entered a period of regression of capitalism – this process of deterioration explains the specific characteristics of the mental attitudes of the late Renaissance.

According to Borkenau's book, this conception was evidently unfounded. That shift in the axis of world trade plays no role in his attempt to interpret the Renaissance; he does not even mention it. In this he is quite consistent. Having asserted that Renaissance Italy had only advanced as far as monetary capitalism, and that productive capitalism never existed there, he cannot discern any setback in the development of industrial capitalism. He clearly finds it superfluous to adduce the phenomenon of the revolution in the world market to the end of the fifteenth century, in order to explain the material and intellectual situation of the Renaissance.

7. The conception of the genesis of capitalist production in the other West European countries is just as new as that of Italy's development. This applies firstly to the question pertaining to the initial plant structure in capitalist production. Borkenau adopts Sombart's erroneous interpretation of Marxist theory, according to which Marx had designated manufacture as the first stage of capitalist plant structure,[30] and he even places the thesis that "manufacture is the first period of capitalist industry" at

---

[27] Cf. H. Grossmann, *Das Akkumulations- und Zusammenbruchsgesetz*, Leipzig, 1929, p. 48.

[28] *Die Cultur der Renaissance*, Leipzig, 1899, Vol. II, p. 81.

[29] H. Sieveking, *Genueser Finanzwesen*, Freiburg i.B., 1899, Vol. II, p. 47.

[30] W. Sombart, *Der moderne Kapitalismus*, 2nd ed., 1917, Vol. II/2, p. 731.

the center of his conception and its substantiation! Here again he pays no heed to the 100–150 years of capitalist development which preceded the period of manufacture, namely the period of decentralized putting-out system.

8. Not quite as new as the conceptions just outlined – but perhaps even more interesting – is Borkenau's theory of the genesis of capitalism from the material aspect. Following the publication of *Das Kapital*, theoretical controversy arose about this genesis, in which W. Sombart, M. Weber, H. Sieveking, J. Strieder, G. von Below, Heynen, A. Doren, H. Pirenne, R. Davidsohn, and many others participated either directly or indirectly. One basic question was the following: According to *Das Kapital* the bearers of the emerging capitalism did not originate from among the artisans, and such provenance would have been impossible. This impossibility relates to (a) capital necessary to operate a capitalist enterprise, (b) the new techni- [175] cal processes, (c) the precognition of complicated elements of profitability (cheap sources of raw materials in distant markets, the currency and legal conditions of foreign marketing outlets, costs of transportation, customs, etc.), (d) to the technical and monetary organizational problems of a large enterprise, and finally (e) to the class origin of wage laborers.

The competition which threatens local handicraft comes with the rise of world commerce and international trade fairs in the thirteenth century. In order to neutralize this and prevent any social differentiation within the community or the guild, the rules of the medieval guilds try to block the master's ascent to capitalist status through regulations governing the number of tools he may use, the number of journeymen (*Gesellen*) he may employ, etc. Thus the accumulation of larger, freely disposable amounts of capital within the guild is rendered impossible. At the same time and for the same motives, any technical innovation is discouraged, the established technique becomes rigid routine, and production is adapted to the local market from which the competition is being excluded. The narrow horizon of the production of the local guilds prevented them from surveying distant markets of raw materials; the artisan obtained his raw materials second- or third-hand from the wholesale merchant. Likewise the artisans had no knowledge of foreign outlets for exports, of conditions governing foreign currencies and of customs duties. But above all, the guilds' artisans were lacking all organizational prerequisites for the creation of large-scale undertakings, as well as the ability to rationally calculate a production process extending over longer periods. How could the impoverished artisans, in their process of decay, who respected the spirit of traditionalism and routine, and rejected every innovation, have acted as historical signposts and have opened new horizons? Even in the best of circumstances the accumulation of capital within the framework of local manufacturing production was too slow and did not answer the new commercial requirements of the world market. It was also incapable of creating a new class of industrial entrepreneurs.[31]

The new capitalistic plant structures gradually emerged "outside the control of the ancient urban system and its constitution of guilds" – be it in the rural areas

[31] Marx, op. cit., I, p. 776. [*Marx-Engels-Werke*, Vol. 23, pp. 777f; *Capital*, Vol. I, p. 750.]

or in the trading ports by the sea where, for specific reasons, the structure of the guilds became relaxed. Yet the bearer of this new revolutionary development was [176] naturally not the artisan who belonged to a guild but the large-scale merchant, i.e. the trading and usury capital. The first big capital funds in circulation accumulated through banking and usury, before they could be used in the sphere of production. "Usury centralizes money wealth where the means of production are dispersed."[32] The large-scale merchant possessed larger capital means, knowledge of sources of raw material supplies and of buyers' markets for finished merchandise in which he traded. He was accustomed to doing business on credit – in short, he had all the prerequisites necessary for the new plant structures. The latter was not created all at once, rather it developed gradually in the course of a lengthy historical process. The large-scale merchant bought finished products from the artisans who originally had been working directly for the consumer, and thus, as he cut them off from their sales outlets, made them dependent on him. As he was advanced money and soon also supplied with raw materials for processing, the craftsman became even more dependent; and finally, despite his formal autonomy, he sank to the status of a wage laborer, while the production process, manual technique, did not change. In this manner the large-scale merchant provided work for numerous artisans who worked separately in their own homes with their own tools, formally independent but in fact totally dependent on him. Thus originated the putting-out system, the first capitalistic, albeit decentralized, large-scale enterprise. In view of the relatively small amount of capital funds accumulated, this form of enterprise was the most appropriate and rational, since the entrepreneur saved capital expenditure for factory buildings, lighting, heating, taxes, etc. We encounter the first beginnings of capitalistic production in this form of putting-out systems in fourteenth-century Italy and even as early as in thirteenth-century Flanders.

The next stage in the process of subordinating production to capital was that the large-scale merchant, hitherto only the organizer of other peoples' production, proceeded to take over the production under his own management. But this change was also gradual and extended over long periods. At first the merchant takes over single stages in manufacture, e.g. dyeing and dressing, while the other processes (for instance from spinning to weaving) continue in the usual way. The centralization of [177] the workers in closed factories, manufacture, is only the last stage of this lengthy historical development and itself constitutes the beginning of a new evolution of manufacture which takes place gradually – of a new process to which we shall later revert (see no. 10 below).

This is not the place for delving into this theory's details. Numerous historians have brilliantly demonstrated its validity using the historical material.[33] Especially

---

[32] Op. cit., Vol. III/2, p. 136. [*Marx-Engels-Werke*, Vol. 25, p. 610; *Capital*, Vol. III, p. 596.]

[33] H. Sieveking, "Die kapitalistische Entwicklung in den italienischen Städten des Mittelalters," in: *Vierteljahrsschrift für Sozial- und Wirtschaftsgeschichte*, Vol. VIII, 1909, pp. 73, 80. Cf. also Adolf Schaube's criticism of Sombart on the basis of English historical material: Die Wollausfuhr Englands vom Jahre 1273, in *Vierteljahrsschrift für Sozial- und Wirtschaftsgeschichte*, Vol. VII, 1908. Heynen, *Zur Entwicklungsgeschichte des Kapitalismus in Venedig*, 1905, pp. 121ff. Broglio

with regard to Italy, Doren has proved with the aid of an abundance of factual material the correctness of Marx's conception.[34] The same proof has been supplied, just as convincingly and also on the basis of ample source material, by H. Pirenne[35] for thirteenth-century Flanders and the Netherlands, by W. Cunningham, W. J. Ashley and G. Brodnitz[36] for fifteenth- and sixteenth-century England, and by Baasch[37] for sixteenth-century Holland. Other authors have shown via exhaustive historical research that in fifteenth- and sixteenth-century France the erosion of the artisans, their narrow horizon and their adherence to routine were too strong for new plant structures and techniques to emerge from their midst – and that in France, just as in England, it was the monetary and commercial capital which pioneered capitalist production – the putting-out system.[38] One can say that this theory of the historical genesis of capitalism has become the predominant one; it has already been introduced into textbooks of general economic history, such as those by H. Sée and J. Kulischer.[39]

Such a genesis of capitalism does not, however, fit into Borkenau's "structural" [178] scheme of development. He regards mechanics as the immediate prerequisite for the rise of the mechanistic philosophy in the first half of the seventeenth century, and the beginning of the analysis of the labor process into its constituent phases and of quantitative working methods as the immediate prerequisite for mechanics. According to him, the beginnings of capitalism are here and not in the thirteenth and fourteenth centuries. The large-scale merchant as the bearer of capitalist development does not fit very well into this scheme. Borkenau does not have capitalism emerge from monetary and commercial capital, but from the guild's craftsmen and through the rationalization of the methods of artisanship by analysis of the labor process – and he shifts its beginning in one leap across centuries into chronological proximity with the mechanistic philosophy, in the late sixteenth century! "It is," expounds Borkenau, "one of the most important insights resulting from all of Max Weber's research, that the main body (*Grundstock*) of manufacturing entrepreneurs, the first to introduce systematically capitalist methods into the production process, does not originate from the moneyed and trading bourgeois classes but from the ascending craftsmen" (p. 155). "The new manufacturing technique is not employed by religiously indifferent capitalists but by Calvinist, ambitious little men . . .." It is

---

D'Ajano, *Die Venetianer Seidenindustrie bis zum Ausgang des Mittelalters*, Stuttgart, 1893. R. Davidsohn, *Forschungen zur Geschichte von Florenz*, Vol. IV, Berlin, 1922, pp. 268ff.

[34] A. Doren, *Studien aus der Florentiner Wirtschaftsgeschichte*, Vol. I, Stuttgart, 1909, p. 23.

[35] Henri Pirenne, *Les anciennes démocraties des Pays-Bas*, Paris, 1910.

[36] W. Cunningham, *The Growth of English Industry and Commerce*, London, 1890, Vol. I. W. J. Ashley, *Englische Wirtschaftsgeschichte*, Vol. II: *Vom 14. bis zum 16. Jahrhundert*, Leipzig, 1896. G. Brodnitz, *Englische Wirtschaftsgeschichte*, Jena, 1918.

[37] Baasch, *Holländische Wirtschaftsgeschichte*, Jena, 1927, pp. 86, 156.

[38] E. Levasseur, *Histoire des classes ouvrières et de l'industrie en France avant 1780*, Paris, 1901, Vol. II: "Au XVIIe siècle les corporations opposaient un obstacle presque insurmontable à la création de la grande industrie et même de procédés nouveaux dans l'industrie" (p. 174). "La grande industrie ne pouvait pas naître dans le sein de la corporation" (pp. 271, 154). Similarly Henri Hauser, *Les débuts du capitalisme*, Paris, 1927, pp. 22ff.

[39] Henri Sée, *Les origines du capitalisme moderne*, Paris, 1930, pp. 13, 15. J. Kulischer, *Allgemeine Wirtschaftsgeschichte*, Munich, 1929, Vol. II, p. 110.

generated "by the efforts toward rationalization of craftsmanship" (p. 90). Manufacturing capitalism has *everywhere* "been recruited from the higher strata of artisans and from aristocrats who had turned bourgeois" (p. 157).

Borkenau does not notice that Max Weber's views on the origin of capitalism, to which he refers, were criticized and superseded in the discussion mentioned, and he is not aware that Weber himself had become unsure and doubtful of his theory.[40] Elsewhere Borkenau refers to Boissonade with regard to French manufacture.[41] [179]

Apart from these, of all historians of economy only J. Kulischer is mentioned once. Boissonade is the source of Borkenau's information! An "exemplary treatment of material," through which all other works on the origin of French capitalism are supposed to have become "obsolete"! The seminal works of Fagniez, E. Levasseur, Germain Martin, E. Tarlé, J. Godart, Henri Hauser, Henri Sée et al., each of whom provides deeper insight than Boissonade into the essence of the historical processes, are supposed to be obsolete!

Actually, Boissonade's book does not at all constitute a revolution in French historiography of economy. In 1899 Boissonade presented the first results of his research, providing archival documentation on 582 manufactures. In 1901 E. Levasseur already treated the findings of Boissonade's research with critical irony.[42] Since that time, in almost thirty years of untiring research in archives, Boissonade has considerably enlarged the number of known manufactures. But our knowledge about the origin of capitalism was not advanced through this investigation but rather became even more obscure. Due to its methodological insufficiency and ignorance of capitalistic forms of enterprise, his work was already outdated on publication (1927) and fell short of the results of earlier findings of French research.[43] Thus e.g. J. Kulischer blames Boissonade for having overlooked the putting-out system as the first capitalist form of undertaking and having mistaken it for artisanship! Here Tarlé's criticism had a clarifying effect. "Sée too stresses in a

---

[40] "In the occident the early capitalist putting-out system did not always, and not even usually, develop from within craftsmanship, but it originated very often beside the artisans ..." (M. Weber, *Wirtschaftsgeschichte*, Munich, 1923, p. 145). [Max Weber, *General Economic History*, transl. by Frank H. Knight, Illinois: Free Press, 1950 (1927), p. 158.] "To sum up, one should always be aware that the factory did not originate from the workshop nor at its expense, but initially emerged alongside it (Weber identifies the factory with manufacture and criticizes the distinction made between these two concepts by the 'early science, also Karl Marx,' op. cit., p. 149. – H. G. [op. cit., p. 162f.]). Above all, it seized upon new forms of production and new products, e.g. cotton, chinaware, gold brocade or surrogates – none of which were manufactured by the craftsmen organized in guilds" (op. cit., p. 157 [English: p. 173]).

[41] "On the entire development of manufacture and of commerce protected by the state from Louis XI to Louis XIII, we now obtain very comprehensive information from P. Boissonade, *Le socialisme d'état*, Paris, 1927. It is theoretically insufficient and inadequate for the history of the relations of production, yet it is exemplary in its treatment of the material for the history of the productive forces. Since its recent publication, and despite the imperfections, all other works on the genesis of French capitalism have become obsolete" (p. 173).

[42] E. Levasseur, *Histoire des classes ouvrières ...*, op. cit., Vol. II, p. 239.

[43] Boissonade's confusion of concepts is evident already in the title of the book, which calls the mercantile policies of the French governments in the sixteenth and seventeenth centuries "Le Socialisme d'État."

number of his writings that in France, just as in England, the industrial capital was preceded by the commercial capital, which tried to dominate the production of the small craftsmen."[44]

9. In view of the central role assigned in Borkenau's thinking to the origins of capitalism, we have tried to clarify the question as to the time of its first appearance, the monetary and commercial capital as its bearers, and finally the putting-out system as its first form of enterprise. [180] Now it would be thinkable that capitalism, though not when it first appeared, but at some later stage of its development, viz. in the transitional phase from the decentralized putting-out system to the centralized manufacturing plant, did develop in the manner claimed by Borkenau. However, even if understood this way, this theory of the genesis of capitalism proves untenable.

In view of the problem's importance we would like to cite some proof for this statement. It can be ascertained from the sources that the overwhelming majority of the first manufacturing entrepreneurs in seventeenth-century France were monied people, capitalists, merchants, speculators, high officials, in short, anything but "little men with high aspirations."

Some typical examples from the whole period between Henry IV and Louis XIV will show who were the carriers of manufacturing. In Troyes, under the reign of Henry IV, the manufacture for satin and damask is founded by J. Schier, a wealthy merchant. (Marjépol, in Lavisse, Vol. VI/2, p. 78.) Thomas Robin, "maître de requêts" of Queen Marguerite, founds the "manufactures royales des Toiles fines et des Toiles de coton" in Rouen and Nantes (1604–1609). (Boissonade, loc. cit., p. 255.) The merchants J. Wolf and Lambert founded in 1606 the "manufacture des toiles fines de Hollande" at St. Sévère near Rouen. (Levasseur, op. cit., Vol. II, p. 171.) The first big manufacture "des industries des lainages" of the firm of Cadeau, arising at Sedan under state sponsorship, was founded by three Parisian merchants. (Boissonade, loc. cit., p. 254.) The manufacture of wallpaper was started under Colbert, and a factory was erected at Beauvais by Hinard, a Parisian merchant, while the first mirror factory was built in 1663 in Orleans by Denoyer, "receveur de tailles."[45]

These are not solitary selected examples. As was always the case with undertakings sponsored "from above," there soon appeared speculators and adventurers wishing to exploit this chance. E. Levasseur states about the time of Henry IV: "Pierre Sainctot, de Paris, membre de la Commission du Commerce; Claude Parfait, sellier, riche marchand de Troyes, étaient des capitalistes. Dans ces affaires d'argent, il se glissaient déjà des spéculateurs suspects, comme Moisset de Montauban . . . et des habiles, comme Nicolas Le Camus qui, arrivé à Paris avec 24 livres, passa pour avoir laissé à sa mort une fortune de 9 millions." (Further examples, Levasseur, op. cit., Vol. 1, p. 175; Vol. 11, pp. 200, 258. Lavisse, op. cit., Vol. VII/1, p. 220.)

Colbert, the actual initiator of the manufacturing system, surrounded himself with a team of agents who – always traveling throughout the country in the factories'

---

[44] J. Kulischer, op. cit., Vol. II, p. 110.

[45] E. Levasseur, Vol. II, p. 258. Lavisse, Vol. VII/1, p. 220.

interest and being partners in Colbert's foundations – represented a mixture of fortune hunters, speculators, and apostles of the new capitalist creed. "Pour fonder des manufactures, Colbert employa un certain nombre d'agents pris dans le commerce ou dans la banque, qui furent en quelque sorte les missi dominici de la réforme." [181] The principal agent was Bellinzone, an Italian naturalized under Mazarin. He was appointed "inspecteur général des manufactures" with a salary of L. 4000 and was imprisoned for "malversation" at Vincennes after Colbert's death. Another agent, the banker Jabach, appointed director of the wallpaper manufacture of Aubusson, participated as a capitalist in a series of undertakings. The team included the merchant Camuzet of Paris, founder of innumerable manufactories for silk stockings, and finally the two brothers Poquelin, Parisian merchants, who had offices in Genoa and Venice and likewise participated in a number of manufactories, e.g. the mirror plant in the Faubourg St. Antoine. (Levasseur, op. cit., Vol. II, p. 238.)

Sagnac emphasizes that in Colbert's times the foundation of manufactories mainly took the form of capitalist joint-stock and similar companies, so that their original basis was not the "extreme effort" of the little man, but the participation of capital. "Sociétés en nom personnel, sociétés en commandite, sociétés anonymes surtout, recueillent des capitaux des marchands, des magistrats et des nobles euxmêmes, s'efforçant de draîner vers les grandes affaires une partie de la richesse, d'habitude employée en achat de rentes sur l'Hôtel de Ville ou d'offices royaux."[46]

The form of the joint-stock company or limited partnership enabled the merchants and magistrates to invest capital without having to leave their offices to personally look after the business. "Colbert pressait ... les gens riches qui étaient sous sa main, bourgeois et marchands de Paris, de Lyon, de Rouen, de Troyes, courtisans, magistrats, banquiers, officiers de finances et traitants d'apporter leur contingent" to the capitalists of the newly emerging joint-stock companies.[47]

Not only the capitals, the stock owners, and other suppliers of money came from the circles of commerce, finance and the magistrates – but also the managers, i.e. the practical directors, were usually taken from the estate of traders. "C'est parmi les marchands," says Sée, "que se recrutent ordinairement les directeurs de manufactures ... Ces marchands-manufacturiers n'appartiennent plus en aucune façon à la classe de maîtres des métiers; ils échappent à l'organisation corporative."[48] For Jacques Savary, the famous author of *Le parfait négociant* (1673) and [182] Colbert's counsellor in all legislative matters of manufacturing organization, it is a matter of course that the big merchants are those who establish manufactories. Thus he provides instructions for "Négociants qui voudroient établir des manufactures."[49]

---

[46] In Lavisse, op. cit., Vol. VIII/1, p. 230.

[47] Levasseur, Vol. II, p. 241; cf. Lavisse, Vol. VII/1, p. 222.

[48] *Esquisse d'une histoire économique de France*, Paris, 1929, pp. 300/301. This statement by Sée is in agreement with Levasseur's, op. cit., Vol. II, p. 402.

[49] Jacques Savary, *Le parfait négociant*, Vol. II, Chaps. 6 and 7, quoted from the fifth edition, Lyon, 1700.

We see that Borkenau's historical conception that capitalism in general and manufacture in particular "were not created by monetary capitalists, but by upward-striving little men" does not correspond to historical reality. It is a theory which presents the origins of capitalism, i.e. of the original accumulation, as an "idyll" according to which the "work," the "unlimited effort" (p. 176) serve to create the "solid" manufacturing capital (p. 155), and the "ascent to the capitalist class through strict rationalization of work" (p. 157) is achieved.

10. In the above we have shown how Borkenau simply disregards the development of capitalism in the premanufacturing period. Now we will examine manufacture and its division of labor.

In his opinion, manufacture's span of life extended from "the beginning of the sixteenth century" (p. 13) until the last third of the eighteenth century, i.e. over a period of almost 300 years. It is clear to anybody who has studied history that manufacture cannot have remained unchanged over such a long period. Borkenau does not take this into account. The problem of the "period of manufacture" is for him a simple and unambiguous matter. He speaks of "the manufacturing bourgeoisie" (pp. 13, 162) and of "manufacturing mentality" (p. 404), as if these were always concerned with absolutely fixed and unequivocal categories. "As is well known, the manufacturing technique consists of nothing but an extremely developed division of labor, while entirely retaining the foundations of the production process in craftsmanship" (p. 2). Manufacture abolishes the qualification for work, it replaces the skilled artisan with the unskilled laborer whose work consists of "the performance [183] of a perfectly simple manipulation which is accomplished with precision and which "should be feasible even to a child, even to an imbecile" (p. 7). Thereby all special training becomes superfluous, manufacturing work loses all particular quality and "becomes pure quantity." Thus at the turn of the seventeenth century, manufacturing has replaced qualified work by "general human" or "abstract" work, therefore developing that concept which is the basis of modern mechanics. Thus the emergence of scientific mechanics in the beginning of the seventeenth century presupposes the prior development of manufacturing.

This presentation of the character of manufactorial work at the turn of the sixteenth and seventeenth centuries is sheer fantasy. It contains an inner logical contradiction. The term "craftsmanship" already implies a skilled type of work for which one is qualified. Work that can be performed by unskilled laborers, by anybody – also by children and imbeciles – for which no training is required, ceases to be "craftsmanship." Borkenau's generalized conception of manufacture is evidently based on the description in the first chapter of *Wealth of Nations*, illustrated by the far-reaching division of labor and dissection of the work process into simple manipulations of the production of metal pins. He transfers the situation and conceptions described by A. Smith which apply to the conditions of the second half of the eighteenth century to those prevailing in the sixteenth century, without giving a thought to the question as to whether the "manufacture" of the sixteenth century can be identified with that of the eighteenth century.

Borkenau has overlooked the various stages of development in manufacturing. Manufacturing has undergone various successive phases of development in its over two hundred years of existence. The characteristics are clearly identifiable. (1) In the beginning, manufacturing appears in the form of simple cooperation between workers in a spacious workshop, without any trace of division of labor. Although laborers assembled in a workshop is a precondition for the subsequent division of labor, at first – during the extensive period of cooperative manufacture – this division of labor does not yet exist. At the end of the sixteenth and at the beginning of the seventeenth centuries there hardly existed any division of labor in the most advanced manufacture, the Dutch one; even less existed in the relatively backward French one. Cooperative manufacturing was followed by the (2) heterogeneous and (3) serial manufacture; these are not only two different basic modes but also two consecutive phases in the development of manufacturing history. Finally there arises [184] the fourth and last phase, the "combination of manufactures" which, although not universally accepted, did exist as a tendency: the combination of different manufactories into an "overall manufacture." The highest stage of technical development is represented by the "organic" manufacture which subdivides the work process into the simplest, repetitive manipulations performed with virtuosity, where the end product of one worker is the starting point for his successor's labour. This "organic" phase represents the "finished form," the "perfected form" of the development of manufacture.[50]

It is a blatant anachronism to apply the division of labor in "organic" manufacture of the eighteenth century to the "cooperative" and "heterogeneous" manufacture of the end of the sixteenth and the start of the seventeenth century. In the second half of the seventeenth century, in England, William Petty only knows the "heterogeneous" manufacture, i.e. a plant structure in which several independent artisans work in one workshop under the same capitalist, and nevertheless fashion their products entirely by the traditional method, without division of labor into simple manipulations, where the final product, e.g. a clock or a carriage, results from "simple, mechanical assembly of separate partial products." Almost until the end of the seventeenth century, the division of labor into simple manipulations is out of the question, as is the replacement of skilled workers by unskilled ones, children, and imbeciles. The manufacture is based upon specialized and highly qualified craftsmanship; once specialized, the participants' separate tasks in the total complex are frozen, and a hierarchy of qualified partial specialists is formed.

Beside the hierarchical pyramid of differently trained and specialized workers there appears a new "class of so-called unskilled laborers," for, within specialized work, there are also "certain simple operations of which everybody is capable." In the latter class, which is an exception within the general specialization, "the cost of apprenticeship vanishes"; the lack of specialization is thus also turned into a [185] specialty within the hierarchical specialization of manufacture.[51]

---

[50] Cf. Marx, *Kapital*, op. cit., Vol. I, pp. 342–348. [*Marx-Engels-Werke*, Vol. 23, pp. 362–371; *Capital*, Vol. I, pp. 342–350.]

[51] Op. cit., Vol. I, p. 351. [*Marx-Engels-Werke*, Vol. 23, p. 370; *Capital*, Vol. I, p. 350.]

Because of the qualified character of the work, manufacture was dependent on the workers, as they could not easily be replaced. This is also the reason for the struggle and efforts of governments to attract foreign workers (e.g. Colbert's demand for glassblowers from Venice, for tinplate workers from Germany, etc.), whereas on the other hand the emigration of specialized workers was forbidden and was threatened by heavy prison sentences.

Nothing is more characteristic of manufacturing work's qualified character than the conditions in the first mirror factory, established in 1663. Italian workers, brought from Murano in Venice at great expense, difficulty, and danger through the intermediacy of the French ambassador, earned 3–4 ducats daily. They were to train a certain number of French workers annually, yet they strictly preserved their professional secrets, so that the manufactories with their precious equipment "dépendaient absolument du caprice des étrangers." Once, when one of those Italian workers "celui qui gouverne les glaces sur les grandes pelles" had broken his leg, the manufacture had to be closed for ten days, but the workers had to be paid and the fires in the big furnaces maintained, because the other workers "ne savent faire sa fonction et n'ont pas même voulu y essayer, disant que c'est la plus difficile et qu'il faut l'avoir appris dès l'âge de 12 ans" (G. Martin, *La grande industrie sous le règne de Louis XIV*, Paris, 1899, pp. 77, 78).

Nowhere is the arbitrariness of Borkenau's construction better illustrated than in this question. With the progress of the division of labor, each partial procedure did not become simpler nor did qualified work become superfluous and replaceable by unskilled work. Parallel with the development of the division of labor one can observe a strengthening of the role played by qualified work rather than a weakening thereof. At the end of the fifteenth century – earlier in some countries and later in others – parallel with the development of the division of labor, a process of diversification of production began. Formerly only few and simple types of cloth were produced in England, so that one and the same clothmaker could master the spinning, weaving, and dyeing; at the end of the fifteenth century new types of cloth appeared: ordinary and fine cloth, straights and kerseys, were now made; the statute [186] of 1484 contains half a dozen varieties in addition to the aforementioned ones. With the diversification of products came greater demands on the skills of the artisans, weavers, dyers, etc. – a development which would accelerate in the future.[52] The weaver had to learn to weave ten to fifteen different kinds of cloth, the ribbonmaker had to produce twenty or thirty kinds of ribbon, etc. We see a similar diversification in Holland. By the end of the sixteenth century, new branches of production, new raw materials, new techniques appear, and all these innovations demand higher qualifications; in Leyden, e.g., begins the weaving of fustian (1586), of serge (1597), and of "draps changeants."[53]

[52] W. Cunningham, *The Growth of English Industry*, German translation, Halle, 1912, Vol. I, p. 508.
[53] Baasch, *Holländische Wirtschaftsgeschichte*, Jena, 1927, p. 84.

New dyes, such as cochenille and later indigo, caused a revolution in the technique of dyeing. The smallest error could spoil large quantities of material. In Haarlem, in addition to delicate tablecloths, there were the famous "Bontjes" (linen mixed with cotton). In Amsterdam there was the production of ribbons and of velvet, in Rotterdam there was plush and "Bombasin." The same development occurred in France. With the rise of the wealthy bourgeoisie in the sixteenth century, luxury became more widespread (up to the fifteenth century this was limited to the nobility and the clergy), while cheaper, "lighter" luxury materials such as satin de Bruges, crêpe de soie, serges, étamines, caddis, etc. now came into demand.

11. We have seen how, according to Borkenau, "rational technology" was impossible during the period of "predatory capitalism" and how it only arose with the "solid" manufacturing capitalism, because the industrial bourgeois which developed from artisanship "needed a rational structuring of operations" (p. 9). Manufacture, thus rationalized, therefore represented a superior plant structure which soon replaced the previous forms of production. "The displacement of handicrafts by manufacture, though it had its beginnings already in the sixteenth century, nevertheless became general only in the seventeenth century, and introduced sophisticated manufacturing techniques" (p. 2). Alongside with this fundamental view, we find elsewhere another remark which evidently contradicts the former. There we learn with regard to the first half of the seventeenth century that in France "the emerging manufacturing bourgeoisie ... had to rely on government support in every respect" and that "without the direct protection of the government it could not exist at all" (p. 171). And this despite the "rational technique" and despite the great "sophistication" of that technique! [187]

The "displacement of the handicrafts" by manufacture, which according to Borkenau has "become even more general," is a pure illusion. Let us take the example of France to examine the character of manufacture and the truth of Borkenau's statement. In general the handicrafts were not replaced by manufacture; in the seventeenth and even in the eighteenth century the workshop remained the predominant plant structure; even though there existed undertakings which in everyday and administrative language were called "manufactures," there was no manufacture in the sense of A. Smith up to the end of the seventeenth century, i.e. as a basis of far-reaching division of labor; the capitalist forms of undertakings, as far as there were such forms, were almost exclusively represented by the system of home industry.[54]

When after the civil war the state under Henry IV (1589–1610), the "créateur" and "père" of the mercantile economic policy, began to sponsor the manufacturing system, it endeavored to keep in the country the money which was payable abroad for luxury articles. Therefore "manufactures" of luxury goods – silk and

[54] Thus e.g. J. Kulischer says about the French silk industry in the seventeenth century: "... The flourishing silk, velvet and brocade industry of Lyon (including also the use of gold and silver threads for braids, lace, fringes, bows, etc.) was exclusively a home industry; there were no manufactures. About half of all French silk goods were produced in Lyon" (J. Kulischer, *Allgemeine Wirtschaftsgeschichte*, Vol. II, p. 171).

wallpaper plants, the manufacture of tapestries, crystal, and mirrors – were founded in the country. Since the luxury industry was never a field for division of labor and employment of unskilled workers, but always used highly qualified, artistically and technically trained craftsmen, these were imported at great expense and with difficulty from abroad: from Milan, Venice, and even from the Levant, and this despite the fact that France itself suffered from severe general unemployment (in Paris in 1595 there were more than 14,000 unemployed, in Amiens in 1587 almost 6,000, in Troyes in 1585 nearly 3,000). In these establishments "rational techniques" [188] were out of the question. The system of official support and premiums was bound to encourage uneconomical, speculative undertakings, even when conditions for normal profitability did not exist. Despite the monopolies and financial subsidies granted by the king, these manufactories could not hold out. "La plupart de ces créations avaient succombé de son vivant ou après sa mort."[55]

In the next half-century there was no improvement, but a deterioration in the industrial development sphere. After Henry IV's death (1610), the king's creations went bankrupt under Maria di Medici's rule. Fresh creations were, of course, out of the question. Until Richelieu became minister, there followed "quatorze années de mauvaise administration et de désordre qui arrêtèrent de nouveau le progrès de la nation," – in short, it was a "période de stérilité."[56]

The eighteen years of Richelieu's ministry (1624–1642) were a period of general decline and exhaustion in the country "peu favorable à l'industrie."[57] Richelieu was too strongly occupied with higher politics, with the struggle against the Habsburgs, to devote his attention to industry. His most important creation is the Imprimerie Royale (1640); not even Borkenau would wish to claim that this was a special area for manufactured analysis of the labor process. Then came the time of Mazarin and the Fronde. Before Louis XIV came of age, France again went through a period of civil war. "La Fronde (1648–1652) ... porta un grand préjudice aux affaires industrielles et commerciales."[58] It was "the time of France's total ruin. How then could one find industries?"[59]

And Levasseur's judgment is not different: "Quand Louis XIV prit la direction de l'État . . . l'industrie et le commerce paraissaient languissants." "Le nom de Mazarin . . . en réalité ne mérite pas une place dans l'histoire économique."[60]

Our analysis has shown that the "période semi-séculaire de 1610–1660 a été plus agitée par les troubles à l'intérieur et par la guerre avec l'étranger. La classe industrielle souffrit."[61] This half century which, according to Borkenau, was the [189] period in which modern mechanics emerged, was not a period of technical progress

---

[55] E. Levasseur, op. cit., Vol. II, p. 176; cf. p. 170.

[56] E. Levasseur, op. cit., p. 187.

[57] Op. cit., p. 188.

[58] Op. cit., p. 199.

[59] C. Hugo, "Die Industrie im 16. und 17. Jahrhundert," in: *Der Sozialismus in Frankreich*. Stuttgart, 1895, p. 814.

[60] Op. cit., p. 201.

[61] Op. cit, p. 410.

but one of general economic decay and of sterility in industrial development in particular, in which there could be no question of "sophistication of techniques" and progressive division of labor. The decay was so complete that Colbert had to start the reconstruction of industries anew. Thus he himself wrote about his efforts for the establishment of manufactures: "La grande manufacture étant chose presque nouvelle, hasardeuse. . ."[62]

In France, even under Colbert and up to the end of the seventeenth century, there existed no manufactures in A. Smith's sense with extensive division of labor. Most of the manufactures established with government subsidies and privileges operated too expensively and therefore found few customers – which exposes the techniques they were based on! For instance, in Berri as well as in some other provinces "les marchants aimaient mieux acheter comme par le passé, aux petits fabricants qu'à la manufacture," since the small artisans were cheaper. How, then, might the "rational" division of labor of these manufactures have looked?[63]

In addition, due to Colbert's strict official regulation of industry (Règlements généraux of 1666 and the subsequent special regulations for individual sectors of industry), all technical procedures were precisely prescribed by law, which impeded all technical progress! All the historians, such as Mosnier, Sée, G. Martin, Sagnac, Levasseur, and Kulischer, agree on this point. Thus H. Sée says about the control: "Elle a pour effet de maintenir l'industrie dans l'immobilité, d'empêcher toute innovation."[64]

Despite the generous government subsidization, the "manufactures" went broke in France. This was not as a consequence of external coincidence; their ruin was the necessary result of the internal shortcomings of the Colbertian system of protection. They were an artificial product of the royal administration; they could thrive under [190] the wings of royal protection and not by virtue of a rationalization of production processes. Rationalization as well as division of labor in particular are a necessity for the entrepreneur, imposed on him by the struggle of competition: a reaction to the difficulties of marketing. Through technological progress and division of labor, production should become cheaper, and through the drop in prices an advantage should be gained over the competitors. But the "manufacture," privileged by the state institutions, need not be afraid of competition, for it relies on royal subsidies, import restrictions, and monopolistic privileges. Instead of developing and becoming efficient in the competitive struggle, it loses its fighting strength in the unhealthy atmosphere of monopolistic protectionism. Borkenau himself admits that the emerging

[62] Lavisse, op. cit., Vol. VII/1, p. 221.

[63] Levasseur, op. cit., Vol. II, p. 274; Mosnier, op. cit., p. 127.

[64] H. Sée, *Esquisse d'une histoire*, p. 295. Cf. Mosnier, op. cit., p. 140; Sagnac, op. cit., p. 210; Levasseur, op. cit., Vol. II, pp. 339, 341; J. Kulischer, op. cit., Vol. II, p. 107. For Borkenau's construction, which speaks of the "displacement of handicraft by manufacture, the assessment of the decline of manufacture by both contemporary writers (Vauban, Boisguillebert, Fénelon) and by present historians is a fatal fact. According to him, the decline of manufacture was only a decline in quotation marks, a result of intentional blows directed by the monarchy against capitalism! (p. 263).

French bourgeoisie "was in every respect dependent on government support," and that "without direct governmental protection it could not exist" (p. 171).

12. However, the manufactural work was by no means made redundant by unskilled labor -on the contrary, it always remained quality work. Also and especially for this reason, its effects on scientific mechanics were and had to be different from those stated by Borkenau! The highly sophisticated nature of the manufactural work makes it impossible for it to give impetus to the development of that "general human" and "abstract" kind of work which is the basis of scientific mechanics. On the contrary, manufactural work was fundamentally unsuitable for this. The most important characteristic of every mechanical labor is its homogeneity; the work done is always identical qualitatively and is only different quantitatively, and these differences can be exactly measured. (Descartes, in the preface to his *Traité de la Mécanique* (1637) presupposes such homogeneity of performance as a condition for measurability.)

It is just this characteristic of homogeneity which every labor of man or animal is lacking. The manufactural worker's performance is not "general human," i.e. qualitatively always uniform, but is dependent on the worker's strength and skill, and therefore individually different, subjective – therefore not *homogeneous*, not *uniform*. In the long run man performs uniform movements only very imperfectly.

This individual, subjective character of human labor precludes, according to [191] Marx, "truly scientific analysis," viz. exact quantitative methods are not applicable to it. Borkenau makes an effort to formally agree with Marx's standpoint (p. 2), but then to prove the contrary, namely that manufactural work had excluded qualification thereby becoming "general human" work; it had thus founded the basis for exact scientific analysis, for exact quantitative methods in mechanics!

If the far-reaching division of labor sufficed for the development of a "general human" labor, then scientific mechanics would have already emerged in the fourteenth century. Borkenau says repeatedly that the manufactural technique of the seventeenth century consists of a "division of labor to the utmost degree," yet in this matter, which is of decisive importance for his conception, he does not adduce a single example, not even a source. If one compares the division of labor in England and France of the sixteenth and seventeenth centuries with that practised in Italy during the fourteenth century, one will see that the former had a rather miserable appearance, whereas e.g. in the silk industry of Lucca and Venice a total of sixteen separate processes of labor are mentioned, including winding, twining, boiling (*cocitori*), dyeing, rolling bobbins (*incannaresse*), warping, weaving, etc.[65]

Because of human labor's aforementioned subjective, heterogeneous character it could not serve – divided or undivided – as the basis of scientific analysis; therefore

---

[65] Broglio d'Ajano, *Die Venetianische Seidenindustrie*, pp. 21/23. In the Florentine cloth industry at the beginning of the fifteenth century, one distinguished between the following processes: sorting, washing, beating, combing, scraping and carding of wool, weaving, dyeing, shearing, weaving, degreasing, fulling, roughing, stretching, smoothing, pressing, rolling, etc. of cloth – altogether up to thirty different partial processes: Doren, *Studien*, op. cit., Vol. I, p. 43.

the impetus to theoretical mechanics was not given by human labor but by the material means of labor, the machine, i.e. only to the extent that this narrow subjective barrier of human labor was overcome! In the manufacture "the process was previously made suitable to the worker"; thus "the organization of the social labor process is purely subjective." "This subjective principle of division of labor no longer exists in production by machinery. Here, the process as a whole is examined objectively" and therefore open to scientific analysis, to quantitative methods. "The implements of labor, in the form of machinery, necessitate the substitution of natural forces for human force, and the conscious application of science, instead of rule of thumb."[66]

Thus we arrive at the decisive point: In the course of their development since the middle of the fifteenth century, the mechanistic thinking and the progress of scientific mechanics show no trace of a closer relationship to manufactural division of labor, but are always and everywhere closely related to the use of machines! It is typical that Borkenau suppresses all traces of the use of machines over a period of some three hundred years, thereby deterring the reader from thinking that modern scientific mechanics have anything to do with machines! Thus, for instance, he speaks of "the technique of the artisan, which is almost exclusive to the period of manufacture" (p. 8).[67] Thus he does not mention Descartes' *Traité de la Mécanique* of 1637, although he discusses all his other works. [192]

As a matter of fact, manufacture has never been a form of production in which artisanship "is almost exclusive." From the very beginning, machines were used in manufacture – and even before – and for two purposes:

(1) As motor mechanisms, where human labor was replaced e.g. by water power, as in mills and other water-driven machines. This in particular was the strongest incentive for going deeper into theoretical mechanics. Namely, when attempts were made to achieve an increased performance (e.g. driving two milling processes or two stamps by means of one water wheel), the overstrained mechanism of transmission became incompatible with the insufficient water power, which led to research into the laws of friction.
(2) As working machines – wherever there was a matter of crude, undivided, largescale processes requiring the application of brute force: crushing ore in metallurgy, so-called stamping mills in pits and mines, grinding rags in paper mills, etc.

Water power was instrumental in one of the greatest upheavals of technology, the revolutionizing of the iron and mining industries. Since Roman times, iron was obtained from ore in the smithies' primitive furnaces in the woods. Farmers usually did this as a sideline. The invention of casting iron and the transition to blast furnaces [193]

---

[66] Marx, *Kapital*, op. cit., Vol. 1, pp. 383, 390. [*Marx-Engels-Werke*, Vol. 23, pp. 401 and 407; *Capital*, Vol, III, pp. 380, 386.]

[67] Thus he already contradicts himself on the following page, where he says that the seventeenth century was a century of water, while the nineteenth was a century of fire. But it could become a "century of water" only through the natural force of water applied as the driving power for machines which replaced artisan's labor.

and to the indirect production of crude iron came with the beginning of the fifteenth century. The technical starting point of this upheaval was the use of water as driving force in the production of iron, viz. the water-powered hammers in smelting and in moving the bellows when melting and forging. This technical upheaval, itself connected with the upheaval in the technology of warfare and the greater need for iron, soon led to a social upheaval, to the relocation of the iron industry from the heights of the mountains and woods to the river valleys. There, the numerous small furnaces were replaced by large-scale enterprises with mass production: impressive blast furnaces with foundry buildings, water wheels, bellows, stamping works, and heavy water hammers operated on a capitalistic basis with wage labor and rational bookkeeping.

Furthermore, water power caused the upheaval in the mining industry from the second half of the fifteenth century. The use of water as another mechanism for powerful pumping works and conveyor systems enabled the first really deep excavations, the building of deep mines and shafts. In general the exploitation of the natural forces (water in the mines, machines for crashing ore, etc.) enabled the application of concentrated power which transcended human power, thus rendering mankind independent of the latter and placing it before new tasks. This was the beginning of the technological age.[68]

[194] It is evident that man, in all these technological upheavals, acquired new, important material for observing and contemplating the actions of forces. In the machines, in the turning of the water wheels of a mill or of an iron mine, in the movement of the arms of a bellows, in the lifting of the stamps of an iron works, we see the simplest mechanical operations, those simple quantitative relations between the homogeneous power of water-driven machines and their output, viz. those relations from which modern mechanics derived its basic concepts. Leonardo da Vinci's mechanical conceptions and views are only the result and reflection of the experiences and the machine technology of his time, when one new technical invention follows the other or the previous inventions are improved and rationalized.[69]

---

[68] The technical revolution in mining brought about a thorough social upheaval. With the extension of mining, the need for more capital to finance the building of shafts, ventilation, ore-lifting, and water storage systems caused a *thorough change in ownership and concentration of capital*: on German soil and in adjacent regions, in the middle of the fifteenth century the small medieval (communal) enterprises became dependent on a few financially powerful putting-out capitalists, usually wholesale ore dealers (as e.g. the Fuggers in Augsburg), who granted them advances, took possession of their shareholdings (Kuxe), while the original members of the miners' union, deprived of their ownership, were reduced to wage laborers. In this manner *industrial capitalism in the German, Tyrolean, and Hungarian mining industry became a major power long before the Reformation*. The financial support of the Fuggers was not only instrumental in 1519 in the election of Charles V as emperor; this big power, as we know from Ranke, was even capable of thwarting the strengthening of the central government within the empire, so as to safeguard the interests of its own price monopoly and unrestricted profits.

[69] Since the middle of the fifteenth century a technical literature emerges. The oldest printed publication on technical matters, with numerous descriptions of machines, is the book of Valturio Roberto of Rimini, written about 1460 and printed at Verona in 1472. Vanuccio Biringuccio of Siena, the originator of modern metallurgy, mathematician, engineer, and practical director of

Here in the case of machines we see the tendency toward the replacement of qualified work by unskilled labor at an early stage – which Borkenau ascribes to the division of labor in manufacture. Yet for Borkenau the mechanical aspect of manufacture does not exist; he does not even mention it. Even though during the period of manufacture the work of machines was quantitatively less important than the work of human beings, it was most significant for theoretical mechanics. Marx has demonstrated that the sporadic use of machines in the seventeenth century was extremely important and inspired the great mathematicians of the time to initiate modern mechanics. Research in economic history has since revealed much new material; chronologically the use of machines began much earlier and their sophistication and frequency was greater than was assumed only sixty years ago. But Borkenau wants the basic concepts of theoretical mechanics to be derived from the manufactural division of labor, which is why the history of machines and their use must be obliterated from the horizon.

13. According to Borkenau, the manufacturing period at the turn of the sixteenth and seventeenth centuries *put capitalist accounting into practice*, thereby [195] also enabling the observation of nature according to quantitative methods. It must be stated against this: capitalistic calculation has nothing to do with any work processes. As Max Weber correctly remarked, it is a formal procedure of comparing the monetary value of expenses (costs) with income (prices) for the sake of maximal profitability. Once it had developed in the sphere of trade, capitalist calculation could easily be extended to the sphere of production. Exact accounting, like the general partiality for exact methods of measuring in diverse areas of knowledge, was first developed in Italy during the thirteenth and fourteenth centuries.[70] This development culminated in the first scientific system of double-entry bookkeeping in Fra Luca Paccioli's book (1494), in which Paccioli theoretically formulated a practice in use for a hundred years, viz. since the second half of the fourteenth century (Sombart, loc. cit., p. 312). The oldest well-kept Italian account books originated in the fourteenth and fifteenth centuries. In Italy, the period's leading capitalist country, says Sombart, "the general spirit of rationalization and mechanization was most advanced" (loc. cit., p. 325). Double-entry bookkeeping "organizes the phenomena in an intricate system which one can call the first cosmos based on the principle of mechanical thinking .... It is the consistent application of the basic idea of quantification which entails the endeavor to conceive all phenomena merely as quantities,

mines and iron works, describes in his *Pirotechnia* (1540) the mechanical system for the better exploitation of water power, which he invented and introduced in northern Italian iron works: a large bucket-wheel, which set in motion a number of bellows and could serve four fires at the same time, for which otherwise four water wheels had been needed. – Georg Agricola shows in book VIII of his work *De re metallica*, written around 1550 (Basel 1556), the construction of the crushing machines which were already used in Germany in the fifteenth century for the crushing of iron ore. The water wheel moves at first one, and later three or four crushing stamps, which entailed a considerable *rationalization* of the work and a saving in manpower. (Cf. Ludwig Beck, *Geschichte des Eisens*, Braunschweig, 1893, Vol. II, p. 87.)

[70] W. Sombart, "Die Entstehung der kapitalistischen Unternehmung," *Archiv für Sozialwissenschaft*, Vol. 41, 1915, pp. 311, 325.

an idea which brought to light all the wonders of the cognition of nature." In short, the "double-entry bookkeeping, developed in the fourteenth century, originated from the same spirit as the systems of Galileo and Newton" (loc. cit., p. 318). Here again, Borkenau has eliminated from history two hundred years of capitalist methods of calculation.

The Substantiation of Borkenau's Conception

Until now we have outlined Borkenau's conception and have confronted it with [196] reality; now the question arises of how he substantiates his conception. To him there are only two ways of considering historical facts: the descriptive presentation, which he scorns, and the emphasis on "structural" moments, i.e. their arrangement into a structured scheme. We have seen how he neglected the historical development of natural sciences and presented the historical change in the concept of natural laws in its place. We see the same disdain for facts in his principal conception of the connection between mechanistic philosophy and manufactural division of labor. Here, too, a proof is replaced by an assertion. Borkenau himself comments on his thesis: "if this conception is valid, then the actual scientific research of the time had to be done at [sic!] the manufactural production process itself" (p. 6). This can only mean that scientific research had to frame its basic concepts according to the manufactural division of labor which presented the material for scientific analysis. Now Borkenau himself establishes that three different technical procedures existed side by side during the period of manufacturing: (1) the traditional artisanship, (2) the division of labor in manufacture, and finally (3) "the factory which was emancipated to a large extent from artisanship," i.e. if we express this phrase more clearly – mechanical production by machines. And Borkenau finds: "It is striking that the science of the period allows itself to be led exclusively by the methods of manufacture" (p. 4). In the face of this "striking exclusiveness" it ought not be difficult to adduce the necessary evidence. Yet no such evidence is produced.

Borkenau tries to illustrate his thesis by the example of physiology: "At the very beginning of the seventeenth century, physiology obtains its scientific foundation through Harvey's discovery of blood circulation which he explains with the analogy of a pump mechanism" (p. 5). One asks with surprise: what does a pump mechanism have to do with the manufacturing methods based on division of labor? After all, the pump is a machine. Thus, instead of demonstrating the connection of mechanically conceived physiology with the division of labor in manufacture, Borkenau demonstrates its orientation towards machines. Elsewhere he says with regard to the seventeenth century that the "manufacturing period" was simultaneously "the century of water" (p. 9), that is, a century which built machines driven by water. But what have water-driven machines to do with the division of labor in manufacture? Finally, on a third occasion he asserts that this connection is "evident in Simon Stevin, the field engineer of Moritz of Nassau," the founder of modern mechanics in Holland (p. 6). And again we ask in wonder: what does the practice of field [197] engineers have to do with the method of division of labor in manufacture? These are the only examples given for the historical proof of the alleged connection. Galileo's mechanics and those of his time are said to be nothing but the scientific treatment of the process of manufacturing production. "According to the latest state of research, this thesis can now be critically confirmed by the recourse to the sources, which was until recently impossible" (p. 6). Yet Borkenau does without this attestation from sources. To console us he says that this can be found in another author's book. Olschki, he says, "in his excellent research on Galileo and his time," has proven that what is innovative in Galileo's quest is the rejection of theoretical tradition and the reference to the active technicians' practice.[71] The same interconnection with practice, Borkenau says, was also "self-evident" for Simon Stevin, the field engineer (p. 6). But we can only repeat our question: what does the connection with the praxis of technicians have to do with the scientific treatment of the division of labor in the manufactural process of production? After all, we know that three different procedures existed side by side in the practice of the manufacturing period. The "connection with the praxis" does not yet indicate with which praxis the connection was established – the artisan's, the manufacturer's or the praxis with machines. Thus Borkenau thinks with regard to Francis Bacon that it was "precisely Bacon's close ties with the most highly developed (i.e. mechanical, H. G.) forms of industrial praxis" which impeded his access to those basic forms of technique which became the foundations of the mechanistic world view (p. 90). Therefore, if Borkenau's thesis is to make sense at all, then proof should be provided not only of the connection with some kind of praxis but of manufacture based on division of labor. For this is the "thema probandi" of Borkenau's book. He does not provide the proof, and Olschki, whom he cites, does not either.

In addition to the historical evidence in the sources, Borkenau wants to provide a second, theoretical proof: "The new mechanistic world view's dependence on the technique of manufacture can also be easily shown from their respective contents." And now the reasoning we already know follows – by the division of work into simple manipulations, the skilled workers are replaced by unskilled ones, whereby all work is reduced to uniform, "general human," and thus quantitatively measurable labor. Only thereby do the quantitative methods which are the foundations of mechanics become possible. We have already shown what this reasoning is worth. Where and how the argument by "content" is supported by evidence is essential here. This is already provided in the introductory remarks on p. 7 of the book, before the start of the research and before any material has been presented. In the book itself, especially in the section on Descartes, no further proof is brought; the previously developed trend of thought is simply repeated (p. 357). [198]

[71] Everyone familiar with Alberti's and Leonardo da Vinci's achievements, knows that Galileo's rejection of the traditional academic science and his reference to practice is not "innovative." One hundred and fifty years before Galileo, Alberti, this "truly universal Titan" – as Burckhardt calls him – studied all possible sciences and arts; "he went into physics and mathematics and simulataneously learned all the skills of the world, asking artists, scholars, and craftsmen of all sorts, including shoemakers, about their secrets and experiences" (Jakob Burckhardt, *Die Cultur der Renaissance in Italien*, op. cit., Vol. I, p. 150).

We are told of Descartes' whole family's civic history, the professions of his father and grandfather, of his mother's father and grandfather, of the grand-uncles and other ancestors; we are given a lengthy interpretation of the dreams in the mystical crisis of Descartes' youth, from which – after all the biographers' earlier interpretations – nothing substantially different or better emerges; we find many other superfluous details – however, what is really essential for the thesis is missing, viz. positive evidence of the connection between mechanistic philosophy and division of labor in manufacture. Within the system of categories actually used, the reduction of the elements of the mechanistic world image to the division of labor in manufacture proves to be decorative, "materialistically" adorning the genesis of mechanistic philosophy, but by no means serving as a means of analysis. In the book itself this technique of division of labor in manufacture is inconsequential in the analysis of individual thinkers' actual ideas or of concrete intellectual trends.

Only when one bears this in mind does Borkenau's attitude toward a series of phenomena become comprehensible – e.g. toward the inventions of the Renaissance: there were many, and some were "of the greatest importance," but were made only accidentally, by practitioners, without a possibility for perfecting them systematically. Again, it is enough to mention Leonardo da Vinci to see this assertion's baselessness. All his inventions – and there were dozens of them – emerge from the theoretical cognition of the relevant subject matters. Leonardo himself writes: [199] "The practice must always be based upon good theory."[72] "Science is the captain, practice the soldiers."[73] The research on air and air pressure laws led him to construct the parachute, invent the pluviometer (which measures the humidity in the air), the pendulum of the anemometer (which measures the wind force) and to his systematic, long-lasting endeavors to construct a flying machine.[74] The discovery of the most important laws of mechanics, of the law of the lever, of the inclined plane, the screw, etc., all of which he traces back to the pulley, leads him to the construction of various pulleys and combinations of pulleys, winches and various lifting machines. The discovery of the laws of hydrostatics leads him to the idea of the artesian well, for which he also constructs the suitable drilling equipment.

For Borkenau the inventions of the Renaissance are purely "accidental." Had he really applied the thesis of the connection between mechanics and division of labor, he would soon have encountered factual connections which would have induced him to revise his thesis. He would immediately have grasped the connection of Renaissance inventions with the situation prevailing in Italian industry. But he did not pay any attention to the Italian economy's development at that time. He made do with the empty formula of the incipient monetary capitalism as a general explanation.

Due to lack of space it is impossible to describe in greater detail Italy's state of affairs in those days. Let us only recall that, as a consequence of the shift of the international trade axis from the Mediterranean to the Atlantic coast of Europe,

---

[72] M. Herzfeld, *Leonardo da Vinci*, op. cit., p. xvii.

[73] G. Séailles, *Leonardo da Vinci, l'artiste et le savant*, Paris, 1906, p. 353.

[74] Op. cit., p. 231. Cf. F. M. Feldhaus, *Die Technik*, Leipzig 1914, and idem, *Leonardo da Vinci, der Techniker und Erfinder*, Jena, 1913.

Italian capitalism – which had been flourishing for almost two hundred years – experienced a sudden recession. This was aggravated by the wage increases caused by the best manpower's move from the cities to the country – into gardening. In order to compete with the world markets, efforts were made to reduce production costs. Hence the trend toward replacing expensive human labor with cheap natural power – water power – a context which clearly emerges from Leonardo's[75] writings.[76] Do we here encounter the problem of capitalist "rationalization"? By no means, according [200] to Borkenau; for him it is an established fact that rational technique can only "be put into practice for the first time during the manufacturing period," that it only "develops from the endeavors to rationalize crafts," and that its bearers could not be the religiously indifferent Renaissance men, but only the "Calvinist ambitious little men" (p. 90).

Not only the general connections between industrial development of Italy and the invention of industrial machinery become clear with an economic analysis. One can go further and recognize that certain inventions are determined by the social situation of a particular stratum. Since the late fifteenth century, Venice, the maritime power, had been using, on its war galleys propelled by rowing slaves, mitrailleuses of a special structure consisting of twenty barrels arranged in two circles, with the ten interior barrels longer than the ten outer ones. During normal service on the galleys the whip was enough to impose obedience; in face of the enemy in battle the situation was different. In this connection we learn about the purpose of the mitrailleuses: "elles servaient à tenir les rameurs en respect pendant l'action, quand le fouet des surveillants n'y suffisait pas." The salvo from the shorter barrels was intended for the more closely placed slaves, that from the longer barrels for the more distant side of the ship.[77]

The Sources of Descartes' Mechanistic Conception According to His Texts

Since in Descartes there is no reference to division of crafts labor, the question arises: what do his texts reveal regarding the sources of his mechanistic inspiration? In all his principal works we find numerous explicit references to machines. These are not just occasional remarks but are the foundations of his mechanistic conception. His concept of the world and its parts being a mechanism is demonstrated in decisive passages of his argumentation with the example of machines. None of these numerous passages, which are at the core of the Cartesian arguments, is [201] ever mentioned by Borkenau! And there is even more. He also negates the practical importance that Descartes attributes to the machines as a way of reducing human

[75] Grothe, op. cit., p. 10.

[76] G. Cardano, in his book *De subtilitate* (1550) stresses the most important advantages of the use of machines: (1) the savings in manpower, (2) the possibility of employing unskilled, and therefore cheaper, workers, (3) less waste of material, thereby making production even cheaper, (4) general advantages of hygiene, thus saving cleaning expenses.

[77] E. Hardy, *Les Français en Italie* de *1494 à 1559*, Paris, 1880, p. 37.

labor, in short, as productive forces. "By the way, there is no doubt that, in regard to the practical usefulness of knowledge, he was much less interested in the development of productive forces than in medicine. His mechanical inventions were limited to appliances for polishing lenses" (p. 274). Indeed, Descartes was a physician and not an engineer. His interest in the development of productive power was not based upon the utilization of his own inventions but on his conviction that science can be generally useful for practical tasks of life. Although "there is no doubt" that Descartes was less interested in the practical applicability of knowledge and in the development of productive power – a statement not substantiated by a single word – there is evidence to the contrary. It is precisely the development of productive forces, the practical usefulness of knowledge for dominance over nature and alleviation of human toil and labor which Descartes – in contrast to the speculative philosophy of earlier times – posits as the main tasks of science. In this respect he does not differ in any way from Bacon.

In the first part of the "Discours" of 1637 he already says, "... que les mathématiques ont des inventions trés subtiles, et qui peuvent ... faciliter tous les arts et diminuer le travail des hommes."[78] And he pursues that same thought even more consistently at the close of the "Discours": "Les notions générales touchant la physique" – which Descartes acquired – "diffèrent des principes dont on s'est servi jusqu'à présent" ... "car elles m'ont fait voir qu'il est possible de parvenir à des connoissances qui soient fort utiles à la vie, et qu'au lieu de cette philosophie spéculative qu'on enseigne dans les écoles on en peut trouver une pratique par laquelle, connoissant la force et les actions du feu, de l'eau, de l'air, des astres, des cieux, et de tous les autres corps qui nous environnent, aussi distinctement que nous connaissons les divers métiers de nos artisans, nous les pourrions employer en même façon à tous les usages auxquels ils sont propres, et ainsi nous rendre comme maître et possesseurs de la nature." To leave no doubt as to how this domination of nature should be achieved through increased knowledge of nature, he adds that [202] this is desirable "pour l'invention d'une infinité d'artifices (i.e. artificial machines, H. G.) qui feroient qu'on jouiroit sans aucune peine des fruits de la terre et de toutes les commodités qui s'y trouvent."[79]

Here – in Descartes' own words – the source which provided the initiative for working out the mechanical concept of work is in the machines, and not connected with the division of labor in manufacture, as Borkenau claims. The practical aim of easing human labor through the work of machinery presupposes the comparison

[78] *Oeuvres*, Adam and Tannery, eds., Paris, 1897, Vol. VI, p. 6.

[79] This practical function of Descartes' philosophy is so evident that J. H. von Kirchmann could write sixty years ago: "Here the same tendency is evident in Descartes as in Bacon. Both were so enthusiastic about the new discoveries that they emphasized above all the need for inventions of methods and machines that were to prove beneficial in practical life" (*R. Descartes' philosophische Werke*, Berlin, 1870, part I, p. 70). Ten years after the "Discours," at the end of his preface to the "Principes," Descartes stresses the importance of science for improving the quality of practical life and shows "combien il est important de continuer en la recherche dc ces vérités et jusques ... à quelle perfection de vie, à quelle félicité elles peuvent conduire" (*Oeuvres*, Vol. IX, p. 20).

between both types of work, their reduction to general mechanical concepts of work and the quantification of the work done. Only thus can it be ascertained whether the machine really does reduce human labor.

For Descartes, *mechanics is first of all a theory of machines*, whose principles are subsequently extended to physics and to the whole universe. Thus in his work one finds lifting machines, which were used for centuries in architecture and in loading ships' cargoes, mentioned as the first group of mechanisms. In the "Traité de la Mécanique" which he develops in his letter to Constantin Huygens (5 October 1637), using the examples of the *poulie* (pulley), the *plan incliné* (inclined plane), the *coin* (wedge), the *tour* (winch), the *vis* (screw) and the *levier* (lever) – the simplest elements to which every machine can be reduced – he gives the "explication des machines et engins par l'aide desquels on peut avec une petite force lever un fardeau fort pesant."[80] In his letter to Mersenne (13 July 1638) Descartes develops the same thoughts; he deduces the principles of mechanics from the contemplation of machines and at the same time, like Leonardo da Vinci 140 years before him, reduces all machines to the inclined plane, as their basic form.[81] Descartes gives us the theory of the pulley (*mouffle* or *poulie*), the *plan incliné*, and the *levier* one after the other. The latter, he says, is nothing but "un plan circulairement incliné." Likewise, "le coin et la vis ne sont que des plans inclinés, et les roues dont on compose diverses machines ne sont que des leviers multipliés, et enfin la balance n'est rien qu'un levier qui est soutenu par le milieu." [203]

The relations between machines and the principles of mechanics are just as clear and close in Descartes as the relations between machines and his mechanistic philosophy. A short synopsis of his major works will confirm this.

In his early work, the "Cogitationes Privatae" (December 1618), he is already interested in the mechanical motions of the machine, and he describes a statue with pieces of iron in its arms and legs.[82] Immediately afterwards he mentions the artificial mechanical dove of Archytas of Tarent: "Columba Architae molas vento versatiles inter alias habebit, ut motum rectum deflectat."

This is not the place for a close investigation of Descartes' experiences with artillery. Suffice it to say that he was entirely familiar with cannons, which after all are machines, with the specifics of the motions they produce, with the trajectory and speed of the projectiles, and with all the factors on which the performance of this machine, namely the trajectory and speed, depends;[83] visible and important

[80] *Oeuvres*, op. cit., Vol. I, p. 435.

[81] Clerselier, *Lettres de Descartes*, Paris, 1657, Vol. I, letter LXXIII. Cf. *Oeuvres*, op. cit., Vol. II, pp. 236–223

[82] *Oeuvres*, op. cit., Vol. X, p. 231.

[83] In the Jesuit college at La Flèche, Descartes received an education which included, among other subjects, instruction in "l'art des fortifications et l'emploi des machines"; this education was "orientée vers la pratique militaire et orientée à former . . . un officier d'artillerie ou du génie" (P. Mouy, *Le Développement de la physique cartésienne*, Paris, 1934, p. 2).

traces of this remained in his thinking. In the second essay, "De la Refraction"[84] of the "Dioptrique" (1637), which was published simultaneously with the "Discours," he tries to explore the laws of optical reflection by establishing first an analogy between the unknown connections of optical phenomena and the laws of ballistics with which he is familiar. He compares the ray of light and the laws of its refraction with the trajectory of an artillery missile and its laws. A cannon ball, shot into a river at a certain angle, will not penetrate the water surface but will be refracted [reflected] at the same angle to the other side, as if it had hit a solid object. "Ce qu'on [204] a quelquefois expérimenté avec regret, lorsque, faisant tirer pour plaisir des pièces d'artillerie vers le fond d'une rivière, on a blessé ceux qui étaient de l'autre coté sur le rivage."[85] Descartes' great discovery, the law of refraction, is also determined by his experiences with artillery pieces!

As a former artillery officer he also obviously knew all other factors determining the trajectory and speed of the missile (such as the length and elevation of the barrel, the amount and chemical composition of the gunpowder), which are elaborated at length in the *Principes de la Philosophie* (1647).[86]

Elsewhere in the "Principes" he mentions the cannons which are based on the principle of compressed air: "ce qui a servi de fondement à l'invention de diverses machines, dont ... des petits canons, qui n'étant chargés que d'air, poussent des balles ou des flèches presque aussi fort ... que s'ils étaient chargés de poudre."[87]

But aside from the wide area of lifting machines and machines for military use, mention is also made of other machines which were equally important for the development of mechanics: the clock and the motor mechanism in industry, the "machines mouvantes." These represented the real origin for Descartes' mechanistic conception.

Borkenau quotes the following sentence from the fifth part of the "Discours": "les règles des mécaniques, qui sont les mêmes que celles de la nature" and states: Descartes "demonstrates it there with the famous example of blood circulation, mainly adopted from Harvey" and derives the very laws of nature from mechanics (p. 356). But how did Descartes arrive at his mechanics? Borkenau reiterates the well-known conception: Up to Descartes, scientific mechanics were impossible, because "the world was a sum total of static systems" and therefore the manner of observation had to be a qualitative one. Only when the social world is set in motion, a movement which dissolves all traditional stratified orders, does "the qualitative mode of observation fall away, and what replaces it must be at the same time quantitative, mathematical and dynamic" (p. 357). This quantitative mode of observation is once again connected with the division of labor in manufacture, since [205] "in manufacturing work, the quantified performance, the computable movement belongs with the quantified material" (loc. cit).

---

[84] *Oeuvres*, op. cit., Vol. VI, p. 93.
[85] *Oeuvres*, op. cit., Vol. VI, p. 99.
[86] *Oeuvres*, op. cit., Vol. IX, p. 262.
[87] *Oeuvres*, op. cit., Vol. IX, p. 227.

The arbitrariness of such an assertion shows itself precisely in the quotation from the "Discours" – if only one reads it till the end! Only in the fifth part, and referring to the description of the heart's and blood's movements, does Descartes say: "que ce mouvement que je viens d'expliquer suit aussi nécessairement de la seule disposition des organes . . . qu'on peut connaître par expérience, que fait celui d'une horloge, de la force, de la situation et de la figure de ces contrepoids et de ses roues."[88] There is no allusion to the division of labor in manufacture, but there is a comparison with a machine, with the clock; the movements of the heart and blood are just as much conditioned by the disposition of the bodily organs as the movement of a clock is conditioned by the disposition of its weights and wheels! After the description of the blood circulation, in which Descartes says that the rules of mechanics are the rules of nature – which in this context can only mean that the movements in nature take place according to the same principle as the mechanical movements of a clock – he presents the problem of the automatics of the movements of single organs of the body, e.g. of the muscles, and believes that these, by virtue of their disposition, "se puissent mouvoir sans que la volonté les conduise."[89] How does he illustrate the possibility of such automatic movement of muscles, to make it understandable to his contemporary reader? Not by the division of labor in manufacture, but by motor mechanisms! The possibility of automatic body movements, says Descartes, will not surprise anybody, "qui sachant combien de divers automates ou machines mouvantes l'industrie des hommes peut faire, sans y employer que fort peu de pièces." Every animal body, because it is more complex, and consists of a large number of components, is more perfect, compared to these machines and we can consider the body "comme une machine qui, ayant été faite des mains de Dieu, est incomparablement mieux ordonée et a en soi des mouvements plus admirables qu'aucune de celles qui peuvent être inventées par les hommes."[90] Thus the human body's functions are of the same kind as mechanical movements, yet machines, as compared with man, show a limitation by which they can be clearly distinguished. In order to demonstrate this, Descartes uses the fiction of a perfect man-machine which is capable of moving and uttering words. Even if this is of the best possible [206] external likeness, such a machine will be different from man in principle, since it can only speak a few exactly prescribed sentences and execute only certain movements, whereas man can react in varied ways to all possible situations by means of his reason, because "la raison est un instrument universel."[91] In this way, the simplest man differs from the highest animal or the best machine. For animals possess no reason; even if they execute certain functions better than man, they are only acting mechanically: "C'est la nature qui agit en eux selon la disposition de leurs organes: ainsi qu'on voit qu'un horloge, qui n'est composé que de roues et de ressorts, peut compter les heures et mesurer le temps, plus justement que nous avec toute

[88] *Oeuvres*, op. cit., Vol. VI, p. 50.
[89] *Oeuvres*, op. cit. Vol. VI, p. 55.
[90] *Oeuvres*, op. cit., Vol. VI, p. 56.
[91] *Oeuvres*, op. cit., Vol. VI, p. 57.

notre prudence."[92] In the third meditation of the "Méditations" (1641), the famous proof of God's existence is based on the reality of the idea of God, and in answer to Caterus' "first objections" (1647): the essential argumentation is based on the example of machines. Just as the idea of the machine is based upon its builder's knowledge of mechanics, thus the idea of God must originate from God.[93]

This conception is even more strongly expressed in the seventeenth century by Descartes' successors, whose education was based on his teachings and who all use the example of the clock.

Like Descartes, Robert Boyle (1626–1691) regards the human organism "tamquam machinam, e partibus certis sibi adunitis consistentem." In his endeavor to reconcile religion with science within a unified world picture, the watchmaker's relation to the clock (which he illustrates with the Cathedral of Strasbourg's famous clock) serves as the model for the teleological conception of nature on a mechanical basis.[94] It is not different with Newton (1642–1727). His God appears as a watchmaker who – according to an ironical remark of Leibniz (1715)[95] – needs to rewind the clock of the world from time to time. Voltaire – a Newtonian – still writes in a letter dated 26 August 1768: "Les athées n'ont jamais répondu a cette difficulté qu'une horloge prouve un horloger."

[207]

Descartes, in his last work *Les passions de l'âme* (1649) again reverts to the comparison with the clock in order to make the difference between a living and an inanimate body comprehensible: "Le corps d'un homme vivant diffère autant de celui d'un homme mort que fait une montre, ou autre automate (c'est-à-dire, autre machine qui se meut de soi-même), lorsqu'elle est montée, et qu'elle a en soi le principe corporel des mouvements pour lesquels elle est instituée, avec tout ce qui est requis pour son action, et la même montre ou autre machine, lorsque'elle est rompue et que le principe de son mouvement cesse d'agir."[96]

An even more important role than that of the clock is played by various other types of motor mechanisms, "machines mouvantes," in particular water-driven machines, which were then originally invented for the Italian industry, and subsequently used for purposes of domestic convenience and for the embellishment of the affluent's palaces and gardens. In "Discours VIII," entitled "De l'arc-en-ciel," in his work *Les Météores* (1637), Descartes describes his experiences, which were probably gathered in Rome, with the artificial creation of rainbows whose form could be varied according to different arrangements of the holes in the fountain, in

---

[92] *Oeuvres*, op. cit., Vol. VI, p. 59.

[93] "Ce que j'ai éclairci dans ces responses par la comparaison d'une machine fort artificielle, dont l'idée se rencontre dans l'esprit de quelque ouvrier; car, comme l'artifice objectif de cette idée doit avoir quelque cause, a savoir la science de l'ouvrier ... de même il est impossible que l'idée de Dieu qui est en nous, n'ait pas Dieu même pour sa cause" (Abrégé de la troisième Méditation, *Oeuvres*, op. cit., Vol. IX, p. 11. Cf. pp. 83–84).

[94] *The Works of Robert Boyle*, London 1772, Vol. II, "Of the Usefulness of Natural Philosophy," p. 39.

[95] *Hauptschriften zur Grundlegung der Philosophie*, E. Cassirer (ed.), Leipzig, 1903, Vol. I, pp. 120, 126 [Leibniz's first and second letter to S. Clarke].

[96] *Oeuvres*, op. cit., Vol. XI, p. 331, art. 6. Cf. also arts. 7 and 16.

which he saw an experimental confirmation of his mechanical theory of refraction.[97] Finally, in his *Traité de l'Homme* (1644) he regards man as a machine, composed of various partial mechanisms which function in the same manner as clocks, water mills, carillons, organs, etc.: "... Je suppose que le corps n'est autre chose qu'une statue ou machine ..."[98] He uses stereotypically the expression: "cette machine"[99] to characterize all organs, such as the tongue with the sense of taste, the nose with the sense of smell, the respiratory organs, the heart, the eyes, the stomach, etc. He wants to clarify the movements of all body parts through muscles and nerves, and [208] the latter through the "esprits animaux" in a purely mechanical way by comparing them with the driving force of water,[100] because "la seule force dont l'eau se meut en sortant de la source, est suffisante pour y mouvoir diverses machines, et même pour les y faire jouer de quelques instruments, ou prononcer quelques paroles, selon la diverse disposition des tuyaux qui la conduisent," ... "ainsi que vous pouvez avoir vu dans les grottes et les fontaines qui sont aux jardins de nos Roys."

The nerves are compared with the "tuyaux des machines de ces fontaines," the muscles and sinews with the "divers engins et ressorts," and respiration and other natural functions with the "mouvements d'une horloge ou d'un moulin, que le cours ordinaire de l'eau peut rendre continus."[101] The heart's and arteries' functions are compared with the "Orgues de nos Églises," that is, with their bellows (*soufflets*).[102] The external world thus acts upon our senses and causes their movements in a purely mechanical way, "comme des étrangers qui, entrant dans quelques-unes des grottes de ces fontaines, causent eux-mémes sans y penser les mouvements qui s'y font en leur presence ... selon le caprice des Ingénieurs qui les ont faites." The reasonable soul's role is comparable with the water engineer's function (*le fontenier*) who, by rearranging the machines' pipes, changes their pattern of movement.[103] At the end of the work, he says: "Je désire que vous considériez que ces fonctions suivent toutes naturellement en cette machine, de la seule disposition de ses organes, ni plus ni moins que font les mouvements d'une horloge, ou autre automate, de celle de ses contrepoids et de ses roues."[104]

Elsewhere, in order to explain Harvey's theory of blood circulation, Descartes says that the veins and arteries are like ducts through which the blood flows incessantly into the chambers of the heart "en sorte que ces deux cavités sont comme

---

[97] *Oeuvres*, op. cit., Vol. VI, pp. 343ff.
[98] *Oeuvres*, op. cit., Vol. XI, p. 120.
[99] *Oeuvres*, op. cit., Vol. XI, pp. 125, 138, 145, 148, 163, 173, etc.
[100] Op. cit., Vol. XI, p. 130.
[101] Op. cit., p. 131.
[102] Op. cit., p. 165.
[103] Op. cit., p. 131.
[104] Op. cit., p. 202.

des écluses par chacune desquelles passe tout le sang."[105] Thus the chambers of the [209] heart are compared to sluices.[106]

Appendix: Galileo, Hobbes

What we have shown so far regarding Descartes is also demonstrable for all the other representatives of mechanistic philosophy in his time. For lack of space we must restrict ourselves to some brief references. According to Borkenau, the "basic concepts of mechanics, which Galileo and his contemporaries were the first to develop comprehensively, were nothing but the exact formulae of the relations emerging between human labor and the object of their work in manufacture's extremely divided manual production process. Mechanics, i.e. the science of the manufacturing period, is the scientific exploration of the process of manufacturing production" (p. 6). This assertion, made with certainty and acknowledging no possible doubt, arouses the suspicion that Borkenau knew very little of Galileo's mechanics. In the first chapter of Galileo's *Mechanics*, first published in French by Mersenne,[107] he shows quite clearly from where he derived his mechanical concepts. He does not base himself on the division of labor in manufacture but on the machines, and lifting machines in particular! In the first chapter "qui montre l'utilité des machines" he already mentions machines for the transportation of heavy loads, machines for bringing up water from the depths of wells, pumps for removing water from the hulls of ships, and finally water mills and other machines with wheels, which replace and reduce human and animal labor costs.[108] Having thus circumscribed the purpose of mechanics and his research's objectives, he in Chapter II provides the definition "afin d'en tirer les raisons de *tout ce qui arrive aux Machines, dont il faut expliquer les effets* ..." "Or, puisque les Machines servent ordinairement pour *transporter les choses pesantes*, nous commençons par la définition de la pesanteur, que l'on peut aussi nommer gravité" (op. cit., p. 6). Then in Chapters VI-X he demonstrates the mechanical principle of the asymmetrical lever, the scales, the winch and the crane, the pulley, the screw and its uses for drawing up water, the syphon and the pump, those ordinary machines which during the almost two hundred years from L. B. Alberti and Leonardo da Vinci to Descartes, had always been objects of contemplation for theoretical mechanics.

---

[105] Op. cit., p. 332.

[106] Finally, it should be mentioned that Descartes' disciples had the same conception of mechanics as their teacher. In the "Traité de la Mécanique," published by N. Poisson in 1668, mechanics is first a theory of machines, whose principles are subsequently extended to physics and the whole universe. "De même aussi on peut considérer le corps humain comme un automate ou machine" (P. Mouy, "Le Développement de la Physique Cartésienne," op. cit., p. 63).

[107] G. Galilei, *Les Méchaniques*, transl. from Italian by Mersenne, Paris, 1634.

[108] "La troisième utilité des machines est très grande, parce que l'on évite les grands frais et le coût en usant d'une force inanimée, ou sans raison, qui fait les mesmes choses que la force des hommes animés ... comme il arrive lorsque l'on fait moudre les *moulins* avec l'eau des estangs, ou des fleuves, ou un cheval, qui suplée la force de 5 ou 6 hommes... par le moyen des *roues et des autres Machines* qui sont ébranlées par la force du cheval, et qui remplissent et transportent le vaisseau d'un lieu à l'autre, et qui le vident suivant le dessin de l'ingénieur" (op. cit., p. 5).

In our context Th. Hobbes is of special importance, since he first applied the mechanical conceptions to the social sphere, whereas previously they referred to natural phenomena only. In the foreword of his principal work[109] he already depicts the state and its citizens as a huge machine whose essence can only be grasped if one analyzes in thought the constituent elements which originate in human nature: "for as in a watch, or some such engine, the matter, figure, and motion of the wheeles, cannot well be known, except it be taken in sunder, and viewed in parts." [210]

On the Genesis of Theoretical Mechanics

What can be proven by the survey presented in the preceding paragraph? Perhaps the fact that Borkenau disregarded several passages in Descartes, or that he evidently distorts Descartes' conception by connecting it arbitrarily with the division of labor in manufacture? Not at all. More is actually at stake: the history of the origins of a whole science, viz. modern mechanics! The machines mentioned by Descartes, which can be divided into four categories: artillery, clock, water and lifting machines, also represent the most important areas of practical mechanics, by which the basic concepts and laws of theoretical mechanics could be developed. Mechanics was only slowly created by the struggle of human ratio with the empirical material. For nearly two centuries – from the middle of the fifteenth to the beginning of the seventeenth century – all those who took part in the struggle (L. B. Alberti, Leonardo da Vinci, Nicolo Tartaglia, Girolamo Cardano – to name only the most important scholars) derived their mechanical concepts and theorems not from the division of labor in manufacture, but from the analysis and observation of machines and their performance.

Anybody who traces the history of the genesis of theoretical mechanics will immediately encounter the four aforementioned categories of machines; we will briefly look at them:

I. *Firearms*. The discovery of gunpowder and firearms – not even mentioned by Borkenau in this context – constitutes an epoch-making turning point in the history of scientific mechanics. Not only did it break the nobility's monopoly in martial skills, with warfare becoming a bourgeois affair, but also because "the skills of the engineer, founder of cannons and the gunner, acquired in a bourgeois manner, became prominent"[110] and educated bourgeois elements participated in war, impetus was provided to fruitful mass observations which served both the perfection of firearms and the development of the theory of mechanics. [211]

Through the accumulated observations of missile trajectories, the old Aristotelian "milieu theory" of motion, which maintained that the progression of

[109] Th. Hobbes, *Elements*, Part III, *De Cive*. German translation by M. Frischeisen-Köhler, Leipzig, 1917, p. 72. [English quotation according to: Thomas Hobbes, *Philosophical Rudiments Concerning Government and Society* (1651); *Clarendon Edition of the Philosophical Works of Thomas Hobbes*, Vol. III, Oxford, 1983, p. 32].

[110] J. Burckhardt, *Die Kultur der Renaissance in Italien*, op. cit., Vol. I, p. 103. Cf. Max Jähns, *Handbuch einer Geschichte des Kriegswesens*, Leipzig, 1880, p. 831.

the projectile was caused by the movement of air, was totally undermined, for the air's obstructing effect was empirically recognized. With Aristotelian doctrine abolished, the road was opened for new observations and new attempts at theoretical explanations. The oldest work on artillery, the *Livre des faits d'armes* was written by Christine de Pisane around 1400. From Leonardo da Vinci[111] via Tartaglia and Girolamo Cardano an uninterrupted chain of scientific endeavors leads to the establishment of a theory of motion by the experience gathered from firearms. One only need open Nicolo Tartaglia's book *Nuova Scientia* (1537), written almost one hundred years before Galileo,[112] in order to become convinced that the laws of motion were studied above all on the example of artillery projectiles' trajectories.

Anybody familiar with the genesis of mechanics must know that the discovery of the law of free fall is closely connected with the history of firearms, with the observations on the projectiles of guns.[113] Here, as in so many other areas, the economic aspect provided the impetus for the continuation of research, while striving, [212] through rational construction of guns, to attain the same effect with smaller calibers, to achieve better transport capacity and lower construction costs.[114]

Just as Borkenau is silent on the history of firearms as a source of theoretical mechanics, his attitude is the same vis-à-vis the other principal sources and spheres of observation: the construction of water-driven machines, the lifting mechanisms, and finally the clock mechanisms.

II. *The mechanism of clocks*. Nowadays we can hardly imagine the intellectual upheavals connected with the discovery and perfection of the mechanical clock. The close connection established among the Arabs between the construction of clocks and astronomy is well known. Scientific chronometry, i.e, the exact quantification of time, is the precondition of exact observations in all spheres of science. In thirteenth and fourteenth-century Italy the astronomers were often also watchmakers and mechanics. But in mechanics the clock is the first and most important machine, having a uniform motion produced automatically by a system of weights. At first the

---

[111] Numerous texts by Leonardo, which illuminate his problematic show how theoretical mechanics tried to derive its concepts from the flight of projectiles. Just one example: "Si une bombarde avec 4 livres de poudre jette 4 livres de boulet à sa plus grande puissance, à 2 milles, de combien faut-il augmenter la charge de poudre pour qu'elle tire à 4 milles? La puissance du boulet dépend-elle de sa vitesse initiale?" (G. Séailles, op. cit., p. 353).

[112] *Inventione de Nicolo Tartaglia, Brisciano, intitolate Scientia Nova*, divisa in V libri, 2nd ed., Venice, 1550.

[113] Also Tartaglia's other work, *Quesiti et Inventione diverse* (1546), the first volume of which is devoted to the study of the motion of cannonballs; and this, according to the testimony of P. Duhem, had a strong influence on the development of mechanics in the sixteenth century. It was, therefore, of basic significance for the history of dynamics (P. Duhem, *Les origines de la statique*, Vol. I, p. 197).

[114] "On peut restreindre beaucoup de la mesure commune et faire l'artillerie de moindre poids; chose qui rend très grande facilité a la conduire et si espargne beaucoup à celui qui la fait forger" (Vanuccio Bringuccio, *La pyrotechnie ou art du feu*, X livres, 1st ed., 1540, quoted from the French edition, Paris, 1556, p. 142).

automatic motion was of even greater interest than the indication of the time.[115] The public tower clocks in the towns of Italy and Flanders were enormous gear works, combining the mechanism for the measuring of time with the bell-ringing mechanism.[116] At the threshold of the fourteenth century, two astronomers who were also mechanics, the Dondi brothers of Padova, constructed a planetarium (described in a book of this title): a complicated gear work driven by weights, which visualized the observable movements of the sun, the moon and the planets.[117]

A field of observation for scientific mechanics was thus created, which was to stimulate the investigation of the elementary laws of motion: a vertical movement [213] of the slowly descending weight was transformed by a mechanism of wheels into a circular motion. The automation of circular motions in the planetarium had to be adapted to the speed of the individual heavenly bodies' movements, according to astronomical calculations. One single weight set in motion several wheels which had to move at various speeds in different orbits, and this necessarily led to systematic contemplation on the causes for this difference in the space-time relation. The experimental imitation of the structure of heavenly mechanics removed the latter's mystic veil and suggested the idea that the heavenly bodies' movement was governed by principles similar to those of the mechanics of the planetarium. The orreries, which were constructed as increasingly complicated mechanisms, are important for the histories of both mechanics and astronomy; their construction simultaneously reflected the actual state of practical mechanics and of astronomical knowledge – first on the basis of the Ptolemaic system and subsequently of the Copernican system.[118]

Aside from these clockworks whose purpose it was to measure time and illustrate astronomic processes, the clockwork mechanism was also employed in Italy for the production of power: in Milan, around the middle of the fourteenth century, there were mills driven by clockworks.[119]

III. *Lifting mechanisms*. Due to lack of space we will not go into the details of the lifting machines used in medieval architecture and shipping, with which considerable loads, such as church bells and blocks of marble, were raised to great heights. We only mention that in 1466 the tower Della Magione in Bologna, with its foundations, was moved a substantial distance without the slightest damage![120]

---

[115] Mathieu Planchon, *L'évolution du mécanisme de l'horlogerie depuis son origine.* Bourges, 1918, p. 4.

[116] Pierre Dubois, *Horlogerie, iconographie des instruments horaires du XVIe siècle*, Paris, 1858, p. 25.

[117] G. Libri, op. cit., Vol. II, p. 220.

[118] Here I would briefly mention two of the most prominent planetaria of the sixteenth century: the one built in Paris in 1546–1553 by the mathematician and astronomer D'Oronce Finé, and the famous astronomical clock in the cathedral of Strasbourg, built in 1571–1574 by Conrad Dasypodius, professor of mathematics at Strasbourg University. *Conradi Dasypodii, Horologii astronomici Argentorati descriptio*, Argentorati, 1580. Cf. P. Dubois, *Horlogerie*, Paris, 1858, pp. 44–48.

[119] G. Libri, op. cit., Vol. II, p. 232.

[120] G. Libri, op. cit., Vol. II, p. 217.

As Libri rightly said: the technique of building in those times had at its disposal instruments "qui pouvaient conduire à des puissants effets dynamiques."

IV. *Waterworks*. The water structures and the water mechanisms should be mentioned in brief. In twelfth-century Italy, canals were already built for irrigation, since the thirteenth century canals for shipping were built in Lombardy (e.g. the Guastalla [214] canal in 1203). In Venice, hydraulics had also reached a high stage of development in the installations of the lagoons. Since the fifteenth century, locks were installed in rivers, to enable the passage of ships through canals of varying levels.

Water as a driving power for working machines was used in Italy in an early era. In the eleventh century (1044) a mill situated in the lagoons is already mentioned; it is driven by high and low tides and it changed direction every six hours. In the fourteenth century water is used as the driving force for industrial machinery. "Dès l'année 1341, il y avait à Bologne de grandes fileries (spinning mills) mues par la force de l'eau, et elles produisent un effet évalué à quatre mille fileuses."[121] We have already pointed out the upheavals caused by the utilization of water power in the iron works and in mining, as well as the momentous and continuous rationalization of the working process, which was thus attained already in the fifteenth century.

In Italy we see the development of theoretical mechanics parallel to the development of practical mechanics. The latter first circumscribes its terms of reference in L. B. Alberti's book on architecture, written around 1450, and tentatively posits the problems,[122] reaching its first culmination about 1500 in the works of Leonardo da Vinci.

Modernity is already foretold in Alberti's enthusiastic paean to technology: it enables us to "trencher les rochers, percer des montaignes, combler les valées, résister aux débordements de la mer et des fleuves, nettoyer les paluz ou marais, bastir des navires" (preface). Then follows a treatise on a series of important problems in statics and dynamics: practical problems of balance, building sound foundations and arches, the calculation of their load-carrying capacity and their resistance. Volume VI deals with the various methods of transporting loads and the requisite machines; finally, important problems of dynamics are raised: "De [215] deux fardeaux pareils l'un aide l'autre. – Pratique des ouvriers. – Moyens pour le mouvement de grands fardeaux" (p. 111). In the technician's mechanical praxis the struggle begins for theoretical insight; it is important to understand the genesis of theoretical mechanics, even though it did not yet lead to exact results. This next step, likewise concerned with the same matter, the machines, was taken (as already demonstrated) by Leonardo da Vinci. He was followed by Tartaglia and a large number of sixteenth-century theoreticians. Here we shall only mention – from the aspect which is relevant to us – that besides the trajectories of artillery

[121] G. Libri, op. cit., Vol. II, p. 233. This Bolognese machine for spinning silk and cotton thread, with its several thousands of components, cogwheels, axles, etc., was famous and was still repeatedly described in the seventeenth century, e.g. by A. Alidosi, *Instruttione delle cose notabili di Bologna*, 1621 and by J. J. Becher, *Närrische Weisheit*, 1686.

[122] L. B. Alberti, *De re aedificatoria*, Florence 1485 (posthumous), here quoted from the French edition, Paris, 1553.

projectiles the mechanisms of clocks were also the subjects of theoretical study and became points of issue for numerous theoretical treatises. Thus, for instance, the well-known theoretician of mechanics, Maurolycus of Messina (1494–1575) wrote a “treatise on clocks.”[123] In G. Cardano’s work *De rerum varietate* (1557), Book IX (“De motibus”), different kinds of motion are also discussed and the general rule of acceleration changes is established; these motions are studied based on experience gathered in the art of watchmaking. In his work *De Subtilitate* (1550) Cardano sees the importance of the machines primarily in the saving and replacement of human labor. And similarly Conrad Dasypodius, professor of mathematics and builder of the astronomical clock in Strasbourg; the essential task of mechanics as realized in machines consists of saving labor: “quod maxima pondera minimis moveantur viribus et quibusnam talis motus fiat machinis.”[124] As shown above, the same applies to Galileo and Descartes.

This short synopsis also shows that theoretical mechanics derived its concepts from experience with machines, and that these machines have been the subject of discussions since the mid-fifteenth century, while division of labor in manufacture is never mentioned then or in later literature. All these facts, so important for the development of practical and scientific mechanics, are not mentioned by Borkenau. His history of the birth of scientific mechanics lets a ready-made form of mechanics emanate from the heads of Stevin, Galileo and Descartes in the first decades of the seventeenth century. Thus the great mathematicians and mechanists turned into its founders and pioneers who in reality only perfected classical mechanics. [216]

Borkenau’s Method and Its Metamorphoses

We now want to demonstrate that Borkenau’s failure is due to his method. In contrast to the isolating way of viewing history, such as Max Weber’s, who “only knows separate factors in historical events which determine the course of history” (p. 158), Borkenau acknowledges “the dialectic materialism which is based on the categories of totality and objective tendency” (p. 159). He strongly emphasizes that “two inseparably linked determinants, forces and relations of production, determine the whole ideology” (p. 118). However, in his work he neglects to explain the mechanistic world picture via the forces and relations of production at the time of its emergence; in vain we search for a description of productive forces at work – during the epoch discussed in his book – whether in France, Holland, or England. Over and above this: he completely negates the effect of productive forces on the genesis of that epoch’s ideology. He replaces them in these countries with the manufacturing technique! “The mechanistic world-picture,” we read in his book, “is a transposition of the manufacturing process to the cosmos as a whole . . . this transposition can have nothing to do with the development of the productive forces” (p. 127). In transferring the role of productive forces to the technique, to the manufactural technique

[123] G. Libri, op. cit., Vol. III, p. 108.

[124] Conrad Dasypodius, *Heron Mechanicus, seu de mechanicis artibus*, Argentorati 1580, p. E2r.

only, he immediately confronts us with a problem that is difficult to understand: "Mechanics, i.e. the science of the manufacturing period, is the scientific treatment of the process of manufacturing production" (p. 6); but why only of the process of manufacturing production? Is the technique of manufacture the only one existing at that time? By no means. We know that concurrently there existed three different technical methods: next to the traditionally undivided craftsmanship and manufacturing, there is "also the modern ... factory" (p. 4). Is perhaps the manufacturing technique the most advanced? It is not, for besides manufacture there existed the "most highly developed forms of industrial praxis," in navigation, in warfare, in the art of the printer (p. 90). And yet: scientific research, "more strongly based on the observational material supplied by industry," does not consider all these three techniques, not even the most highly developed one, the industrial praxis based on machines; it is based "not on the material of all processes of production, but particularly on the manufactural ones" (p. 5). Borkenau himself finds this "striking"; and actually it is more than striking, especially if one considers that, according to him, the science of that period was led not only exceptionally or mainly, but exclusively by the methods of manufacture (p. 4)! The first question is: why does not Borkenau take into account the category of totality? Why should not all the productive forces be relevant here, instead of only a particular technique which was not even the most progressive one? A second question: Does that epoch's science really allow itself to be led "exclusively" by the methods of manufacture? Borkenau asserts that it does, but – as we have already discovered – does not adduce a single example in support of this thesis.

217]

Moreover, despite the fact that Borkenau so emphatically underlines the importance of manufacturing technique, for him this is not the ultimate causal element of the mechanistic world-picture: the manufactural production also "contains ... very few incentives for the creation of this world picture" (p. 13). The point in time, in which insights into manufactural technique were transformed into the mechanistic world picture "was not decisively determined by the development of manufacture" (loc. cit.).

But if manufactural production lacks the drive to create the mechanistic world-picture, the question remains: what else does create it? "How, then, did we reach this immense generalization of the experiences of manufactural technique?" (p. 3). And further on: "This generalization would never have developed, unless simultaneously forces were active toward a conception of man as a merely mechanically functioning being" (p. 13). What sort of secret "forces" were those? We learn: "As in all periods, thus also in the period of manufacture it is the relations of production which cause the generalization of that which in technology at first exists as mere subject matter for thought" (p. 14).

Through this new methodological twist, through the decisive role now ascribed by Borkenau to the relations of production, it is not clearer by what and how the new world-picture was determined. Rather, new difficulties arise. On the one hand, the role of the relations of production is thought to consist only in [determining] the theoretical generalization of the "material for thought" provided by technique viz. in a rather receptive auxiliary function. On the other hand, however, he assures us that the relations of production are those forces "which urged to perceive of man as a merely mechanically (why must it be mechanically? H. G.) functioning being" (p. 13). Here, then, the relations of production are understood as active, independent forces which are not limited to the auxiliary role in the generalization of the "material for thought" supplied by technology, but of their own accord, press us to perceive man as a merely mechanically functioning being. [218]

Borkenau's method is presented as a truly Protean method, secretly undergoing continual metamorphoses. At the beginning it stated that the new world-picture was determined by two inseparably linked factors – forces of production and relations of production. Then the effectiveness of the productive forces for the genesis of the mechanistic world-picture was negated and was replaced by a particular technique. Finally it emerges that this technique alone is incapable of creating that world-picture, and that it only supplied "material for thought." This was then theoretically generalized "on the part of the relations of production." But a generalized "material" for thought still remains only material and not a world-picture. Thus we conclude that it is the relations of production which create – in an unexplained way – the very mechanistic conception from the "material." The place of the originally "inseparably united" determinants was ultimately taken by the relations of production alone, while the technique of manufacture was reduced to a mere supplier of "material for thought."

Yet we still do not have that "last" element with which the world-picture is explained. It is clear: if one proceeds from the relations of production which are said to "urge" us to perceive of man as a mechanistic being, one is inescapably confronted with the question: why is it just the relations of production at the beginning of the seventeenth century, and not those of the fifteenth or sixteenth centuries which do the "urging"? Relations of production is merely an economic expression for property relations. The property relations of a period are as such static. That the relations of production of the seventeenth century urge toward the mechanistic conception of man, while those of the earlier period did not – this can only be explained by the changes which occurred in property relations. It is impossible to understand any changes in the property relations which as such are static without looking into the dynamic element, the forces of production. The changes in the property relations are the results of the respective changes in the forces of production. Since Borkenau, as we have seen, has excluded the forces of production as possible explanations for the mechanistic world picture, he is lacking the dynamic factor which should explain changes in the relations of production. Consistent with his position he discards all the previously mentioned elements of explanation: forces of production, technique of manufacture, relations of production – and allows for a further change in his system of categories. Instead of the totality of relations of production, of manufacture technique or the economic structure of society as a whole, his final explanation in the analysis of the ideological trends is party warfare. [219]

For this purpose, Borkenau adopted a special method which he wants to apply in his first three introductory chapters in an "abbreviated form," but subsequently the more strictly, the nearer he comes to his "principal theme of the genesis of mechanistic philosophy" (p. 21). He starts from the premise that a thinker can "be considered as truly understood only if understood within the context of the struggles of which he was a party" (p. 21). He therefore believes that he gives "a full history of dogmas and their causal derivation" through "a record of all those elements of thought which are imposed on every thinker by virtue of his position in party struggles" (p. 21).

Before we examine the value of such a method, we want to discover to what extent he fulfilled his promise of "a very special analysis of the parties." The description of "all those elements of thought" which are "imposed" on a thinker "by virtue of his position in the party struggles" means a presentation of all the parties in the period concerned, their mutual relations, their contrasts or common features. Only by analyzing all parties can one obtain an insight into the totality of the historical situation in a particular period. There is no trace of such an undertaking in Borkenau's work. "The interplay of social forces which led each thinker to his system is here only adumbrated; in this book's framework, an unsatisfactory but unavoidable abbreviation" (p. 21). The fact that he does not even keep to this minimum is evident e.g. in the chapters on Thomas Aquinas (pp. 23–35), on Cusanus (pp. 40–53) and on Luther (pp. 104–107). "According to our program we refrain from analyzing ... the [220] question ... regarding Thomas' special position in the struggles of his time" (p. 31). "Again, according to our program, we refrain from analyzing the historical element and Cusanus' role in the struggles of his time" (p. 42). We find the same attitude in the presentation of Luther's ideas. There is not a word on the Reformation's social background, Germany's economic situation, its economic structure and the different classes. Parties, party warfare? – no trace of these. Luther's doctrine of the abysmal depravity of human nature, and the conclusion drawn from there that men can only be restrained by force, is only an accommodation of Lutheranism to the demands of absolutism. The unavoidable question is why the Reformation remained unsuccessful in the two greatest absolute monarchies – in Spain and France. Are "the" teachings of Luther really to be regarded as something so immutable as in Borkenau, who undertook the particular task of "examining the changes in the form of thought"?

Instead of an overall picture of the situation, through which the different classes and parties' position, with the thinkers and intellectual trends belonging to them, would become intelligible, we are given a number of single, separate, incoherent, random facets which tear apart the general context. The particular social stratum to which a certain thinker (Bodin, Vanini, Descartes) belongs is isolated and taken out of context when that thinker is discussed. The prevailing economic and political situations in England are disposed of by some remarks when Hobbes's theory of the state is presented (p. 440). The following survey shows how unsystematically Borkenau deals with the grouping of the French parties: in connection with the libertines' moral teachings, the nobility's social situation is described (pp. 207–208); in connection with Luther's teachings in the chapter on "natural law," the party of the divine rights of kings in France is described (p. 106). Then, after having mentioned the moderate royalist party (p. 114) in connection with Bodin, and later touched on the Dutch situation in passing, there is a chapter on the new morals and the new theology which contains the characterization of the French noblesse de robe (pp. 172ff.) and finally there are the moral groupings of Jansenism, which emerged [221] later on (p. 248).

This method's arbitrariness is striking and becomes evident particularly in what is omitted. The social situation of the absolute monarchy, of the state, is not mentioned. Borkenau speaks of the period of the "emergence of absolutism" (p. 100). We do not learn anything about the causes of this "emergence." In one passage we hear that "the aspiring absolute monarchy has domesticated the rebellious nobility" (p. 171), and elsewhere that "for some time absolutism tried ... to maintain a balance between nobility and bourgeoisie," "to defend the former and to advance the latter" (p. 263) etc. Who is this absolute state? Why does it ally itself with one class and fight the other? Is it at all possible to render intelligible Luther and Calvin's reformations without explaining the state's attitude? How could the Reformation prevail against the church of the Pope? The latter had, after all, an immense secular power. "Depuis deux siècles," says Seignobos, "aucune hérésie n'avait échappé à la destruction."[125] If the Reformation was not already nipped in the bud, this was only possible because Luther in Saxony and Calvin in Geneva could organize their churches under the protection of the state.[126] Wherever the state turned against the Reformation – as in France or Spain – it could not succeed. But why do the states in Germany place themselves at the head of the Reformation, but in France and Spain at the head of its opponents? On this there is also no elucidation. Borkenau evidently believes that for the success of the Reformation Calvin's "doctrine of probation" was much more important than the state's attitude as caused by its social situation.

That which is said about the state applies to the church as well. The presentation of the social situation, the differences within the clergy, the situation of the poor clergy of the orders, the situation of the secular clergy with all its prebends and benefices – all factors of great importance for the Reformation's fate – are not mentioned. The church's social significance at that time becomes evident when one considers that in the *États Généraux* of 1614 out of a total of 464 delegates, the clergy alone numbered 140 representatives compared with only 132 representatives of the nobility; the church was the greatest landowner and disposed of the largest resources. Without knowledge of the church's social position one [222] cannot comprehend the attitudes of: the high and low clergy, the religious parties of the Reformation and the Counter-Reformation, or the other propertied classes – the nobility and bourgeoisie and their parties. For these classes basically lived on the church's benefices which were disposed of by the court! Just as the nobility in France was turned into a court nobility, most French prelates became courtiers, who lived in Paris, far from their dioceses. The bishops' income, no matter how large, did not suffice to cover the cost of their "grand-seigneurial" life style. Therefore, they were dependent on benefices and incomes which the king controlled and thus were dependent on him. Some bishops accumulated the revenues from six large abbeys. The nobility too was dependent of the king. The abbeys were not reserved for the

[125] Ch. Seignobos, *Histoire sincère de la nation française*, Paris, 1933, p. 238.

[126] On the developments in Germany, Ranke says: "The new churches were founded under the protection, the immediate influence, of the reigning powers. It is only natural that thus their shaping was also determined." (*Deutsche Geschichte im Zeitalter der Reformation*, 5th ed., Leipzig, 1873, Vol. II, p. 308).

clergy, they were distributed among the nobility, to women and even to children as additional sources of income. The nobility grew accustomed to regarding the church estates as their sources of income and the church as a career, especially for younger sons. The unmarried daughters viewed ecclesiastic posts as an honorable refuge which enabled them to lead an elegant style of life. The upper strata of the bourgeoisie had similar aspirations. The king presented abbeys to Protestants, to poets such as Desportes and Brantôme for their love songs or daring stories about amorous ladies. After the civil war (1596) the majority of the abbeys were bestowed on lay members. Can one comprehend in such circumstances any party groupings and party struggles in France, when not a single word is mentioned about the church and clergy's role which so deeply influenced all aspects of life?

Borkenau uses an easy method of analysis. If he possesses factual material, e.g. regarding the Catholic Counter-Reformation in France during the second quarter of the seventeenth century, then he expounds this fraction of material concerning the church in full detail, in order to show "by which social stratum the movement was set in motion" (p. 210). The much more important material on the church's social situation as a whole is not presented, and ad hoc he makes a virtue of necessity: "The social stratification of the clergy in that period and its attitude to Jansenism have not yet been sufficiently researched. These belong in a sociological (*sic!* H. G.) history of the church, not in our study of the changes in forms of thought" (p. 265).

But he also omits any description of the peasant's situation and the proletarian elements, although in Germany for instance the peasants' revolution played an [223] important part in the party struggles at the time of the Reformation. Ranke has shown how the parties in the fights between the churches accused one another of responsibility for the outbreak of the revolution, the Protestants pointing at the exploitation of the people by the Catholic clergy, and the Catholics accusing the Protestant preachers of demagogy.

The party struggles of the propertied classes, which were fought at the expense of the broad masses, must remain incomprehensible if these masses are not taken into account. Borkenau describes the programmatic "passionlessness" of Neo-stoicism (p. 187), a philosophy of the magistrates who want to keep aloof from the troubles of the times to preserve their equanimity. In reality, this simulated passionlessness is only a mask which cannot be recognized as such without considering the peasants' revolts in the rural areas against the background of the upper classes' political struggles and without the urban proletariat's revolts. The magistrates in particular experienced the violent outbreaks of a people driven to depression; the outbreaks were directed in the first place against the lower officialdom of the fiscal and judicial administration in the provinces. The rigorous fiscal policy pursued throughout the ministry of Richelieu was the cause of continuous uprisings of the poor: 1630 in Dijon, 1631 in the Provence and in Paris, 1632 in Lyon, 1635 in Bordeaux. Similar outbreaks occurred in smaller towns such as Agen, La Réole, Condom, Périgueux. In Montferand, the citizens beat the priest to death because he preached the duty of paying taxes. In 1636, the peasants revolted in Limousin, Poitou, Angoumois. Mobs of seven- to eight-thousand people roamed the land and manhandled the officials of the fiscal administration. In 1637, insurrections of the "frondeurs" with battles on barricades took place in Gascony and Périgord. In 1639, there was a peasants' revolt in Normandy, led by the "discalced," in which several tax collectors with their personnel were beaten to death. Similar upheavals occurred in Caen and Rouen; these took a particularly violent form, and several tax officials were torn to pieces by wheels with nails affixed. A punitive expedition of four thousand roamed the country murderously. It occupied Rouen in 1640, dispersed the parliament in order to reinstate the king's authority, replaced it by royal commissars, and abrogated all municipal privileges; the municipality (*mairie*) was abolished, and the leaders of the revolt executed by the military. [224]

Such a situation must influence the manner of thinking of the most directly affected magistrates. This officialdom which, as Borkenau asserts, had created for itself a philosophy of "sages" through "Neo-stoicism," a program consisting of "dispassionateness" and "aloofness" from the suffering of the "outside world" (p. 187), was surely not as neutral in its attitude toward the social upheavals as Borkenau claims. Insofar as it remained neutral, this was only vis-à-vis the political fights among the propertied classes; but it stood with all its passion in the midst of the battle when not only party struggles within the propertied classes were concerned but also their common interest vis-à-vis the mass of the people. Then Neo-stoicism completely forgets its "dispassionateness," and Borkenau himself has to admit that "Neo-stoicism was imbued in all directions with the pride of a ruling class" (p. 189).

Whereas so far we have examined the "very special analysis of the parties," promised by Borkenau, from the aspect which he failed to take into account, we shall now look into his perception of those parties with which he deals and to which he ascribes a decisive role in his theoretical construction: the party of the manufacturing bourgeoisie and that of the officialdom or what he calls the "gentry." We only learn about the first that "the divine right is the political doctrine of the mercantile monopolistic bourgeoisie, or of those parts of the bureaucracy and the nobility that are connected with it" (pp. 106f). It is not explained any further, and is clearly considered self-evident why the capitalist manufacturers who, according to Borkenau, emerged from the artisans, elevated "divine right" to their doctrine of the state. Neither does he say why certain strata of the bureaucracy and the upper nobility – who as a whole represent other specific interests – should be "allied" with the monopolistic bourgeoisie, and what is the basis of the alliance: blood relationship, religious ties, or economic interests.

The "very special analysis of the parties" furthermore arrives at the conclusion that divine right, the theory of the state of the monopolistic bourgeoisie, was nothing but "the adaptation of Lutheranism" (or the Anglicanism and the French Catholicism related to it) "to the requirements of absolutism" (p. 105); this would seem reasonable if one considered that the French manufacturing bourgeoisie was "dependent on government subsidies" and "could not exist without government support" (p. 171). But it also becomes evident that "insofar as absolutism sought an understanding with the bourgeoisie," "it had to renounce (!) the doctrines of the divine right" (p. 115). Borkenau does not explain why an "understanding" should be needed here, when the divine right was to be regarded as adaptation by this bourgeoisie to the requirements of absolutism; the contradiction is evident. [225]

Absolutism, "insofar as it sought an understanding with the bourgeoisie," had to renounce its divine right, which means, according to Borkenau, dissociating itself from the monopolistic bourgeoisie.

The core of the manufacturing capitalists, whose origins were in craftsmanship, stood (according to Borkenau) in sharp contrast to the financial capitalists (p. 155). On the following page we are already told that just these manufacturers had "swept along with themselves part of the bankocracy (i.e. of the financial capitalists. H. G.) and together they led the communities" (p. 156). He does not explain how this is compatible with the "sharp contrast." There remains a contradiction; this, like many other contradictions, is the consequence of Borkenau's distortion, to benefit his construct of the historic role played by the parties. It shows up most clearly in the example of the "gentry," that stratum of officialdom from which Descartes was born.

"Descartes was the first to attempt the construction of a unified world-picture from the categories which determined the capitalist individual's life" (p. 268). – "His origins rendered him most suitable for this task. Descartes' family is like an extract from all the important strata of French bourgeoisie, whose center was the *noblesse de robe*. His father was a counsellor in the parliament of Rennes" (p. 269). – In short Descartes "belonged to the gentry" (p. 271). And what was that gentry? "Gentry is . . . the strongest, most independent, politically and intellectually most active class of absolutist France, the noblesse de robe" (p. 172). – "By virtue of its privileges it represented the class interests of the bourgeoisie vis-à-vis the monarchy (p. 174). – Taken all in all, Borkenau presents the "gentry" as the bourgeoisie's hero. At the convention of the Estates General in 1614 it had the absolute majority in the Third Estate's delegation; "there it was the official leader of the tiers état" (p. 175). Although Borkenau sees that the gentry "in its own interest" often acts in league with the nobles, he thinks "one should not be led astray" by such "tactical maneuvers" and constellations (p. 176, note). Rather, "on this gentry devolves the exclusive ideological leadership in the struggle for the new capitalistic way of life in France" (p. 172).

[226] Was the gentry, which represented "overwhelmingly the revenue capital," especially the provincial magistrates – from which Descartes was a descendant – really the leader in the strife for the implementation of bourgeois interests? We have heard before that because the Renaissance revenue capitalists were detached from the process of work, they lacked any motivation for the rationalization of the technique, and became the carrier of Renaissance aestheticism which despised the common people. Although the French gentry is also a revenue capitalist class, aloof from the working processes – it is here supposed to be almost revolutionary, exercising the ideological leadership in the struggle against the monarchy and for the interests of the bourgeoisie – the gentry and not the manufacturing bourgeoisie which is close to the work processes and which strives for rationalization!

"In the fifteenth and above all in the sixteenth century the monied classes invested their capital mainly in real estate, and now they often gave up land in favor of official position" (p. 174). "The social position of the class was ambiguous" (p. 176) in that "a formally bourgeois class, by virtue of its capital power, attains in fact a position of nobility within the bourgeoisie" (p. 172) and is therefore "a stranger in the capitalist world." The magistrates are "almost unaffected by the need for unlimited effort with uncertain success in the capitalist competition ... The magistrates were the only ones whose economic life could take its course in feudal traditional security" (p. 176). And it was just this class, which does not, and does not want to, know anything of the "rigid rationality of the work process," this "noblesse de robe is ... the protagonist of the bourgeoisie" (loc. cit.), despite its independent wealth, despite its remoteness from work, despite the fact that it is "alien" to capitalism! This time neither independent wealth nor monetary and commercial capital are leading to an "aesthetic world view," they fulfill another task: "The propertied class that emerged from monetary and commercial capitalism had already been the principal standard bearer of courtly humanism. Now it was the bearer of ... the new philosophical development" (p. 174).

What was the role of the parliamentary bourgeoisie in historic reality? No other factor has contributed more to the weakening and demoralization of the bourgeoisie than the venality of office. Therefore, in the Estates General of 1560, not only the nobility and the clergy but also the representative of the Third Estate protested against the purchase of offices. It was abolished by the ordinances of 1560 and 1566, but was reintroduced already in 1567 – in the interest of the monarchy which was always in need of money. "A partir de cette époque, on peut suivre pas à pas le travail de décomposition qui s'opère dans les rangs du Tiers."[127] [227]

When in 1604 the purchase of offices was finally authorized by Henry IV, the moral collapse of the Third Estate and the development of the magistrate into a self-seeking class was unavoidable. Perhaps, says Normand, Henry IV had anticipated and intended these consequences of the purchase of offices, perhaps he had "prévu qu'il briserait ainsi toute opposition de la part de ses parlements et des tribunaux inférieurs."[128] In the Estates General of 1614, which Borkenau extols because the magistrates prevailed among the representatives of the Third Estate, it was evident that: "Sur 192 députés 131 étaient titulaires d'offices. La bourgeoisie laborieuse et commerçante avait été presque partout remplacée par la nouvelle noblesse de robe, ignorante ou insoucieuse des besoins du peuple."[129] The representation of a class had become the representation of a caste! This stands to reason. The purchase of offices had to have a paralysing effect on the industrial accumulation and the productive activity of industry, thus delaying the progress of the bourgeoisie, since large amounts of money were withdrawn from industry in order to be used for the unproductive purposes of the court.

During the eighteen years of Richelieu's ministry alone, more than two billion Gold Francs (in today's currency) accrued to the state treasury from the purchase of offices,[130] without taking into account enormous ancillary expenses that went into the pockets of high court officials. An immense, superfluous and idle bureaucratic

---

[127] Charles Normand, *La Bourgeoisie française au XVII'ème siècle*, Paris, 1908, p. 30.
[128] Normand, op. cit., p. 18.
[129] Op. cit., p. 17.
[130] G. D'Avenel, *Découvertes d'histoire sociale 1200–1910*, Paris, 1910, p. 26.

machinery, whose only raison d'être was the exploitation of the people: under Colbert 45,780 financial and justiciary offices were sold – according to Forbonnais' opinion 40,000 too many – whose monetary value was 417,630,842 livres (also billions according to the present currency).[131] "Economiquement, cette puissance d'attraction de l'Etat eut une influence fâcheuse .... En ouvrant ce débouché à la richesse acquise, on lui faisait une retraite au lieu de l'obliger à travailler. Les capitaux à peine formés, sortaient des affaires pour n'y plus rentrer ...." "Si la France, beaucoup plus avancée que l'Angleterre au début du règne de Henri IV (1589), était fort depassée par elle au moment de la Revolution ... cela pouvait tenir à la manière française de placer son argent en valeurs inproductives."[132]

But not only did the "paulette" mean the economic disintegration and weakening [228] of the bourgeoisie; its moral and intellectual consequences were even worse. Why should the industrialist or the merchant send his son to study for many years, when the latter, through the purchase of a judiciary or financial office, could become a "Monsieur" and elevate himself from the ranks of the disdained Third Estate to the nobility? Instead of conquering rights for his whole class in the struggle against the powers-that-be, everybody with money at his disposal strove individually to avoid the struggle by purchasing rights for himself and his descendants. The result for the class as a whole was "l'insuffisance de la volonté pour la lutte."[133] Just as the buying of offices resulted in a regression in the development of the bourgeoisie, it also led to the incapsulation of the "gentry" as a caste. It goes without saying that among the magistrates there were some individuals who excelled in their education and wider political horizon. As a whole, however, due to the simony, the magistrates were venal and incapable of representing the interest of the class as a whole beyond their own narrow interests.

Thus Sée, in a retrospective view of the seventeenth century, says: "Souvent les membres des cours exercent leurs fonctions à un âge où ils ne possèdent ni l'instruction, ni la pratique nécessaires. Dans les Universités, ils ont souvent acquis à prix d'argent un diplôme qui ne prouve, en aucune façon, qui'ils aient étudié le droit .... En somme, beaucoup de parlementaires sont ignorants ou incapables."[134] The "gentry," this alleged protagonist of the Third Estate, was therefore detested by the "philosophers" of the Enlightenment, as well as by all the real protagonists of the revolution.[135] And rightly so. The parliaments opposed all, even the most useful reforms, which were in the bourgeoisie's interest, if the interests of their own

[131] Normand, op. cit., p. 41.
[132] D'Avenel, op. cit., pp. 270–277
[133] Normand, op. cit., p. 43.
[134] H. Sée, *La France économique et sociale au XVIIIe siècle*, Paris, 1933, p. 95.
[135] Thus Diderot passes judgment on the parliaments: 'Intolérant, bigot, stupide, conservant ses usages gothiques et vandales..., ardent à se mêler de tout, de religion, de gouvernement, de police, de finance, d'art et de sciences, et toujours brouillant tout d'après son ignorance, son intérêt et ses préjugés'. And even more damning is Voltaire's (1774) judgment: "Il était digne de notre nation de singes de regarder nos assassins comme nos protecteurs; nous sommes des mouches qui prenons le parti des araignées." (Sée, op. cit.)

castes were affected. They opposed reductions in judicature costs, reforms in the outdated procedure of the penal law with its system of torture, and were against the unification of local common law: "Ils réprouvaient la liberté de la presse; ils condamnaient et faisaient brûler une foule d'ouvrages, comme irrespectueux des vérités religieuses ou des institutions existantes. Ils combattirent la déclaration qui accordait l'état civil aux protestants," in short, Sée speaks of the "esprit conservateur des parlementaires."[136] When they made a stand against the "lettres de cachet," they only did so because they were often affected, and they regarded them as limiting their judicial prerogatives. But "les Parlements se firent les défenseurs de tous les privilèges sociaux et se dressèrent contre toutes les réformes qui s'efforçaient de les atténuer."[137] This was what the "protagonist" of the bourgeois interests looked like! [229]

All the contradictions described above, in which Borkenau gets entangled, are not incidental, but are the unavoidable result of his method, which takes the struggles of the parties as its point of departure for the analysis of the ideologies. It attempts to understand the architectural basic law of a building by explaining the structure of the sixth floor from the character of the fifth, disregarding the foundations and the intermediate storeys. Only the historian of today, looking back at the available historical material, and analyzing methodically the productive forces and the relations of production of the epoch, can grasp the totality of their social situation, and only from such reconstruction of the overall situation (e.g., Italy's situation after the shift of the axis of world trade from the Mediterranean to the oceanic coasts of Western Europe) can he properly understand the various parties or thinkers of that period (e.g., Machiavelli's program for the unification of Italy). In contrast, this situation is reflected only in distorted form as if in a convex mirror in the mere party struggles of contemporaries. Could Machiavelli's contemporaries realize that, when the dynamical and centralizing power of rising Italian capitalism was broken, this spelled the end of the program for Italy's unification as well? The party struggles of that time, the interests defended or opposed by the parties, do not so much express the real situation of the period as the conscious or unconscious illusions entertained by the parties regarding this situation. Therefore, if one adopts as his point of departure the party struggles as such, the ground is cut from under his feet and one forms his judgment not according to the essence of things but according to its more or less shadowy distortions. [230]

Social Origins of Mechanistic Thought [Original English Summary]

Franz Borkenau's book "The Transition from Feudal to Modern Thought" (*Der Übergang vom feudalen zum bürgerlichen Weltbild*), serves as background for Grossmann's study. The objective of this book was to trace the sociological origins [231]

[136] H. Sée, op. cit., p. 96.
[137] H. Sée, op. cit., p. 96.

of the mechanistic categories of modern thought as developed in the philosophy of Descartes and his successors. In the beginning of the 17th century, according to Borkenau, mechanistic thinking triumphed over mediaeval philosophy which emphasized qualitative, not quantitative considerations. This transition from mediaeval and feudal methods of thought to modern principles is the general theme of Borkenau's book, and is traced to the social changes of this time. According to this work, the essential economic change that marked the transition from mediaeval to modern times was the destruction of the handicraft system and the organization of labor under one roof and under one management. The roots of the change in thought are to be sought here. With the dismemberment of the handicraft system and the division of labor into relatively unskilled, uniform, and therefore comparable activities, the conception of abstract homogeneous social labor arises. The division of the labor process into simple, repeated movements permits a comparison of hours of labor. Calculation with such abstract social unities, according to Borkenau, was the source from which modern mechanistic thinking in general derived its origin.

Grossmann, although he considers Borkenau's work a valuable and important contribution, does not believe that the author has achieved his purpose. First of all, he contends that the period that Borkenau describes as the period of the triumph of modern thought over mediaeval should not be placed at the beginning of the 17th century, but in the Renaissance, and that not Descartes and Hobbes but Leonardo da Vinci was the initiator of modern thought. Leonardo's theories, evolved from a study of machines, were the source of the mechanistic categories that culminated in modern thought.

If Borkenau's conception as to the historical origin of these categories is incorrect in regard to time, Grossmann claims it follows that it is incorrect also in regard to the social sources to which it is ascribed. In the beginning, the factory system did not involve a division of labor into comparable homogeneous processes, but in general only united skilled handicraftsmen under one roof. The development of machinery, not the calculation with abstract hours of labor, is the immediate source of modern scientific mechanics. This goes back to the Renaissance and has relatively little to do with the original factory system that was finally superseded by the Industrial Revolution.

While Borkenau, in tracing the social background of the thought of the period, relies chiefly on the conflicts and strife of political parties, Grossmann regards this as one element only in the formation of the general social situation, which in its entirety and in the interaction of its elements explains the development of modern thought.

Descartes and the Social Origins of the Mechanistic Concept of the World*

Henryk Grossmann

I

While the purely mathematical and logical aspects of the Cartesian Algebra or "Science Universelle" have been masterfully expounded by L. Brunschvicg[1] and its further development into Leibniz' "General Science" has been excellently treated by L. Couturat, [2] who also confined himself to the purely mathematical and logical aspects of Leibniz' work, the sociological aspects of Descartes' "Science Universelle" have not been examined by these authors or any others. It may be asked whether such a sociological problem exists at all. In order to understand that it does, we must realize that all Descartes' works are sharply marked by the imprint of mechanical principles; that Descartes derived these principles from his mechanics, that is to say, from the study of machines, and then extended these principles to physics and finally to the whole universe.

Even before the publication of the *Discourse on Method*, Galileo had shown in his *Treatise on Mechanics*,[3] first published in French, that he had derived the mechanical principles and concepts which he used for the explanation of physics and the universe from the analysis of machines. He based his ideas, as he shows in Chapter I, principally upon lifting-engines, machines for the transportation of heavy [2] loads, machines for lifting water from deep wells, pumps for removing water from

* Grossman's working title for many years was "Universal Science versus Science of an Elite. Descartes' New Ideal of Science," which remained the title of the project in the final report to the Institute for Social Research in 1946 although it no longer had much connection to the text itself as it had developed. In the later manuscripts Grossmann changed the title so as to fit the actual content of the work. We have adopted the later title as best reflecting the author's intentions (G.F./P.M.).

[1] Léon Brunschvicg, *Les étapes de la philosophie mathématique*, 3rd ed. (Paris, 1929).

[2] Louis Couturat, *La logique de Leibniz* (Paris, 1901) Chapter VI, "La Science Générale."

[3] Galileo Galilei, *Les méchaniques*, translated from the Italian by Marin Mersenne (Paris, 1634); see also Antonio Favaro (ed.) *Delle meccaniche lette in Padova l'anno 1594 da Galileo Galilei* (Venice: R. Istituto nel Palazzo Loredan, 1899) *Memorie del Reale Instituto Veneto di Scienze*, vol. 26, no. 5.

G. Freudenthal, P. McLaughlin (eds.), *The Social and Economic Roots of the Scientific Revolution,* Boston Studies in the Philosophy of Science 278,
DOI 10.1007/978-1-4020-9604-4_4,

ships, water mills and other machines with wheels intended to save human labor and thus to cheapen it. In Chapter II of his treatise Galileo gives his "Definitions" of the mechanical concepts "*in order to derive from them the causes and reasons of everything that happens to the machines, the effects of which are to be explained* ... And since the machines usually serve to *transport heavy things*, we begin with the definition of heaviness that can also be called gravity."[4]

The same is true of Descartes who in his "Treatise on Mechanics or an explanation of the engines with the help of which a very heavy load can be lifted by a small force,"[5] derives his mechanical notions and principles from an analysis of the six simplest machines: the pulley, the inclined plane, the wedge, the turning-lathe, the screw and the lever, and reduces all machines, including the most complicated ones, to the *inclined plane* as the elementary basic mechanical form.[6]

Descartes interprets not only the outside world as a machine; in his *Passions of the Soul* (1649) he applies the same mechanical principles in the field of psychic processes. He explains these processes mechanically, by impulses which are created through the mere arrangement and the motion of the bodily organs and external objects, thus initiating mechanical psychology. [3]

In view of Descartes' general tendency to give a mechanistic interpretation of all fields of science, of all phenomena of physical, organic and psychic nature, it is natural to suppose – and this surmise constitutes the basic idea of the present essay – that Descartes also applied the same mechanistic principles to his method, to the very structure and functions of his scientific apparatus, and conceived them upon the model of machines. This is true – as will be shown – particularly of Algebra, the "Cartesian science" (in the phrase of Couturat).[7]

What is the meaning of Descartes' universal science? In his eyes, it was a *universal method* applicable to all the sciences. Originally Descartes planned to call his essays published in 1637 not *Discours de la méthode*, but *Le projet d'une science universelle* (project for a universal science), and he emphasizes as a particular feature of this universal science that the matters dealt with "are explained in such a way that *even those who have never studied can understand them*."[8] [4]

Here we encounter the expression of an extremely important intellectual current in the field of science. While in that same France Michel Montaigne in his "Essais" (1579), in the chapter "On the inequality there is among us" had acridly stressed that he thought "there was more distance between one man and another

[4] Ibid., p. 6. Here, as throughout this essay, the italics are mine (H.G.) unless otherwise indicated. See Henryk Grossmann, "Die Gesellschaftlichen Grundlagen der mechanistischen Philosophie und die Manufaktur," in *Zeitschrift für Sozialforschung 4* (Paris, 1935) 200–210.

[5] This treatise is contained in a letter from Descartes to Constantin Huygens, Oct. 5, 1637. See *Oeuvres de Descartes*, ed. Ch. Adam und P. Tannery: AT I, 435–447.

[6] AT II, 236–237, and Grossmann, op.cit., pp. 202–203.

[7] Leibniz (1666) calls Descartes "the inventor of analysis, that is to say, of algebra." See Couturat, op. cit., p. 180.

[8] Descartes, letter to Mersenne, March 1636, AT I, 339: "sont expliquées en telle sorte, que ceux mesmes qui n'ont point estudié les peuvent entendre" (AT I, 339).

than between some men and some beasts"[9] and while later, barely twenty years before Descartes, Montchrétien, full of the pride of the ruling class, had nearly exactly repeated the words of Montaigne and asserted that "very often there is as much difference between one man and another as between man and beast,"[10] and while even thirty years after Descartes, Spinoza was to say that "the people are not capable of understanding high matters,"[11] we find Descartes appealing not to the upper strata of scientists and specialists, but to *the great mass of the unlearned*. The universal science was to be universal not only in that it would be applicable in all the fields of science, but also in that it would be accessible to *all the people*, including the large masses.[12] Here the philosopher's rebellious spirit manifests itself, making known his intention of breaking through the narrow boundaries of the erudite world and making the corporate discipline of philosophy a science not confined within the limits of a corporation.[13] [5]

To understand the real meaning of the Cartesian universal science it is necessary briefly to examine Descartes' relation to mathematics. In analyzing the part played by mathematics in Descartes' thinking, Pierre Boutroux reaches the conclusion that although Descartes' occasionally praised mathematics he did not estimate it particularly highly.[14] Brunschvicg even finds in Descartes "some disdain for the investigations of abstract mathematics."[15] Descartes did not value his own achievements in the field of mathematics any differently,[16] and Boutroux agrees with those who, like Brunschvicg, for instance, believe that in Descartes' philosophical career

[9] *Les Essais de Montaigne* (edited from the 1588 edition by H. Motheau and D. Jouaust) Paris 1886–1889, vol. 2, p. 206. Although Montaigne himself warns that external appearances distort our judgment, he continues: "Comparez . . . la tourbe de nos hommes, stupide, basse, servile, instable, et continuellement flotante en l'orage des passions diverses, qui la poussent" (op. cit. pp. 208–89).

[10] Antonyne de Montchrétien, *Traicté de l'oeconomie politique* [1615] (ed. Th. Funck-Brentano) Paris, 1889, p. 37. See also Fr. Bouillier, *Histoire de la philosophie cartésienne*, 3rd ed. (Paris, 1868) vol. 1, p. 149 [not in 1854 ed.].

[11] Spinoza's letter to Blyenbergh, January 1665. See *The Correspondence of Spinoza*. trans. by A. Wolf (New York, 1928) p. 149. Spinoza also opposed his friends' intention of publishing the *Theologico-Political Treatise* in the popular language, Dutch: he said that he had written his treatise not for the great mass, but for a small number of educated men. See Jacob Freudenthal, *Spinoza, Leben und Lehre* (Heidelberg, 1927) pp. 190, 196 (*Bibiliotheca Spinoziana*, vol. 5).

[12] While the universal science in the first meaning drew great attention, the problems connected with "science universelle" in the wider meaning were neglected!

[13] In the political and social fields Descartes rejected all "the turbulent and unrestful spirits" and their radical reforms and preferred, as he insists, the "most moderate" opinions (see *Discourse on Method*, part II, and the beginning of part III, *Philosophical Works of Descartes*, trans. by E. S. Haldane and G. R. Ross, Cambridge, 1931, vol. 1, pp. 89, 95). He was on the other hand convinced that in the sphere of thinking his radical reform would not meet any insurmountable difficulties (see *Discourse* II). That his philosophical radicalism would later bear fruit in the social and political spheres, and actually did bear fruit in France, is well known.

[14] "Descartes has little esteem for pure mathematics." See Pierre Boutroux, *L'Idéal scientifique des mathématiciens*, 3rd ed. (Paris, 1920) p. 102.

[15] Ibid., p. 115. See also Descartes' critique of traditional mathematics in the "Rules for the Direction of the Mind," Rule IV (*Philos. Works*, I, 11).

[16] Boutroux, op. cit., p. 102.

his mathematical activities were only an "episode."[17] For instance, this is what [6] Descartes wrote in an early letter (April 15, 1630) to Father Mersenne:[18] "I am so weary of mathematics and at present I think so little of them." Immediately after the publication of his treatise on Geometry Descartes wrote Mersenne that he intended to put an end to his mathematical studies. "Do not expect anything more from me in geometry; for you know that for a long time I have been opposed to continuing the practice of it";[19] and Descartes' criticism, as he explained to Mersenne, was directed against "*abstract* geometry, that is to say, the study of problems that serve only to exercise the mind," while he, Descartes, preferred "to cultivate another kind of geometry, *which attacks the problem of explaining the phenomena of nature*."[20]

In view of this disdainful attitude toward mathematics, how can we explain the [7] great importance that Descartes attached to his Algebra?[21]

The contradiction is only apparent. Descartes' criticism is addressed only to that mathematics which is conceived as a separate science and loses itself in abstract problems of this restricted field. He sharply censures his predecessors in mathematics because they used their minds and their strength for solving a handful of unimportant mathematical problems. Thus he criticizes Fermat for his predilection for particular cases which he treats in great detail, for having applied his theory of curves only to the special case of parabolas without attempting to give it a *general* form, so that it was not possible to see at once that it is applicable to *all* curves and that it can be expressed in a general form in the formulas of analytical geometry.[22] Descartes' criticism of Viète has a similar point: he, Descartes, "determines *generally*, for all equations," the laws of which Viète "gave only a few

---

[17] Ibid., p. 104, and L. Brunschvicg, op. cit., p. 124.

[18] AT I, 139.

[19] September 12, 1638. AT II, 361–62.

[20] July 27, 1638, AT II, 268. In the *Regulae* Descartes also criticizes the purely abstract character of the new algebra, foreign to every practical application. "For I should not think much of these rules [of algebra], if they had *no utility* save for the solution of the empty problems with which Logicians and Geometers have been wont to beguile their leisure" (Rule IV, *Philos. Works*, I, 10). We find the same criticism in the *Discourse on the Method*, II: "As to ... the Algebra of the moderns, besides the fact that [it] embrace[s] only matters the most abstract, such as appear to have *no actual use* ... one is so subjected to certain rules and formulas that the result is the construction of an art which is confused and obscure, and which embarrasses the mind" (*Philos. Works*, I, 91–92).

[21] "I am convinced that it is a more powerful instrument of knowledge than any other that has been bequeathed to us by human agency, as being the source of all others" (Rule IV, *Philos. Works*, I, 11).

[22] P. Boutroux, op. cit., p. 116. On this subject it is particularly interesting to analyze the famous mathematical duel between Fermat and Descartes, AT XII, 260–67. Boutroux points out that Fermat, the "restitutor" of Apollonius, did not consider himself, like Descartes, an adversary of ancient geometry, but rather as its successor who only developed it and generalized its results. (Pierre Boutroux, "La Signification historique de la géométrie de Descartes," in *Revue de métaphysique et de morale 22* (1914) 818).

particular instances" thus "showing that he [Viète] could not determine them in general."[23]

Descartes' relation to algebra is quite different; in his eyes it is not a particular traditional science with a fixed, limited, *separate object*, not a "numerous collection of results"[24] *but a general method of investigation*, which, thanks *to its general character, becomes applicable to ever new objects and problems* that neither Viète nor his disciples have ever considered.[25] In brief, Descartes' *mathématique universelle* is an extension of geometrical methods *to all the problems* of mechanics, physics, biology and psycho-physiology,[26] and as such a *general* method, algebra constitutes the core of the "universal science" and is Descartes' great discovery to which he points with special pride. [8]

Wherein lies the *significance of the new method*? With Thomas More, Francis Bacon and Descartes a new historical epoch begins. Under the influence of the needs and interests of the growing middle class, the new generation rebelled against the traditional world of disintegrating scholasticism that thought it could achieve an adequate image of reality with the help of speculative concepts and the play of syllogisms. The new generation, on the basis of observation and experience, wanted not only to understand this reality, but also to *shape it rationally*. In the realization of this goal a decisive role is assigned to science: it must not be practised as it was in the Middle Ages and antiquity, for the sake of the satisfaction derived from *contemplative* thinking;[27] in the eyes of the new thinkers, the goal of science, in addition to *knowledge* was above all the *domination* of nature for the purpose of the practical improvement and rational shaping of human life. In so far as knowledge was in question, *the idea of cognition* had now changed. In the Middle Ages, when nature was to be influenced, this took place through the personal procedure of each individual scientist, through a secret art known *only to him*, and scientists did not hesitate to invoke the help of supernatural powers; each of them tried to achieve his goal empirically, gropingly, *without the help of a general method*. Now a new conception of knowledge had emerged. People were convinced that science was able [9]

---

[23] Descartes, letter to Mersenne, December 1637, AT I, 479. See also Ch. Adam, AT XII, 217.

[24] Boutroux, "La signification historique," op. cit., p. 816. In the Rule IV, Descartes states: "Nothing is less in my mind than ordinary Mathematics, and I am expounding quite another science." (*Philos. Works*, I, 11).

[25] "Not having restricted this Method to any particular matter, I promised myself to apply it ... to the difficulties of other sciences" (*Discourse*, II, *Philos. Works*, I, 94). The disregard of this fact explains both the accusation of Baugard (a disciple of Viète's) that Descartes in Book III of his *Géométrie* plagiarized several rules published by Viète as early as 1615, and the unjustified criticism directed by a modern writer against Descartes – we refer to Ritter's fight for the priority of Viète's merits in algebra. According to Ritter, Viète was the first to construct the edifice of algebraic science which later was only improved by Descartes and others. See Frédéric Ritter, *François Viète, inventeur de l'algèbre moderne, 1540–1603* (Paris, 1895) p. 94 sqs.

[26] Léon Brunschvicg, op. cit., p. 113.

[27] This contemplative character of contemporary science alien to the practical purposes of life is illustrated in Bacon's *Novum Organum* (1620) by the example of "some monk studying in his cell or some nobleman in his villa" (Book I, 80; Works I, 187).

to prescribe *universally binding rules* for human research work, *methods* that made this work simpler, more fertile, and, as Adam formulated it, "*accessible to all*, by [10] dint of a little study."[28]

The representative of the new current also expected knowledge to be *practically applicable*. They wanted not only man's salvation in the next world, but also his happiness in this world; they appealed to people not with a Bible in their hand, but by means of reasonable arguments. To the existing conditions Thomas More opposed the "rational institutions" of the Utopians or of other peoples, as for instance, the Polylerites, under which people live "in happiness." Francis Bacon ascribed "the great cause of the little advancement of the sciences" to the very fact that in former times there was no clear awareness about the above-mentioned goals of science.[29] Therefore he strove to define these goals with particular clarity, saying: "The use of mechanical history is, of all others, the most fundamental towards such a natural philosophy as shall not vanish in the fume of subtle, sublime or pleasing speculations: but be operative to the endowment and benefit of human life." In another passage Bacon emphasized that truth alone is insufficient as the goal of science, "as we regard not only the truth and order, but also the benefits and advantages of [11] mankind." True knowledge must be completed by practical action,[30] *homo sapiens* must be complemented by *homo faber*.

His "New Atlantis" (1629) is only outwardly a "Utopia", in reality it is a magnificent project for a "House of Solomon", that is, a scientific institute promoted by the state, a kind of "academy of sciences". Its task would be *to lead to new inventions*, not through fruitless scholastic discussions and definitions, but through systematically organized investigations in laboratories, and experiments in all fields of natural science: physics, optics, the theory of heat, mechanics and the theory of machines, animal and plant biology, comparative anatomy, etc., through the establishment of technical museums of experimental agricultural stations, through the honoring and rewarding of inventors, etc., etc. This must not be done haphazardly as had been the case heretofore, but on the basis of careful and systematic planning.

This scientific goal must not be achieved by individual scientists who would make the results of their investigations, discoveries and inventions accessible to human society. Bacon felt that this would lead to the division of society into an

---

[28] Charles Adam, AT XII, 228. "For vague and arbitrary experience is mere groping in the dark, and rather confounds men than instructs. But when experience shall proceed in regular order ... by a determined rule, we may entertain better hopes for the sciences" (Bacon, *Novum Organum*, Book I, 100, *Works* I, 203). There is a similar utterance by Descartes, see below.

[29] *Novum Organum*, Book I, 81 (*Works* I, 188): "It is impossible to advance properly in the course when the goal is not properly fixed. But the real and legitimate goal of sciences *is the endowment of human life with new inventions and riches*."

[30] "For information *commences* with the senses. But the whole business *terminates in Works* ... I will proceed therefore with the instances which are pre-eminently useful for the operative part" (*Novum Organum*, Book II, 44; *Works* I, 320).

elite of scientists enjoying the monopoly of science and a large mass of uneducated people. He wanted to eliminate the danger of such a split.[31] [12]

The emergence of the new task that Bacon formulated for science was a reflection in the intellectual field of the social changes in England. The growing use of machines in the industries of England and the most advanced countries of the European continent brought about *a great revolution in* men's thinking. During the predominance of handicraft and the beginning of manufacture, skilled *specialization* and the individual worker's *virtuosity* in a limited trade set the standard; with the emergence of automatically working machines in industry it became clear that these machines, independently of and without any handicraft training or personal talent of the workers, could perform the work better, in greater quantities and with greater speed; and that this work could be done by *anyone* who knew how to handle the machine by simple manipulations, by women and children, indeed even by idiots and cripples, because the automatism of the machines simplified the operations so drastically.

Under the influence of this fundamental technological and social revolution – the transition to large-scale industry and machine production – the conviction grew in the most advanced minds that a new epoch had begun; that *specialization* was a thing of the past that was indispensable only so long as no adequate instruments were available, but that it was no longer necessary in the machine age. This conclusion, drawn from the experience of industrial production, was then extended by a process of generalization to intellectual production, to the sciences, and the idea asserted itself that specialized individual talent was of decisive importance only at [13] the lower level of development of human society when mankind still lacked adequate intellectual instruments, but that later, at higher social levels, even average human intelligence would enable every man actively to participate in the intellectual work of the nation and to attain the knowledge of highest truths if he only knew how to use the proper "instruments." For the intellectual auxiliary means of a systematically conceived procedure, the *method*, was assigned the same part in the intellectual production that the *machine*, that technical auxiliary means played in industrial production.

Thus science would not depend in the least on the achievements of an elite of particularly gifted specialists and intellectual virtuosi; it would rather tend to level differences in talent. "Our method of discovering the sciences," says Bacon, "is such as to leave little to the acuteness and strength of wit, and indeed rather to level wit and intellect. For us in the drawing of a straight line or accurate circle by the hand, much depends on its steadiness and practice: but if a rule or compass be employed,

---

[31] According to Bréhier the very idea of an academy of sciences here shows, "qu'il a compris que le travail scientifique devait être un travail collectif, réparti entre une foule de chercheurs, et il a consacré . . . la *New Atlantis* à la Description d'une sorte de république scientifique, où il assigne à chacun sa tâche" (Emile Bréhier, *Histoire de la philosophie,* vol. 2: *La philosophie moderne*. Paris: Presses Universitaires de France, 1938, p. 42).

there is a little occasion for either: so it is with our method ... The lame in the path [14] outstrip the swift who wander from it."[32]

*****

Descartes developed and deepened Bacon's fundamental idea, which constitutes the real kernel of his *science universelle*, or algebra.[33] But it was not born in Descartes' mind in the definite form in which it is familiar to us. For that reason we shall follow it in its successive stages of development and discover how, from a still vague, imprecise general idea in 1629, it gradually assumed ever more concrete forms, until it found its definitive form in the *Discourse on Method* and the Cartesian algebra [15] of 1637.[34]

---

[32] *Novum Organum*, Book I, 61 (*Works* I, 172). See also I, 122.

[33] The last quoted sentence from Bacon is almost literally repeated by Descartes: "Those who proceed very slowly may, provided they always follow the straight road, really advance much faster than those who, though they run, forsake it." (*Discourse on Method*, I, *Philos. Works*, I, 82). Elsewhere Descartes says: "It were far better never to think of investigating truth at all, than to do so *without a method*. For it is very certain that unregulated inquiries ... only confound the natural light and blind our mental powers" (*Rules for Direction*, Rule IV, op. cit., vol. 1, p. 9, *italics* H.G.). See a similar statement by Bacon, fn 28.

[34] The question may be raised whether it is permissible to connect Descartes' youthful dreams of 1629 with his mature ideas: are there not developmental differences between Descartes' dream of a universal science and the *Regulae* as well as between the latter and the *Discours de la Méthode*? With regard to this question it must be stressed that one is perhaps less justified in speaking of a *gradual evolution* of philosophical conceptions in Descartes than in any other great modern philosopher. During more than two decades of his scientific activity he immensely increased his knowledge of the outer world and changed his views on the facts, but never did he change his fundamental philosophical conceptions once they were formed. This explains why it is so difficult to determine the dates of Descartes' writings when they are not explicitly indicated by himself or mentioned in his correspondence. All his writings start from the *same* philosophical premises so that if we go by their content they can be ascribed to any period of his creative activity. It is for instance characteristic that Ch. Adam, the competent editor of Descartes' complete works, considers the dialogue "The Search after Truth" a youthful essay and places the times of its writing at about 1628, while Cassirer declares that it was written only in 1649, toward the end of Descartes' life; according to Cassirer, this dialogue was an introduction to Descartes' philosophy written for Queen Christina (see Ernst Cassirer, *Descartes*, Stockholm, 1939, p. 140). And it is interesting that neither Adam nor Cassirer base their determination of the date of this dialogue on its philosophical content, for the very reason that this offers no clue as to its date of composition, but upon secondary stylistic factors, the pedagogic intention, or the dialogue form itself. See also the discussion about the date of composition of the dialogue by M. Cantecor, "A quelle date Descartes a-t-il écrit la Recherche de la Vérite?" *Revue d'histoire de la philosophie 2* (1928), 254 sq., and the answer of H. Gouhier, ibid., vol. 3, (1929), 296 sq. This constancy of Descartes' philosophical conceptions can be illustrated by many examples. The view, for instance, that animals are soulless machines, which he expressed in the *Cogitationes privatae*, as early as 1620, even before outlining his philosophical system (see AT X, 219) was maintained by him throughout his life; we find in the *Discourse* of 1637, in his *Replies* to the objections to his *Meditations* of 1641, and finally in 1648/49 when he sharply emphasized it in his correspondence with Henry More (AT V, 267 sqs. and AT V, 340). The same constancy can be found in Descartes' ideas on physics. Gilson states emphatically that all the theories of Descartes' period of maturity are only an elaboration of his early ideas: "*Since 1619–1620 the Cartesian physics had been constituted* in its method and spirit ... Its spirit is the same that will animate *Le monde* (1632) or the *Principia* (1644)" (E. Gilson, *Etudes sur le rôle de*

II

The starting point for the discussion of the problem of the "universal science" was Father Mersenne's letter to Descartes in 1629,[35] in which Mersenne asked him for his opinion of a project put forward by an unnamed person for a "new universal language." Descartes sharply criticized this project and rejected it on the grounds that an artificial language was only a matter of grammar and that such a language, although advantageous in some respects, had important drawbacks.

Nevertheless, Descartes observed, a universal language was desirable but one of an entirely different kind, a universal language not of words but of ideas. The prerequisite of such a language, he added, was "the true philosophy." Just as it is possible to learn all the numbers up to infinity in one day, because the *natural order* of the numbers enables us to survey them easily, so all – actual and possible – ideas could be ordered *according to definite classifications*, and such a universal language [18] of ideas might be learned within *a very short time*.[36]

---

la pensée médiévale dans la formation du système cartésien, Paris, 1930, p. 17). The same is further true of Descartes' methodological views that he had begun to note down as early as 1628/29 in his *Regulae* and later preserved fundamentally unchanged although in a more developed form in his [16] *Discourse* of 1637 (Léon Brunschvicg, *René Descartes*, Paris, 1937, p. 15). Finally, he expounded the same methodological views in the *Méditations* of 1641 and as late as 1644, he summarized them without changing them, on the occasion of the Latin edition of the *Discours* (ibid., p. 28). – The same is true of the problem treated here, Descartes' Algebra conceived as a *Universal Science*, for after all it is only a "sample" of Descartes' general philosophical method. Brunschvicg, in the chapter of his book devoted to the idea of the *mathématique universelle* (*Les étapes*, op. cit., pp. 105 sqs.) rightly uses as equally valid sources the *Regulae* and the *Discours* although they were written at different periods. Moreover, we know that as late as March 1636 Descartes intended to entitle his Discours "*Le projet d'une science universelle*" and that he chose another title only upon Mersenne's advice (see Ch. Adam, AT XII, 227–229). Therefore, J. Sirven justly says: to suppose that the *Universal Science* was only a "dream of Descartes' youth" is to regard it as a concept of an "ideal impossible to be realized." In opposition to such a view, Sirven points out that if the concept of a Universal Science was a dream, then Descartes never abandoned this dream since the first [17] winter days of 1619 (J. Sirven, *Les années d'apprentissage de Descartes 1596–1628*, Paris, 1928, p. 440). More, he was convinced, that he had realized that dream.

[35] Actually this letter was only a pretext; the idea of the *science universelle* goes back farther, to the memorable night of November 10, 1619, when Descartes noted in his *Olympica*: "cum mirabilis scientiae fundamenta reperirem" (AT X, 179 and Adam, AT XII, 50). Then, in 1619, says Gilson, Descartes "conceived the idea of the *unity of the science* and of the absolute unity of mathematical methods. From that moment on, *the Cartesian revolution has been a fact*." Later, Descartes only elaborated the ideas developed at that time (E. Gilson, op. cit., p. 286).

[36] "So that it could be taught in a very little time, and this by means of the order, that is to say, by establishing *an order in all the ideas* that can enter the human mind ... just as there is an [order] *naturally established in the numbers*; and just as one can learn in one day to name all the numbers up to infinity" (Descartes to Mersenne, Nov. 20, 1629, AT I, 80). Even the *Cogitationes Privatae* of 1620 emphasize that because of the interconnection of all the sciences it is easy to keep them all in the mind – like a series of numbers: "Catenam scientarum pervidenti, non difficilius videbitur, eas animo retinere, quam seriem numerorum" (AT X, 215).

More than thirty years later Leibniz had a similar idea and envisaged the classification of all actual and potential judgments, the creation of an "alphabetum cogitationum humanarum," so that man could by analysis derive all principles from a small number of the highest truths. "Scientiam

However, without the aid of a "vraie philosophie" the invention of such a language of ideas was impossible. For in order to establish an "order in all the ideas that can enter the human mind," these ideas, beginning with the simplest, would have to be classified in definite groups according to their properties; for this it was necessary to have a clear notion of each idea and clearly to distinguish it from every other idea; but as things stand today "the words we have, so to speak, have only [19] confused meanings," and for that reason is never completely grasped by people.[37]

From the foregoing it can be seen that such a language of ideas grouped according to classes of ideas and progressing from the simple to the complex would actually be a *philosophic method* enabling us to achieve adequate notions of all things. And of such a *langue de pensées* conceived as a method Descartes says: "I hold that this language is possible and that one can find the science on which it depends, with whose help *the peasants would be able to judge of the truth of things better than the philosophers do now*."[38]

*****

The letter of 1629 quoted above, like the previously quoted letter of 1636, in which Descartes assures us that his universal science will also be accessible to those who have never studied, expresses his dislike of *specialization*, his fear that specialists, [20] because of their preoccupation with details lose their feeling for the essentials of truth, in contrast with men of the people, who, because of their average intelligence

---

generalem dico, quae modum docet, omnes alias scientias ex datis sufficientibus inveniendi et demonstrandi" (*De scientia universali seu calcuo philosophico*; *De natura et usu scientiae generalis*, in G.G. Leibitii *Opera philosophica omnia*, ed. Erdmann, (Berlin, 1840), p. 83, p. 86). See also Louis Couturat, *La logique de Leibniz*, op. cit., p. 35.

The literature of the time shows how widespread such ideas of a philosophical world language were. See Sir Thomas Urquhart (1611–1660) *Logopandecteision: or an Introduction to the Universal Language* (London: Calvert, 1653), George Dalgarno (1626–1687) *Ars signorum: vulgo character universalis et lingua philosophica* (London: Hayes, 1661), and John Wilkins (1614–1672) *Essay towards a Real Character and a Philosophical Language* (London 1668); see Otto Funke, *Zum Weltsprachenproblem in England im 17. Jahrhundert* (Heidelberg 1929) (*Anglistische Forschungen*, No. 69 ed. Dr. J. Hoops). Dalgarno's treatise exhibits a *methodical classification of all possible ideas*, and a selection of characters adapted to this arrangement so as to represent each idea by a specific character without reference to the words of any language. He admits only *seventeen classes of ideas*, and uses letters of the Latin alphabet and also some Greek characters "to denote them." Leibniz, in a letter to Thomas Burnet of Kemney, dated 1697, alludes to the "Ars Signorum."

[37] AT I, 81.

[38] AT I, 81. – If, like Maxime Leroy, we relate the last quoted sentence not to the *langue des pensées* conceived as a method, but to the ordinary linguistic idea of a universal language, the result is pure confusion. Leroy says: "He [Descartes] believed that if a universal language were ever to exist, the peasants could judge of the truth of things better than the philosophers do now." This sounds fantastic; it is impossible to see why the knowledge of one or even a dozen languages should enable peasants to grasp the truth better than philosophers do and why the construction of such a language should depend on "a true philosophy" as yet undiscovered. See Maxime Leroy, *Descartes Social* (Paris, 1931) p. 51.

and "the simple reasoning which a man of common sense can quite naturally carry out" is better prepared for the discovery of the truth.[39]

How can such a far reaching result – the foundations of a universal science – be expected from the above mentioned "theory of order?" And how will such a science enable the man of the people, the uneducated man, to discover the highest truths? Does not the philosopher's trade require long professional training and scientific specialization such as, for instance, the farmer requires and through which he becomes a farmer, not a philosopher?[40] [21]

To grasp the true meaning of the Cartesian "universal science" and the full significance of the problem thus raised for the first time by Descartes, one must recall the dangers threatening modern culture as a result of over-specialization. Descartes foresaw *the great intellectual crisis of today*, which is the necessary result of specialization. The specialist buried in the forest of details of his special discipline loses the ability to understand contemporary social and intellectual life as a whole and therefore loses the ability to judge it correctly; ultimately this must lead to the lowering of the intellectual and cultural level of society as a whole.[41]

The same conflict viewed from another aspect presents itself as a problem of relation between the particular and the universal. For the "specialists" in the time of Descartes, as for the positivists of today, the world was nothing more than a

---

[39] *Discourse on Method*, II (*Philos. Works*, I, 88). In another passage Descartes warns against the "learned" [who] frequently employ distinctions so subtle that the light of nature is dissipated in attending to them, and even those matters of which no *peasant* is ever in doubt become invested in obscurity" (*Rules for the Direction*, Rule XIV, *Philos. Works*, I, 57). Later in his *Traité de la lumière* (1629–1633) Descartes says ironically that "the philosophers are so subtle that they manage to find difficulties in things that seem extremely clear to other humans" (AT XI, 35), and he ridicules their "superfluous subtleties" (ibid., p. 45). Descartes expresses the same belief in the "natural light" independent of any acquired education when he judges himself. According to what he told his friends, he was convinced that he did not arrive at his ideas as a result of his studies; he believed that even if "his father had never made him study, *he would not have failed to write the same ideas, in the same way and perhaps even better* than he did" (see Adrien Baillet, *La vie de Monsieur Des-Cartes*, Paris, 1691, II, 470–71; see also AT XII, 32).

Similarly, in *The Search after Truth* (1629) Eudoxe (who represents Descartes in the dialogue) says to Polyander: "You will come to believe that a man with a healthy mind, *had he been brought up in a desert and never received more than the light of nature* to illuminate him, could not if he carefully weighed all the same reasons, adopt an opinion different from ours" (*Philos. Works*, I, 311).

[40] The originality of Descartes' idea will be revealed more sharply if we compare it with the conceptions of another philosopher, Hegel. Obviously alluding to Descartes' "natural light" Hegel attacks the opinion that "natural reason" alone, without professional preparation could enable one to philosophize, as though philosophy "consisted in that very lack ... of study." Hegel further speaks of the "long road of education" required for the philosopher, in brief, "of educated reason" (G. W. F. Hegel, *Die Phänomenologie des Geistes*, Bamberg, 1807, Preface, pp. lxxxiv–lxxxvi).

[41] Thus a contemporary author rightly says: "Knowledge in its ideas, language, and appeals is forced into corners; it is *over-specialized*, technical, and esoteric, *because of its isolation*. Its lack of intimate connection with social practice leads to an intense and elaborate over-training which increases its own remoteness" (Joseph Ratner, *The Philosophy of John Dewey*, New York, 1928, p. 415).

generalized statement of isolated, precise, factual observations instinctively made in a restricted field of experience; whereas Descartes was convinced that it was [22] impossible to understand the isolated phenomena in a particular field without establishing their relationship to the universality of things as a total object of experience common to man, because the particular phenomenon receives its proper significance only in the frame of that universality. Descartes' Algebra as a "Universal Science" is an instrument to establish this general relationship and to overcome the dangers of isolationism and specialization.

Descartes advances two arguments intended to show first that a new, simplified, and thus for everyone easily accessible method – a real universal science – is not only desirable, but also necessary; and secondly, that the elaboration of such a method is possible.

Students of the Cartesian revolution have above all emphasized Descartes' striving toward the *certainty* of *cognition*. But it must be emphasized that a no less [23] important factor of this revolution is the *simplicity* of the method,[42] the *shortening* of the road to truth, because this factor was destined to counteract over-specialization; [24] for only through the simplification of this road to knowledge can the mediation of professional scholars for the "unlearned" be eliminated, can the attainment of the truth be made *easily accessible to everyone literate* and thereby a *really universal science* established. Actually Descartes reproached contemporary science (in 1628) for burying in a mass of volumes whatever correct knowledge it offered, thus making it practically inaccessible to the people.[43]

For that reason Descartes' fundamental aspiration is, in addition to the certainty of knowledge, to find "*an easier path*" leading to it.[44] Similarly, he says at the end of his Geometry: "It is not my purpose to write a large book. I am trying rather to include much *in a few words*."[45] Therefore, "instead of a great number of precepts of which Logic is composed" he gives his well-known four simple rules that he

42 "By a method I mean *certain* and *simple* rules" (Descartes, Rules for Direction, Rule IV, *Philos. Works*, I, 9). The importance of this requirement, as an expression of the revolt of reason against the absurd subtleties of scholasticism becomes comprehensible when it is compared to the complicated determinations that according to Thomas Aquinas are necessary to grasp the nature of the act of cognition. This act must be dealt with from the following ten points of view: *operatio, actio, motus, mutatio, generatio, formatio, assimilatio, unio, perfectio, vita* (Alfons Hufnagel, "Studien zur Entwicklung des thomistischen Erkenntnisbegriffes in Anschluss an das Correctorium 'Quare'," *Beiträge zur Geschichte der Philosophie und Theologie des Mittelalters 31*, ed. M. Grabmann, No. 4, Chapter III).

43 ". . . even if all the knowledge which we can desire is to be found in books, that which they contain of good is mingled with so many futilities, and confusedly dispersed in such a mass of great volumes, that in order to read them more time would be requisite than human life can supply us with" (Descartes, *The Search After Truth*, *Philos. Works*, I, 306).

44 Ibid., I, 306.

45 *The Geometry of René Descartes*, trans. by David E. Smith and Marcia L. Latham (Chicago, 1925) p. 240.

considers "quite sufficient."[46] Only through the *shortening* and simplification of the road to knowledge, can the knowledge of the truth be made accessible to *all*, including people without long professional training and thus be taken out of the *narrow circle of the professional scientists*. And, anticipating the objections which would be advanced against drawing to philosophy the large untrained popular masses, Descartes expressed the hope that in the world of knowledge it does not make any difference whether an idea comes from Plato or Aristotle or from less prominent men, if only it is sound, that is to say, "has a current value in the world ... just as money which is in nowise of less value when it proceeds from the purse of a peasant than when it comes from the treasury."[47] Descartes makes his idea more pointed by assigning in the *Dialogue* a prominent role to Eudoxe (= Descartes), "a man endowed with *ordinary* mental gifts, but whose judgment *is not spoiled by* any false ideas, and who is in possession of his whole reason in all the purity of its nature."[48]

*****

But to describe the need for such a method and its usefulness was not enough. Was it possible? Descartes answered this question in the affirmative on the basis of the *unity* of all the sciences which makes it possible to realize his dream of a universal science and to overcome the dangers of over-specialization. Descartes dealt with this problem in the *Rules for the Direction of the Mind*, written in 1628 or 1629[49] [25] and in "The Search After Truth,"[50] a dialogue dating from the same period.

In Rule 1 Descartes criticized the point of view of those who in the "sciences" see only one profession among many others, of which a man can learn only one at a time, because according to the opinion of these thinkers every man has a gift for only one *special* trade: the smith who devotes himself to working metals, they say, can hardly be a good farmer. "Hence they have held the same to be true of the sciences also, and distinguishing them from one *another according to their subject matter*, they have imagined that they ought to be studied *separately*, each in isolation from

---

[46] *Discourse on Method* II (*Philos. Works*, I, 92). On this question Gilson says in his "Commentaire": "La méthode se ramène à quatre précepts simples et ces quatre précepts concernent deux opérations de la pensée: l'intuition et la déduction. Or, ces deux opérations ne consistent que dans l'usage spontané de notre lumière naturelle; elles ne peuvent donc être enseignées, et ainsi ... tous les précepts de la Dialectique [i.e. scolastique, G.] deviennent superflus".

The twenty-one rules of the *Regulae* thus add nothing essential to the four of the *Discours*, so that the "Regulae ne contiennent en définitive que les quatre préceptes du Discours, et que ce qu'elles leur ajoutent ne consiste pas en précepts supplémentaires, mais en règles practiques destimées à faciliter leur application. La méthode a donc été dès le début quant à essentielle, ce qu'elle devrait toujours demeurer; Descartes n'a varié que ce qu'il a de moins en moins cru á la possibilité d'en formuler et d'en enseigner le procédés d'application" (Gilson, "Commentaire" pp. 195–196).

[47] *The Search After Truth*, *Philos. Works*, I, 306.

[48] Ibid., p. 307.

[49] Published for the first time in Latin, in 1701.

[50] The first germ of the theory of the unity of all sciences can be found as early as 1620, in the *Cogitationes privatae* and even in his first dream from 1619 (*Olympica*). See above, p. [17], fn 35.

all the rest. But this is certainly wrong ... since the sciences taken altogether are identical with human Intelligence, which always remains one and the same, however applied to different subjects..."[51] As there is only one general human intelligence, that "intelligence or that universal science" can *facilitate for us the simultaneous study of all* the sciences. "For neither does the knowing of one truth have an effect like that of the acquisition of one art and prevent us from finding out another, it [26] rather aids us to do so."[52]

All the branches of "Science" essentially form a unity, because they are nothing but human intelligence in action, and because, for that very reason, there exists only one *façon de comprendre*. Hence there must be *only one method* of taking into account all the data involved in a given problem, that is, to grasp these data according to an intelligible order in such a way that there is only a single chain of simple relations between simple elements.[53] And because there can be only one method, it follows "that all the sciences are so inter-connected, that *it is much easier to study them all together* than to isolate one from all the others." For what is to be discovered is the *order* of the elements, not the individual subject-matters of the individual sciences detached from this order, as we are told by the specialists. The task of philosophy is not to solve "this or that special difficulty of the School," but on the contrary "if ... anyone wishes to search out the truth of things in serious earnest, he ought not to select *one special* science; *for all the sciences* are conjoined with each other and interdependent."[54] What is sought in all the sciences is *always the same thing*, the manner of connecting the elements, not the substantial differences of [27] the elements themselves. "Although the objects of Mathematics are different, they ... take nothing under consideration but the *various relationships* or proportions which are present in these objects."[55]

Because this interdependence of the sciences described by Descartes is a fact he thinks that the method desired by him is actually realizable. Descartes shows this in his "Search After Truth" (1629), where Eudoxe (= Descartes) says: "We shall treat in detail of *all* the sciences, ... and we shall support *a method* whereby they may be carried on much further, and *find of ourselves, with a mind of mediocre ability, what*

---

[51] *Rules for the Direction*, Rule I, *Philos. Works*, I, 1.

[52] Rule I, *Philos. Works*, I, 1–2.

[53] See L. Brunschvicg, *Les étapes*, Book II, Chapter VII: "La mathématique universelle de Descartes et la physique."

[54] *Rules for the Direction*, Rule I, *Philos. Works*, I, 2.

[55] *Discourse*, II (*Philos. Works*, I, 93). "The proper object of universal mathematics ... is *relation*," says L. Brunschvicg (*Les étapes*, op. cit., p. 106). See also Ch. Adam, AT XII, 51. Only the above considerations make it clear why the peasants, and more generally those who have not studied, would be able, with the help of the "language of ideas" envisaged by Descartes to discover the truth more easily than philosophers had before. For a language of ideas presupposes the ordering of these ideas, from the simplest to the most complex, in clear and distinct classes and the establishment of the relations among these classes; such a classification would provide every person of average intelligence with a method which had not existed previously and with the help of which the truth could be discovered with *certainty* and in the *shortest possible way*.

those most subtle can discover."[56] And the discovery of such a method is possible, because "those branches of knowledge ... are, as a matter of fact, united by a bond so marvellous, *they are capable of being deduced from one another* by sequences so necessary, *that it is not essential to possess much art or address* in order do discover them, provided that by commencing with those that are most simple we learn gradually to raise ourselves to the most sublime."[57] [28]

This conception of Descartes – a universal science understood as a general method accessible to *everyone* – is not, as we have seen, an accidental idea peripheral in the Cartesian system; rather, as Descartes himself emphasized, it is the natural result of his philosophy, the doctrine of unity of all the sciences. Like Francis Bacon who advanced the idea that with the help of an instrument or a machine *anyone* could carry out precisely a definite task previously performed by skilled hands, independently of the individual endowment of the worker operating the machine, Descartes put forward the view that as a result of the interdependence of all the sciences everyone, even an individual of average intelligence without much art or address, can – with the appropriate method – discover the highest truths; something that previously, without the help of such a method, only the "most subtle" mind could accomplish.

The first phase of the intellectual development of the Cartesian "science universelle" ends with the clear definition of these goals as the basis of a reform of method. [29]

III

In the second period of this development Descartes turned to the *preparatory measures* for the construction of such a universal method that deals with the relations and proportions among things; in brief, he began to realize a method which he hitherto had conceived only as an ideal, in the form of a "language of ideas" that he did not define more precisely. Here too, the inspirations for the concrete shaping of the method – as we are told in his *Discourse* – derived directly from his conception of all the sciences as a *unity*. Because they are a unity, because they must be treated not separately but *all together*, and further because proportions between phenomena or groups of phenomena must be sought which are proper to the most varied sciences (mechanics, physics, biology, psycho-physiology), these proportions must be characterized by the shortest possible signs, such as *a*, *b*, *c*, so that it will be possible to retain all of them and their mutual relations simultaneously in the memory.[58] [30]

---

[56] *Philos. Works*, I, 311.

[57] Ibid., I, 306. See also Ibid., I, 327.

[58] "...if I only examined these proportions in their general aspect ... Then, having carefully noted that in order to comprehend the proportions I should sometimes require to consider each one in particular, and sometimes merely keep them in mind, or take them in groups ... I considered, however, that in order to keep them in my memory or to *embrace several at once*, it would be essential that I should explain them by means of certain [letters], the shorter the better." (*Discourse* II, *Philos. Works*, I, 93). The word "letters" is put in brackets because I have inserted it. The

In addition to this method of designation, the conceptions of the unity of all the sciences led to another even more important innovation: for the very reason that the phenomena of the most various scientific fields are to be dealt with *simultaneously* and brought into relation with each other, first they must *be made comparable*, that is to say, reduced to a common denominator, before the nature of their relationships and the particular form of their proportions can be found and described.

To achieve this end Descartes started from space and its dimensions, but he soon went beyond spatial notions. Space is a system of figures susceptible of being measured according to the three dimensions; it is, for instance, possible to start from length in order to reconstruct spatial reality. But his method of composition does not exhaust all the elements constituting reality. For reality does not consist of spatial dimensions alone. The spatial magnitudes are only one particular aspect of reality; every other element analogous to length that is measurable can also be regarded as a dimension; in this manner it is possible to introduce as many "dimensions" into a problem as one pleases. Thus Descartes extended the concept of dimension from the three spatial dimensions to all the other measurable properties of reality. The importance of this generalization of the concept of dimension, which constitutes the central point of the *Regulae* can hardly be overestimated; it freed men's minds from dependence upon spatial representations in mathematical thinking and *thus* [31] *made the mathematical method applicable to all fields of science*. "Thus it is not merely the case that length, breadth and depth are dimensions; but weight also is a dimension ... so, too, speed is a dimension of motion ... It clearly follows that there may be an *infinite* number of dimensions."[59] We are justified in speaking of *pandimensionism* in the case of Descartes; only after the development of the problem of dimension in the generalized form he gave it, could the various phenomena of reality enter into the fundamental equations not only of mechanics and physics, but of all the sciences that deal with measurable magnitudes. Only now, after Descartes had created for himself a powerful instrument in the new concept of dimension, could he *undertake the reform of philosophy – its mathematization;*[60] only now, with the help of this new technical apparatus, could he undertake to give concrete shape to the *science universelle*. Thus the Cartesian idea entered upon its third, definitive phase. [32]

IV

The task was to construct a method independent of the individual sciences, but applicable *to all* of them, whether it was a question of numbers, lines, sounds, or any other field of investigation. "There must be some *general science* to explain that element

translators put "formulas" instead, which is false, because not formulas, but designation by letters are in question; the French text says: "*par quelques chiffres*" (AT VI, 20).

[59] *Rules for Direction*, Rule XIV, *Philos. Works*, I, 61 (AT X, 44–48).

[60] L. Brunschvicg, *Les étapes,* op. cit., pp. 111–112. Thus he wrote to Mersenne, "que toute ma Physique n'est autre chose que Géométrie" (Letter from July 27, 1638, AT II, 268).

as a whole which gives rise to problems about order and measurement, restricted as these are to no special subject matter."[61] This general "science of order," this "science of relations" is precisely the Cartesian Algebra. The *Geometry* is, according to Descartes, only a "sample," an illustration of his method, which includes both the general philosophical method, the subject-matter of the *Discourse*, and, in the words of Boutroux, "the mathematical method which is only a particular application ... of the general method."[62] Or as Boutroux says elsewhere: "Pure Algebra must not be regarded *as an objective science*"; it is "a *technique of calculation*, empty of content in itself. *It is a method*,"[63] applicable to all the sciences. For "no science is acquired except by mental intuition or deduction."[64] But the latter is based on the *comparison* of a sought but unknown magnitude with one given and known, "such that we may discover some equation between what is unknown and something known."[65] [33]

Descartes explained here why, in our striving for the discovery of the truth, reasoning must express itself in the form of mathematical equations in order to find the proportions and relations which are the object of our search. The essence of this method consists in the fact that when a number of known elements: a, b, c and unknown elements: x, y, z are given, one disregards the content of these elements; only *magnitudes* and *their ratios* are important.[66] Thus, "if there are several equations, we must use each ... either considering it alone or comparing it with the others ... we must combine them ... until there remains *a single unknown* ... which is equal to some known line, or whose square, cube, fourth power, fifth power, sixth power etc., is equal to the sum or difference of two or more quantities, one of which is known, while the others consist of mean proportionals between unity and this square or cube or fourth power, etc., multiplied by other known lines."[67]

The mathematization of modern science is not a transformation of or a return to the number mysticism of the Pythagoreans; it was achieved by Descartes in the service of modern natural science. Only in this connection can we understand the methodological importance of Cartesian pandimensionism and can *the new role of mathematics* become fully comprehensible. Only the generalization of the concept of dimension explains why the method of mathematical equations and algebraic analysis did not become the object of a *particular mathematical* discipline, but a *general key to all the other sciences* with the help of which the secrets [34]

61 *Rules for the Direction*, Rule IV, *Philos. Works*, I, 13.
62 P. Boutroux, La signification historique de la "Géometrie" de Descartes, op. cit., p. 816.
63 P. Boutroux, *L'Idéal scientifique des mathématiciens* (Paris, 1920) p. 100.
64 Rules for the Direction, Rule IV, *Philos. Works*, I, 10 (AT X, 372).
65 Rule XIV, ibid., I, 61.
66 Rule XIV, ibid., I, 56.
67 Descartes, *Geometry*, op. cit., Book I, p. 9.

of nature can be discovered. According to this conception of Algebra, philosophy is not reduced to mathematics, but conversely, *mathematics is subordinated to philosophy*.[68]

Thus, according to Descartes, Algebra is nothing but a science of combinations in accordance with fixed rules, the "*ars combinatoria*," that enables us to find a desired and unknown magnitude, and that anyone – once the rules of this science of combinations are disclosed – can apply *purely mechanically*, without any special intellectual effort.[69] This science of combination, Algebra, is "*a universal method*."

[35]

Appendix: On Raymond Lully

Boutroux's mentioning of Raymond Lully (1235–1315) in this connection seems to me to be based on a gross misconception, because Lully's *Ars Generalis*, despite the promising titles of his works ("Ars compendiosa inveniendi veritatem seu ars magna et major," "Introductorium magnae artis generalis ad omnes scientias," etc.) was not by any means and was not intended to be a universal method of knowledge and a key to all the other sciences; in fact, it was the *very opposite* of what Descartes was trying to achieve, and there is only an external resemblance between the two because both Lully and Descartes addressed their writings not to the educated few but to the mass of the people.

The fundamental principle of Lully's system is the conviction that *faith is the preliminary condition of all intellectual knowledge*. Faith is a preliminary disposition by means of which reason is capable of deducing *a priori* all truth, whether natural or supernatural. There is no separation between the rational and supra-rational, between natural and revealed truth. Reason can and ought to demonstrate everything, including mysteries. Thus, Lully perverts the relations between philosophy and theology and confuses the former with apologetics. The whole office of philosophy is to prove that Catholicism is true.[70] Lully wrote his *Ars Generalis* for apologetic reasons, to defend Christian faith against Arabism, especially against the influential philosophy of Averroes; he appeals to the mass of the common people (in the contemporary *Vita Coaetanea* we read: "let a general art be made ... in accord with the capacity of the simple [fecit artem generalem ... secundum capacitatem simplicium]") who lived among the infidels in Spain, the Balearic Islands, and North Africa and were exposed to their religious attacks. This mass of people

[36]

68 "It was not the case that ... our method was invented for the purpose of dealing with mathematical problems, but rather mathematics should be studied almost solely for the purpose of training us in this method" (Rule XIV, *Philos. Works*, I, 57).

69 "One must know how to forget the significance of the elements combined and pay attention only to the mechanism of combination..." "Science will be reduced to a work of *mechanical combination*" (P. Boutroux, *L'Idéal scientifique*, op. cit., pp. 86, 126).

70 Maurice de Wulf, *History of Mediaeval Philosophy*, transl. by Ernest C. Messenger (New York, 1926) II, 153–154.

was to be given an instrument for defending themselves against such attacks; and for that purpose they had to learn how to handle the abstract notions that were used in disputes, such as kindness, grandeur, eternity, power, wisdom, will, virtue, glory, perfection, justice, beneficence, pity, humility, domination, patience.[71] Therefore, the *Ars generalis* contains a practical section that gave instructions for answering a number of concrete questions.[72] Keicher observes justly: "Thus the *Ars* is revealed as a kind of heuristic method with the help of which ideas had to be *imposed in a schematic* manner even on the uneducated who were unable independently to decide upon the doubts and objections to which they were exposed."[73] The *Ars* was intended to offer the ready-made results of human thinking and exploring in an easily understandable form to those who were unable in individual cases to put forth arguments in defense of their own faith.[74] Moreover, Lully applied a special *manner of exposition* intended as a further – mnemotechnical – aid for the people; he used six concentric circles divided into compartments, of which one circle rotated while the others remained stationary. On the periphery of the circles six chambers were drawn, each of which comprised nine categories represented by letters of the alphabet. With the help of this technical construction Lully tried to relate the concrete questions that were answered to the above-mentioned abstract notions, and thus to symbolize the content of these notions. The rotation of the circles with regard to one another automatically supplied the possible permutations of notions and answers to such propositions as might be suggested. "Lully's idea seems to have been the invention of a *logical machine* which would constitute the same sort of labor-saving device in a scholastic disputation ... as an adding-machine in a modern ... business office."[75] The situation was further complicated by the fact that every letter – according to its position – symbolized another abstract category. "Just as in the combination *b,c,t,b,* in which *b* before *c* expresses goodness and *c* magnitude, and *b,* if *t* is first, expresses difference and so on concerning the others." [Sicut in camera de b. c. t. b., in qua b. ante c. dicit bonitatem et c. magnitudinem et b. prius t. dicit differentiam, et sic de aliis.][76] What interested Lully is obvious in the nature of the questions he treated: Quaestio b. f. k. "Whether fire has such good instinct and such good delight through Saturn as through Jupiter." Utrum ignis habeat ita bonum instinctum et ita bonam delectationem per saturnum sicut per jovem?" ... Quaestio b.g.t.i. "Whether in an angel goodness and will are equals [Utrum in angelo bonitas

[37]

[38]

[71] Raymundi Lulli *Opera quae ad inventam ab ipse artem universalem scientiarum artiumque omnium pertinent* (Strasbourg, 1651). See especially the table of categories on the front page.

[72] See, for instance, "Decima pars: De Applicatione" (ibid., p. 488), "Undecima pars: De Quaestionibus" (ibid., p. 562 sqq.) etc.

[73] Otto Keicher, *Raymundus Lullus und seine Stellung zur arabischen Philosophie* (Münster, 1909) p. 20.

[74] Ibid., p. 21.

[75] Lynn Thorndike, *A History of Magic and Experimental Science* (New York, 1929) II, 865.

[76] *L'Ars compendiosa de R. Lulle*, ed. Carmelo Ottaviano (Paris: Vrin, 1930) p. 118 (*Etudes de Philosophie Médiévale 12*).

et voluntas sint equales]?" ... Quaestio f.g.t.h. "How much is one angel greater than another [Quantum est unus angelus maior alio]?"[77]

All this has nothing in common with a "universal method," a "key" to all the other sciences and to the discovery of truths, nor does it have to do with the use of mathematics to explore nature. Indeed, Lully's ideas are in sharp contrast to the aspirations of Descartes, who did not write a catechism with ready-made answers, but on the contrary hoped to give every intelligent person, even though uneducated, a method of discovering scientific truth *independently*. Thus Descartes says in the *Regulae*: "I have always found, not the following of the arguments of others, but the discovery of reasons by my own proper efforts, to yield me the highest intellectual satisfaction."[78] [39]

Although Leibniz, by his own admission, was in his youth influenced by Lully's *Ars Magna*, it seems that this influence must be reduced to the rotation of Lully's circles,[79] which suggested to him the idea of his *Ars Combinatoria*; but this has little to do with the philosophical idea of a "universal method" as a key to the discovery of truths. Lully treats every notion as an independent logical entity which can at will be connected with any other notion to whatever genus it may belong. Can a *logical system* be constructed on the basis of such accidental permutations of concepts?[80] For that reason it seems to me that Descartes was not unjustified when he expressed himself disparagingly about the "art of Lully," which served to enable "one to speak without judgment of those things of which one is ignorant."[81] [40]

V

What are, according to Descartes, the great *advantages* of this new universal method? What was actually at stake in the well-known mathematical duels between Descartes and his opponents? This problem has not been clarified in the existing literature on Descartes. One might expect that the struggle was waged about great "principles," previously unknown truths that Descartes discovered and that supposedly remained unknown to his adversaries? What exactly does Descartes reproach his opponents with? Are they unable to solve the problems he submitted to them? But they can solve them. The point is, that they could solve them only with the help of a backward, slow "artisan" mathematical technique, while Descartes used a superior, rapid technique, comparable to the action and speed of a machine.

To understand this fundamental point, one must recall not only the victorious march of the new machine technique at the time of Descartes (see below Section VII), not only the general tendency of all scientists to use "methods" and inventions

---

[77] Ibid., pp. 156, 157, 159.

[78] Rule X, *Philos. Works*, I, 30 (AT X, 403).

[79] *De arte combinatoria*, in Leibnitii Opera, ed. Erdmann, op. cit. I, 28.

[80] *E. Cassirer,* op. cit., p. 46. – On Lully's *Ars Magna*, see especially Louis Couturat, *La logique de Leibniz,* op. cit., pp. 36f.

[81] Discourse, II, *Philos. Works*, I, 91.

or special auxiliary means to gain for intellectual production the same advantages that machines secured in the field of material production;[82] one must also consider Descartes' participation in this general tendency of his time, his activity and experience as a constructor of machines. [41]

On the basis of his long years of study of glass polishing Descartes knew that it was extremely difficult, when cutting and polishing optical glasses by hand, to obtain the exact hyperbolic form indispensable for precision glasses, and that even the highly-skilled *artisan virtuosi* only rarely succeeded in this. As early as the beginning of April 1635 Descartes reported to Huygens about *machines* "for grinding lenses."[83]Descartes points out in a letter to Constantin Huygens (December 8, 1635) the particular difficulties: "which is good for spherical lenses because all the parts of a sphere are equally curved, ... but this is not the case for the hyperbola whose sides are much different from the middle."[84]Huygens reports that on the first attempt of his apprentices to cut a hyperbolic glass, a turner from Amsterdam broke three glasses.[85] The lens that Huygens sent to Descartes was not good, as Descartes reported to Huygens (Dec. 11, 1635). The turner did not stick strictly to the drawing. "This lens could not have the figure of a hyperbola."[86] Descartes gives further specifications for cutting, and on June 17, 1636 he sent Huygens a "chef-d'oeuvre de ma main," the model of a hyperbola.[87] On Juli 11, 1636 Huygens sent Descartes the second glass cut according to his drawing. But already on Juli 13, 1636 Descartes reported to Huygens,[88] that once again the lens was bad. On the one hand, he reports that it seems "on the whole [en gros]" to have been cut right according to the drawing. "But there are infinitely many small irregularities in the form of waves on its surface ..." and Descartes suspected "that the lens wasn't actually done on a lathe because there is nothing regular or circular about them."[89] Therefore, Descartes, convinced that he would not be able acquire precision glasses without machines, wanted to construct a machine for that purpose.[90] After having

[82] Thus, Nicholas of Cusa (mid-fifteenth century) gave his treatise on three kinds of cognition ("tres modi cognoscitivi, scilicet: sensibilis, intellectualis et intelligentialis") the title "On Spectacles" (*De Beryllo*). Just as material eye-glasses make visible what was invisible without them, so Cusanus sought by means of intellectual eye-glasses to make visible things heretofore intellectually invisible. "Beryllus, lapis est lucidus, albus et transparens cui datur forma concave pariter et convexa, peripsum videns attingit prius invisible. Si intellectualibus oculis intellectualis Beryllus ... adaptatur, per eius medium attingitur invisible omnium principum" (*Nicolai De Cusa Cardinalis Opera*, Basle, 1565, vol. 1, *De Beryllo*, Cap. II, 267 and Cap. IV).

[83] (*Correspondence of Descartes and Constantyn Huygens, 1635–1647* ed. Leon Roth, Oxford: Clarendon, 1926, pp. 1 and 3.

[84] Ibid., p. 10.

[85] Ibid., p. 8.

[86] Ibid., p. 12.

[87] Ibid., pp. 13–14, 19.

[88] Ibid., p. 24.

[89] Ibid., p. 24.

[90] Letter to Huygens July 13, 1636, ibid., p. 8.

in his *Geometry* given the *theory* of what have since been called "Cartesian ovals," which are important for optics,[91] he now concentrated on explaining "only those matters which I believe to have the greatest *practical* value taking into considera-
[42] tion the difficulties [of artisans] of cutting."[92] Descartes' preliminary work for the construction of a glass polishing machine goes back as far as 1629.[93] In letters to Jean Ferrier, a maker of scientific instruments, he described the technical details of his construction; he explained that it is adapted not only for cutting glass, but also "for cutting plates of iron or steel." With its help every glass can be exactly cut in *one quarter of an hour*, while previously a very long time was required.[94] The importance Descartes ascribed to his invention is revealed by his hope that with the help of the precision glasses cut by his machine "we might see ... whether there are animals on the moon."[95]

Eight years later, in his *Dioptrique* (1637) Descartes repeated the views he expressed in his correspondence with Ferrier, particularly in Part X entitled "On the manner of cutting glass." Referring to the difficulties involved in obtaining the exact hyperbolic form by hand-cutting, he says: "In polishing glasses by hand ... in the manner that was exclusively used until today, it was impossible to *do anything good except by chance*."[96] And he emphasized that the machine he had invented did not require any *training or skill* in the part of the workers operating it; it was "an invention with the help of which one could draw hyperbolas *in one stroke*, as one
[43] draws circles with a compass."[97]

Descartes' glass-cutting machine – it could also be used for cutting metals or wood – performed quickly and exactly the work that previously could be done by hand only with difficulty and because of the simplicity of its operation, it did not require any effort or *training* on the part of its operators; it worked *automatically* and for that reason *was accessible to all.*

---

[91] Descartes, *Geometry*, op. cit., Book II, pp. 115f.

[92] Ibid., p. 135. The important bracketed words: "of artisans" were omitted in the English translation; but they can be found in the original. Cutting with the help of a machine presents no difficulties.

[93] Descartes, letter to Ferrier, June 18, 1629 (AT I, 13).

[94] Letters to Ferrier, Oct. 8 and Nov. 13, 1629 (AT I, 32 and 53 sqs.).

[95] AT I, 69.

[96] AT VI, 224 ; cf also *Dioptrique*, Disc. VIII, p. 176f.

[97] AT VI, 215. Even before Descartes, attempts had been made to mechanize the cutting of various curve profiles; Leonardo da Vinci had constructed a screw-threading machine that was intended to replace cutting by hand. See Franz Feldhaus, *Leonardo der Techniker und Erfinder* (Jena, 1913) p. 64. Jacques Besson, Leonardo's successor as "royal engineer" in France, invented a machine for cutting spiral coils, such as required in an endless screw, for instance (Jaques Besson, *Théâtre des instruments mathématiques et méchaniques*, Lyon, 1578, Figs. 5, 6, 7, and 8). In the turnery he also constructed, to replace turning by hand, lathes which could mechanically turn with precision every desired irregular shape. De Caus gives the description and picture of a "machine for turning anything at all in oval form." (Salomon de Caus, *Les Raisons des forces mouvantes avec diverses machines*, Frankfurt, 1615, pp. 27–28.

One is surprised at seeing that the properties of the machines enumerated here are the very same that Descartes repeatedly emphasized as being the advantages of his algebraic method. Let us now examine more closely his well-known mathematical "*duels.*"

To grasp the essence of the problem it is necessary to take into account the tendencies of industrial development that began to manifest themselves even then, to wit, the division of functions between theoretical calculation and planning (with the help of mathematics and mechanics) on the one hand, and the work of practical execution, on the other. [44]

This tendency first appeared in architecture. The architects who drew up the plans for the medieval cathedrals had long since been sharply differentiated from the people charged with the execution of the constructions.

In the first half of the sixteenth century similar tendencies can be observed among other industries; for instance, in Venetian shipbuilding. In a field that had hitherto been exclusively dominated by empirics, the traditional shipmasters, the advent of Vittore Fausto at the Arsenal marked the complete separation between the workmen who made the product with their hands and the theoretician and constructor who made the plans[98] with the aid of mathematics and mechanics. For instance, the mechanical problem involved in determining the length of the oars and the location of the oarlocks and the rowers' benches was but an application of the lever principle. Fausto's theoretical training and knowledge of that principle gave him an advantage over the purely empirical craftsmen.[99]

On the other hand, only because the practical work was separated from the intellectual work, could it be reduced to simple, mechanical, rapid, almost automatic functions, whereby the labor processes were extraordinarily accelerated and their efficiency increased.[100] [45]

Descartes refers to these experiences when he discusses the difference between the inventor of algebraic rules and the persons who mechanically applied them.

In his polemics against Roberval and Etienne Pascal, Descartes stressed the fact that in his analyses he had confined himself to setting up fixed rules that could be mechanically applied by others, and therefore this application could made by anyone.[101]

---

[98] Frederic Chapin Lane, *Venetian Ships and Shipbuilders of the Renaissance* (Baltimore, 1934) p. 69.

[99] Ibid., p. 71.

[100] The most famous example of standardization in the 16th century was that of the Arsenal in Venice with its stores of spare-parts which could be fitted to any ship. When the Dutch at the time of Descartes became a great naval power, their shipwrights used similar methods promoting uniformity in the build of merchant ships, and the progress of science helped them by supplying more accurate and handier measuring tools. (See G. N. Clark, *Science and Social Welfare in the Age of Newton*, Oxford, 1937, p. 52.)

[101] "But what confuses them is that I construct them [my rules] as architects construct buildings, only prescribing what must be done and leaving the manual work to the carpenters and masons" (Descartes to Mersenne, March 31, 1638, AT II, 83).

Descartes' adversaries, Fermat, Roberval, etc., although they did not use analysis, were able to solve the problems given them by Descartes. He criticized the backwardness of their methods and opposed to it the *facility*[102] and *rapidity* with which the solution of the problems could be found with the help of his algebraic analysis, as well as the *ease* with which it could be *manipulated* by anyone, including those who were not specially trained.[103] In his dispute with Fermat about tangents (January 1638), Descartes claimed that his method was superior to Fermat's and "proposed [46] a new problem that he undertook to solve very *easily* by his method."[104] Later in 1638, when Fermat took a long time to solve a problem given him by Descartes, the latter gave the same problem to his former servant Gillot (this was particularly irritating to the southern French nobleman, Fermat), in order to show that with the help of his method anyone was able easily to find the proof.[105]

During the dispute with Roberval, when that gentleman submitted to Descartes the problem of the tangents of a certain curve named the "roulette," Descartes quickly replied with a letter seven pages long,[106] in which he gave the required solution "in such a fashion that it will be easy for everyone to judge it," and added: "And what I have put down here at length so as to be understood by those who do not use analysis, *can be found in three strokes of the pen by the calculus*."[107] When during this quarrel Fermat's friends used a rather unimportant problem (the so called "line E-B") for no apparent scientific reason again and again in the course of six months to launch attacks on Descartes, he accused them of using the case as a mere "chicanerie" as a trick "only invented by them to give him (Fermat) time to look for something better to reply to me." And Descartes added with deprecation: "And it is no great wonder that in six months he found a new round-about way to use his rule."[108] A similar incident occurred again when Fermat at the beginning of 1638 gave Descartes the problem of the tangent of a certain curve named the "galand." Roberval later became involved in this dispute, and when the latter delayed the solution, a partisan of Descartes, a "*mathématicien de province*," Florimond Debeaune from Blois sent in his solution found with the help of the Cartesian method (on April 3, 1639). He explained that he needed "*barely one quarter of an hour*" to develop [47] th required equation, and emphasized the *brevity* and *simplicity* of the solution: if the old "géométrie commune" was used, many sheets of paper (*une main de papier*)

[102] Discourse, II, *Philos. Works*, I, 93.

[103] "Because I speak by *a b* ... I often can put *in one line what they fill several pages with*, and for that reason my method is incomparably clearer, easier, and less subject to error than theirs" (AT II, 83).

[104] Descartes to Mersenne, January 1638 (AT I, 490 and Ch. Adam, AT XII, 13ff, 260ff).

[105] Descartes to Mersenne, June 29, 1638, enclosing the "Réponse du sieur Gillot au Théorème..." (AT II, 195, and Adam, AT XII, 264).

[106] Descartes to Mersenne, July 27, 1638, AT II, 257–263.

[107] AT II, 263.

[108] AT II, 273.

were needed; but with "the new method of 'specific' (*spécieux*) analysis, *it was a matter of only a few words*."[109]

We see that the advantages of the Cartesian algebra with regard to the previous procedure are conceived on the model of the machines in their relation to work by hand. By the automatization of the algebraic procedure, that is, by the mechanical applicability of a few fixed rules,[110] the intellectual work was to be reduced to a necessary minimum, and thus the whole procedure simplified and accelerated and made *accessible to all*.[111] It could also be applied to *all* branches of science; thus the universality of the method was secured, and it was not tied to any special field.

*****

This mechanical character of the algebraic method was understood and emphasized by a number of prominent historians of mathematics, such as Boutroux, Maximilien Marie, H.G. Zeuthen, etc., who failed however to study this problem in detail. [48] Marie, for instance, compares the algebraic method to a mechanical mill: the better the milling machine the less intellectual work is required of the miller operating it.[112] Actually every mathematical rule has this mechanical character that spares intellectual work and much calculation. If, for instance, you wish to know the fourth term of a geometric proportion, you can obtain it by multiplying the second by the third and dividing the product by the first. The fourth term of an arithmetic proportion is found by subtracting the first term from the sum of the second and third terms.

Already in the Liber abaci by Leonardo of Pisa (second edition of 1226) we find such artificial mechanical rules (secundum artem) as opposed to the usual longer procedures (secundum vulgarem modum). Section 9 dealing with the exchange of money contains the following problem: If 12 Imperials = 31 Pisans, and 12 Janninans = 23 Pisans, if, further, 12 Tours coins = 13 Janninans, and 12 Barcelonese = 11 Tours coins, how many Barcelonese will be equal to 15 Imperials?

According to the ordinary method, one would have to find that if 12 Imperials = 31 Pisans, 1 Imperial = 31/12 Pisans, hence 15 Imperials = 155/4 Pisans; then an analogous computation would have to be made for the Pisans, Janninans, Tours coins, and Barcelonese. Instead of this long computation, Leonardo gave a simple [49] mechanical rule that spares the student a considerable amount of calculation. First

---

[109] AT XII, 268.

[110] "L'algèbra se présente à nous comme une technique ayant pour objet le calcul ... Grâce à la simplicité et à la fixité de ses procédés, elle prétend, en effet, opérer rapidement, sûrement, mécaniquement, pertinemment." (P. Boutroux, *L'Idéal scientifique*, op. cit., p. 82.)

[111] "L'algèbra, en effect, est essentiellement une Règle (Regula) ... Les règles de l'algèbra visent à devenir méchaniques, c'est-à-dire applicables par tous et toujours, sans intervention de l'intelligence" (Boutroux, ibid., p. 85).

[112] Marie says that "analytical geometry is a mill of which one has only to turn the handle in order to see solutions of problems come out of it," and he adds the following generalization: "It is precisely such mills that science seeks under the name of methods: the more solutions they can grind the less work they leave to the miller, the better they are." (Maximilien Marie, *Histoire des sciences mathématiques et physiques*, vol. III, "De Viète à Descartes," Paris, 1884, p. 5.)

the currency denominations are written down in the same order (from right to left) as they are given in the problem; the series is repeated underneath. Then the given numerical values are written alternately in the upper and lower rows, and multiplication lines in the form of a zigzag are drawn beginning with the 15 Imperials below:

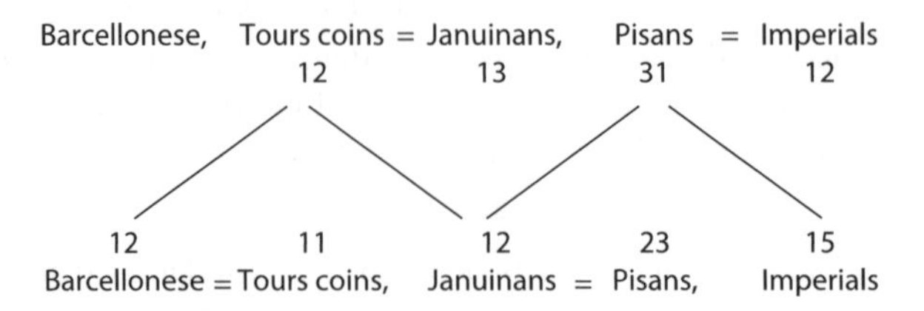

The product of the numbers thus connected ($15 \times 31 \times 12 \times 12 \times 12$) divided by the product of the remaining numbers ($12 \times 23 \times 13 \times 11$) gives the solution: 15 Imperials $= 20\frac{1180}{3289}$ Barcelonese.

Cantor says the following about this rule: "Through this *machine-like procedure*, through the exact prescription as to where the numbers entering the computation should be written, and later, as to how they should be treated, the reader is spared as much of his own intellectual work as possible."[113]

Thus the solution of such problems is made accessible to all, even those not particularly gifted – and that is the very purpose of the rule. [50]

But this machine-like procedure, Cantor goes on to say, was limited by the conditions prevailing in the thirteenth and the two following centuries, when literacy was extremely rare and when mathematical operations were confined to small circles of the clergy and the merchants; it was "*the artisan-like* art of computation" taught and spread by the friars.[114] The situation changed in the sixteenth century and at the beginning of the seventeenth, when people began to make computations *with large* numbers. "The reason for this," says Cantor, "was the fact that the practice of computing penetrated more and more into all classes of people," as the capitalist money economy progressed; further, there was "... the spread of *trigonometric* considerations that made it necessary to compute with trigonometric functions; these were known in the form of *large numbers*, because only a very large radius could provide a sufficient approximation in the end results of the computations." Thus, "it was in accordance with the trend of the times to *facilitate computations with large numbers*."[115]

[51] Special writings aspired to this end, partly through the *invention of new instrumental auxiliary means* and partly through the introduction of *new devices*, the most

---

[113] Moritz Cantor, *Vorlesungen über die Geschichte der Mathematik* (Leipzig, 1892) II, 17–18.

[114] Ibid., II, 95.

[115] Ibid., II, 658. "It was customary to regard sines as lines drawn in a circle of suitable radius, and the necessary accuracy was obtained by making that radius very large, say 10,000,000." (Lord Moulton, "The Invention of Logarithms, Its Genesis and Growth," in: *Napier Tercentenary Memorial Volume*, ed. C. G. Knott, Royal Society of Edinburgh, London, 1915, 8).

important of which was the invention of logarithms by John Napier in 1614. The purpose of these latter was to *shorten and accelerate computations* with large numbers by the application of tables or mechanical rules, as Napier indicates in the very title of his work.[116]

Calculation was further accelerated by the *invention of circular slide rule* by Richard Delamain (1630). Napier – stimulated by the mechanical ideas of his time – arrived at his invention of logarithms by imagining the mutual motion (*fluxus*) of the related arithmetic and geometric progressions.[117] What with Napier was only an intellectual construction, Delamain tried to put into practice with the help of a mechanical device. In the sliding rule the different progressions are marked on lines that actually move in relation to each other. Thus the desired computation – division, multiplication, extraction of square and cubic roots – can be done without the aid of a pen or compass, by instantaneous ocular inspection, that is to say, by simply reading the result on the sliding rule, as Delamain declares in his pamphlet *Grammelogia*, published in London (1630), emphasizing the *ease* with which the most complicated operations can be performed with the help of his device.[118] [51a]

Descartes' reform of Algebra and its fundamental tendency toward automatization, simplification and acceleration of mathematical procedures, were thus in complete accord with the aspirations of his time. However, while the above-mentioned innovators (such as Napier and Delamain) strove to simplify and accelerate mathematical operations by *external mechanical means*. Descartes tried to achieve the same aim by simplifying and automatizing the relevant intellectual process itself in [52] order thus to make the algebraic operations accessible to all intelligent people, to the common man.

A further advantage of algebraic procedure is that it is not tied to a special restricted field but is *applicable generally* to the most varied fields and problems. In his dispute with Fermat about the tangents, Descartes reproached the latter because "his alleged rule is not universal as it seems to him ... For it cannot be applied to

---

116 John Napier, *Logarithmorum canonis descriptio seu Arithmeticarum supputationum mirabilis abbreviatio* (London, 1620). The *Encyclopedia Britanica* says the following about the "Logarithms": "By shortening processes of computation, logarithms have doubled the working speed of astronomers and engineers." (See also Cantor, op. cit., II, 656.)

117 Cantor, op. cit., II, 666, 672.

118 *Grammelogia, Mirifica Logarithmorum Projectio Circularis ... sheving ... how to resolve and work all ordinary operations of Arithmeticke; And those that are most difficult with greatest facilitie, the extraction of Rootes ... etc. And only by an ocular inspection and a Circular motion, Invented and first published by R. Delamain* [n.d.]. (See Florian Cajori, "On the History of Gunther's scale and the slide rule during the seventeenth century", in *University of California Publications in Mathematics*, vol. I (1920) 194, 196.) Other devices were the many trigonometric tables published at the beginning of the seventeenth century, such as M. Bernegger's *Tafeln der Sinus, Tangenten und Sektanten* (Strasbourg, 1612), the tables of Albert Girard (The Hague, 1636) and of Francis van Schooten, Senior (Amsterdam, 1627) (see Cantor, op. cit.). Here also belong the *Tabulae arithmeticae prosthaphaireseos universales*, by [Hans] Georg Herwarth (1610) who tried to facilitate the use of tables (Cantor, op. cit., II, 659). Here the invention of the calculating machine by Pascal (1643) should also be mentioned.

this example nor to other more difficult examples; while my rule *extends generally to all those problems that can fall within the scrutiny of geometry*,"[119] that is to say, to the totality of natural phenomena. More, from this universality of the algebraic method it follows that it extends not only to the totality of the external world of nature, but beyond, to spiritual structures; as a result, the attitude toward reality of the post-Cartesian thinkers is completely different from that of their predecessors. In absolute contrast to the science transmitted from antiquity, which conceived its ideal as the *contemplation* of ideal objects, there arises with Descartes' algebraic method a fundamentally new type of science, a science which is a construction of the mind,[120] which expresses the *active* relation of thinking to reality, and which, [53] with the help of the algebraic method, by progressing from simple to ever more complicated structures, *creates ever new objects, an unlimited world of new combinations* and deduces their properties, like a machine-builder who from the same simple components, such as the inclined plane, the pulley, lever, screw, etc., constructs new machines for ever new uses, and thus constantly extends the world of his practical activity. After the equations of the second degree were solved, more complicated problems were attacked, equations of the third degree. When Niccolò Tartaglia's solution of the third degree (1545) led to the discovery by Ferrari, a pupil of Cardano, of a method of solving equations of the fourth degree, it was hoped that also equations of the fifth and higher degrees could be solved; and the hope seemed justified that it would be possible to move on that path toward the solution of ever [54] more complex problems.[121]

Descartes' pan-dimensionism found its logical conclusion in his pan-algebraicism.[122] Just as in our day, since the invention of the camera, every person of average intelligence, without detailed knowledge of the apparatus or the laws of optics, can take pictures by the purely mechanical application of a few directions, because the inventor of the camera once and for all and for everyone made an intellectual effort that need not be repeated, equations of the second, third and fourth degrees can be solved by the mere application of rules. Once the solution of these equations was found, its further application according to Descartes' way

---

[119] Descartes to Mersenne, January 1638, AT I, 491.

[120] P. Boutroux, op. cit., p. 109.

[121] "For, after having said, for instance, that *straight* lines are the figures defined by polynomial equations of the first degree in x and y (of the form $ax + by + c = 0$); that *conic sections* are the curves defined by polynomial equations of the second degree in x and y (of the form $ax^2 + bxy + cy^2 + dx + ey + f = 0$), nothing can prevent us from adding: 'I name curves of the third order those defined by polynomial equations of the third degree in x and y; curves of the fourth order those defined by polynomial equations of the fourth degree in x and y, etc. . . . and from the equations of these curves *I will deduce their* properties, just as I have done for the conic sections.' *Thus, by the simple play of the algebraic mechanism we call forth a limitless geometric world*." (P. Boutroux, *L'Ideal scientifique*, op. cit., p. 108).

[122] Only early in the nineteenth century was it shown by N.H. Abel that equations of higher degrees are generally insoluble by algebraic methods (see A. Wolf, *A History of Science, Technology and Philosophy in the 16th and 17th centuries*, London, 1935, p. 191).

of thinking "was merely a matter of routine (*simple affaire de métier*) which did not demand any inventive effort on our part."[123] Thanks to this algebraic method, mathematics becomes "a *mechanical* science, which from now on anyone can practice efficiently."[124] Mathematics, which formerly as a science, could show only a limited number of results, became with Descartes a method, leading constantly to an infinite number of ever new results. Zeuthen properly compares the revolution initiated by the Cartesian algebra with the impact of modern large industries upon [55] the old world of handicrafts. "The objects made by hand are produced by efficient individuals. Each piece is something perfect in itself." But cooperation was of no importance; to wit: that "everyone should be able each to perform a part of the same work."[125] Only large industries could employ less skilled workers, through the *cooperation* of many[126] for the common end. By the introduction of a general method with the help of which one could operate *mechanically*, the whole function of mathematical work entered upon a new stage ... From now on it could be applied directly by many *who had no particular talent for* mathematics." In addition, "the greater number of applications caused the emergence of new important problems," and "the discovered results appeared in the shape of formulas that constitute starting points for new mechanical operations."[127] Or – in the words of Boutroux – "Mathematical research had literally become a mechanical combination, a work of manufacture."[128] [56]

VI

Is our interpretation of the genesis and significance of the Cartesian algebra justified? Do not the histories of mathematics generally present *internal, logical* reasons for the advantages of the synthesis of algebra and geometry quite apart from technological and sociological considerations? Before we deal with this point of view in detail, it will be of interest here to point out that similar objections were raised to advancing sociological considerations when explaining the emergence of the Copernican theory: it was maintained that it was the result of an internal, logical

---

[123] P. Boutroux, *L'Idéal scientifique,* op. cit., p. 109.

[124] Ibid., p. 102.

[125] H. G. Zeuthen, *Geschichte der Mathematik im XVI und XVII Jahrhundert* (Leipzig, 1903) p. 216.

[126] Descartes emphasized this cooperative Character of his algebra, which rested on the assistance of many, even in his earliest letter to Beeckman, with the notice of his discovery (March, 26, 1619): "Et certe, ut tibi nude aperiam, quid moliar, non Lulli *Artem brevem*, sed scientiam penitus novam tradere cupio, qua generaliter solvi possint quaestiones omnes ... *Infinitum quidem* opus est, nec unius." AT X, 156–157.

As Gustave Cohen shows in a different context, Descartes emphasizes the "idée de la *collaboration* de tous à l'*oeuvre collective* de la science" (*Écrivains français*, op. cit., p. 470.)

[127] Zeuthen, op. cit., p. 219.

[128] P. Boutroux, *La signification historique de la géométrie de Descartes*, op. cit., p. 820.

development of astronomy brought about by the increased number of observations, and that sociological considerations had nothing to do with it. We think, however, that these objections have been invalid. Burtt showed convincingly that sociological considerations played a decisive part in shaping the Copernican theory.[129] The "internal", "logical" reasons for the correctness of the Copernican theory appear as such only *to us*, that is to the retrospective historical consideration, they did not appear as such to Copernicus' contemporaries. These latter took no note of the internal logical reasons or were even hostile toward them, because their general intellectual orientation was quite different from our own. To the Middle Ages, *man* [57] was in every sense center of the universe, the controlling factor. It was believed that the whole world of nature was teleologically subordinated to him; furthermore, that the *terrestrial* habitat of man was in the center of the astronomical realm.

With the *commercial revolution* of the fifteenth and sixteenth centuries, with the explorers' long voyages and exciting discoveries of unknown continents and civilizations, the business leaders of Europe turned their attention from petty local fairs to the great untapped centers of trade in Asia and America. The realm of man's previous knowledge seemed suddenly small and meager; people's thoughts became *accustomed to a widening horizon. A shift in the point of reference from Europe to other continents* with ancient cultures took place.

What was true of the newly discovered civilizations was also true of the newly discovered religions. If previously Rome had been considered the *religious center* of the world, now there appeared a number of distinct centers of religious life beside Rome. "In this ferment of strange and radical ideas ... it was not so difficult for Copernicus to consider seriously for himself and suggest persuasively to others that a *still greater shift ... of the centre of reference* in astronomy ... must now be made ... from the earth to the sun."[130]

Only under the impact of this general socially conditioned change in thinking were the receptivity of scientists like Copernicus and their interest in the "internal", [58] "logical" reasons in astronomy – that had previously passed unnoticed – aroused; only now did they find the courage to oppose the Biblical tradition, express the new conception and advance arguments in favor of it, as for instance, the explanation of certain peculiarities in the apparent, observable motion of the planets. This example also affords us the possibility of assessing the respective share and contribution of various theological, sociological and logical elements in the formation of the Copernican theory, without confusing them: the first group, the theological elements, was ignored; the second group, the sociological elements, gave the impulse for the shift in the point of reference, and this shift in turn prepared the ground for the acceptance of the internal, logical reasons for the new theory. This example will also facilitate the understanding of our own problem.

---

[129] Edwin H.A. Burtt, *The Metaphysical Foundation of Modern Physical Science* (New York, 1932) pp. 38f.

[130] E. A. Burtt, op. cit., pp. 40f.

The importance of the Cartesian geometry must be evaluated quite differently according to whether we regard it as a *technical reform of mathematics* or whether we consider its *methodological-philosophical aspect*, that is to say, "the extension of the mathematical method to the universality of cosmological problems."[131]

From a purely mathematical point of view, Oresme and Viète laid the groundwork for analytical geometry by their works; "they carried the material of the new science to such a degree of perfection that its advent assumed the character of *logical necessity*."[132] And Fermat's work, *Isagoge ad locos planos et solidos,* published shortly before Descartes' *Geometry,* contained the principle of analytical geometry "stated in the neatest form that could be desired."[133] [59]

Nevertheless, Cartesian geometry represents a *turning point* in the history of modern science; one speaks of the "Cartesian revolution"[134] but one cannot speak in the same way of Fermat's *Isagoge*; the works of Fermat and Descartes, "so *close* in their technical contents, are as *different* as possible if one considers the general orientation of the minds of the inventors."[135]

In order clearly to mark this difference, Brunschvicg asks: What would have been the consequences for science if Descartes had *not* written his *Geometry*? And he thinks that in all probability, "the evolution of *mathematics* would not have been profoundly modified by that fact."[136] But the situation is quite different in the field of philosophy and the reform of natural science by mathematics. Without the Cartesian geometry the development of philosophy in the seventeenth century would have taken a quite different course, in particular it would have lacked "the systematic rigor we find in it." Fermat's *Isagoge* "is the work of a technician."[137]

In contrast to it, Descartes' *Geometry* "is the work of a methodical mind which started from a *universal conception of science* and bequeathed an original notion of scientific truth to its successors."[138] [60]

This new conception of science turned critically against the opinion that scientific truth consists in the knowledge of the substantial difference between phenomena in any isolated field, a knowledge of which the so-called "experts" boast with particular pride. According to Descartes, true science consists in the discovery of the universal relation and connection between the particular phenomena and the universality of things; and he believed that with the help of his Algebra, that is to say, a method accessible to *all* and applicable to *all* scientific fields thanks to its fixed, mechanical,

---

131 L. Brunschvicg, *Les étapes*, op. cit., p. 107.
132 Ibid., p. 105.
133 Ibid., p. 100.
134 Ibid., p. 114.
135 Ibid., p. 99.
136 Ibid., p. 101.
137 Ibid., p. 101.
138 Ibid., p. 101.

and easily applicable rules, the gulf between the specialization of the "experts" and the universal "relatedness" could be bridged.

Not a trace of such a conception of algebra can be found in any of Descartes' forerunners, from Oresme to Viète and Fermat. *Conceived not as a technical instrument of mathematics, but as a philosophical method* specially designed to aid in achieving the universal science, the Algebra is Descartes' exclusive, original creation.[139] Once this difference is clearly realized, it must be inferred that for the understanding of the character of the Cartesian algebra, the "internal," "logical" reasons for its [61] formation, reasons deriving from the internal development of *mathematics*, are of little use, and must be discarded, for the Cartesian Algebra *as a philosophic method is not connected with the previous development of mathematics* and hence cannot be understood on the basis of this development. Other explanatory factors must be sought. But no attempt to find them has as yet been made.

Previously, however, we have referred to two characteristic aspects of this new "*conception universelle de la science*": first that it shows many clearly marked features analogous to those noted in the functioning of machines, alien to the older conception of science, which obviously arose under the impact of *new influences*; secondly, we have seen that a number of historians of mathematics, such as P. Boutroux, M. Marie, H. G. Zeuthen and others have pointed out the *mechanical* character of the algebraic procedures. They stress that the operations of the Cartesian algebra reveal an automatism in the application of fixed rules; that because of the simplification and rapidity of the procedure, they do not require a great effort of intelligence; that they are thus accessible to everyone; that they make mathematical mass production possible – the attacking of great problems beyond the capacity of individuals, problems of which the solution is possible only through the cooperation of many minds. In brief, the above mentioned historians state that the Cartesian algebra objectively shows all the features characteristic [62] of the functioning of machines. It is itself an intellectual machine, an intellectual mill, on which one has only to "turn the handle in order to see the solutions of problems come out of it." Is this resemblance to machines in the functioning of the Cartesian algebra only a coincidence, a mere accident? We prefer another explanation.

In our view, it is perfectly natural that in an epoch when machines were winning their first great victories in England, France, Holland, Italy and Germany, and visibly *demonstrating to everyone the advantages of the automatism of machine labor* over handicraft labor, the enthusiastic admiration for the former, which is expressed in all contemporary accounts, should also influence the thinking of the leading minds – a Bacon and a Descartes. Modern times were heralded by L. B. Alberti's passionate paean to technique.[140] There we read that technique permits us "to cut rocks, pierce mountains, fill valleys, resist the overflow of the seas and rivers, clean out swamps

[139] Descartes could rightly say: "Je commence en cela par où Viète avait fini" (Letter to Mersenne, March 31, 1638, AT II, 82).

[140] Alberti's book "*On Architecture*" was written about 1450 and published only posthumously in Florence in 1485. Here it is quoted from the French edition (Paris, 1553, Preface).

or marshes, build ships." Giovanni da Fontana (1395–1455) of Venice was proud of the mechanical progress of his time and spoke of almost the entire habitable world as full of magnificent fabrics, ingenious machines and organic instruments for carrying on the arts or cooperative sciences.[141] Girolamo Cardano not only spoke with admiration of machines but also, as early as 1550, characterized with full clarity the most important advantages of machine labor in his book *De Subtilitate*.[142] This general admiration for machine technique expressed itself in the early seventeenth century in Bacon's enthusiastic glorification of technology, which he considered to be the determining factor of historical progress and in social transformation. "Founders of cities and empires, legislators, etc.," said Bacon, "extend civil benefits only to particular places." But not so with the authors of mechanical inventions: "For the benefits of discoveries may extend to the whole race of man." The great difference between "the most civilized provinces of Europe and the wildest and most barbarous districts of New India ... *come not from the soil, not from climate, not from race, but from the arts*." The inventions of the last two centuries, such as those of the printing machine, et cetera, "have changed the whole face and state of things throughout the world ... insomuch that no empire, no sect, no star seems to have exerted greater power and influence in human affairs *than those mechanical discoveries*."[143] [63] [64]

Whereas early experience with the Italian republics and in Holland brought to awareness that merchant capital led to wars in the struggle for markets, Bacon emphasized on the contrary that artisanal capital was opposed to military operations.[144] [65]

VII

The question may be raised: Did enough machinery exist in Descartes' time to justify the emphasis I have placed upon it as an influence on Descartes' concepts in mechanics and mathematics?

It is not easy to answer the question because the existing literature on the history of technology – except for a few monographs on selected periods – is little more than an incomplete enumeration of inventions. Even the best books, like Usher's instructive *History of Mechanical Inventions* (1929), treats technology isolated from

---

141 See Lynn Thorndike, op. cit., IV, 167.

142 These are: (1) Saving human labor, (2) usability of unskilled, hence cheap, workers, (3) decrease in the amount of waste, (4) general advantages of sanitary working conditions.

143 *Novum Organum*, Book I, 129 (*Works* I, 221). The discovery of eyeglasses, for instance, brought about the crescendo of intellectual life and an improved standard of education during the fourteenth and fifteenth centuries. Men were able to read more and to read in their maturer years.

144 Bacon: "Certissimum est artes mechanicas sedentarias ... atque manufacturas delicates (quae digitum potius, quam brachium requirunt) sua natura militaribus animis esse contrarias" (*De dignitate et augmentis scientiarum*, Lib. VIII. cap. III. 5, *Works* I, 798).

the social and economic development of the society as a whole. Henri Pirenne's *Economic and Social History of Medieval Europe* (London, 1936) does not even mention in the general index the words: inventions, technique, machines, mechanism, automats, mechanical clocks, water mills, mines, et cetera. A social history of technology showing its evolution, historical conditioning and social effects,[145] still has to be written. [66]

We shall confine ourselves here to a short outline of technological development, stressing particularly the effects of the machine on human thought as the necessary historical background for a correct understanding of Descartes' world.

For the majority of readers the term "machine technique" is associated with the "industrial revolution" conceived as a *sudden* phenomenon which took place at the end of the eighteenth century in England. This view is not in accord with the facts. The revival of technique in Western Europe and the introduction of new machines was a *long process* reaching back to the beginning of the thirteenth century. "Nowhere", says Professor Sarton, "is the growing maturity of the West at the beginning of the thirteenth century more certain than in the field of pure mechanics ... In the course of the century we witness a *true mechanical rebirth*."[146] [67]

The historical causes of this technical revival are not explained in the existing literature. Here we can only briefly indicate them. With a few exceptions antiquity

[145] Lynn White Jr.'s brilliant and suggestive article, "Technology and Invention in the Middle Ages" in *Speculum 15* (1940) is of little help for our purposes. White deals with the earlier period, generally closing with the thirteenth century, whereas, we are interested in the later period from the thirteenth to seventeenth centuries. Analyses are made by White of many and important inventions of that earlier period: the invention of soap, of lateen-sail, compass and gunpowder, of the horseshoe and horsecollar. He analyzes the replacement of the two-field system in agriculture by the three-field system, etc. The *machines*, however, with the exception of a short remark regarding windmills and water-driven mills, are not mentioned.

[146] George Sarton, *Introduction to the History of Science* (Baltimore, 1931) II, 21. Marc Bloch regards A. Vierendeel's assertion (*Esquisse d'une histoire de la technique,* Louvain, 1921, I, 44) "that technique during the Middle Ages was – with the exception of the invention of gunpowder – ... mostly stationary," as "scandalous" already at the time of its publication. "Now," Mr. Bloch says, "we know, beyond a doubt, that at the time the European societies began their conquest of the great oceanic routes they had at their disposal tool equipment immensely superior to that of the Roman Empire at the time of its collapse. But, as yet we have very little information as to just *when* and *how* those various improvements were realized." Marc Bloch, "Les inventions médiévales," in *Annales d'histoire économique et sociale 7* (1935) 634. It is therefore obvious that we can speak only of the acceleration in the introduction of new inventions since the close of the twelfth and beginning of the thirteenth centuries. The new element since this time, however, is the appearance of the machine – the replacement of human by non-human mechanical energy. Concerning the problem of the earlier beginnings of mechanical development in the Moslem world, see Aldo Mieli, *La science arabe et son rôle dans l'évolution scientifique mondiale* (Leiden, 1939). Mohammed ben Musa ben Shaker (825–875) wrote a "Treatise on Machines." (Maximilien Marie, *Histoire des sciences mathématiques et physiques* (Paris, 1883) II, 111). Abu al-Fath Abd al-Rahmân al-Mansur around 1118 wrote various works on mechanics and hydraulics. We find there a theory of the lever and of the balance. Ismail Ibn al-Razzâz Badî al-Zaman, active around 1205, wrote about automata and practical problems of hydraulics (Mieli, op. cit., pp. 154–155).

did not bequeath to us any labor-saving machines.[147] As Ricardo classically formulated it, machines are in competition with labor; where the cost of labor is high, it is replaced by a cheaper machine. In antiquity labor was cheap because of the prevalence of slavery, and it would have been absurd to replace cheap slave labor by expensive machines. The price of a slave was amortized within three or four years at the most; the labor supplied by the slave beyond that period of time cost practically nothing (if we disregard the small cost of sustenance). Thus, slavery can be regarded *as an economic perpetuum mobile*, as a natural machine that continuously supplies energy without costing anything. Hence there was no social need for artificial labor-saving machines. [68]

With the emergence of cities and an urban middle-class in Western Europe in the twelfth and thirteenth centuries, the economic situation changed fundamentally. "City air makes men free." This meant that *every piece of labor performed by an inhabitant of a city had to be paid for*. Labor thus became an important expense item. Only now was it worthwhile to replace expensive labor by cheaper machines.[148] [69]

[147] The ancients had produced many wonderful automatic devices; they were, however, mostly toys used for thaumaturgical purposes, and accessories of priestcraft to mystify the multitude. The water-driven mill is of course of ancient origin; it was invented in the first century, B.C.; but as Marc Bloch points out, it is nevertheless medieval by the very period of its real expansion. Marc Bloch, "Avénement et conquêtes du Moulin à eau," *Annales d'histoire économique et sociale 7* (1935) p. 545; and Ferdinand Lot, *Histoire du moyen age. Tome I. Les destinées de l'empire en occident de 395 à 888*, Paris: Presses Univ. de France, 1928, p. 351 (Collection "Histoires Générale" ed. Gustave Glotz).

[148] Garrett's assertion that machines were discovered by the European mind in a spirit of scientific curiosity, with no clear economic purpose is a myth (Garet Garrett, *Ouroboros or The Mechanical Extension of Mankind*, New York, 1926, p. 6). Numerous documents prove that since the thirteenth century attempts have been made to reduce the high cost of labor by the introduction of machines. In this connection the records of the Cathedral at St. Paul are of special interest. There are references in the account rolls which indicate the presence of a clock keeper as early as 1286. In 1298 stated sums were spent for making "jacks" to strike the hours (see A. P. Usher, op. cit., p. 154). "Jacks" here obviously signifies the same thing as the word "jaquemart" used later in France to denote automatically moving figures striking the hours. Concerning "jaquemarts" see Mathieu Planchon, *L'Horloge, son histoire* (Paris, 1898) Chapter 3. Likewise we know, for instance, that the clock tower, "La Torre della Mangia," in Siena (built around 1348) originally had a watchman who lived in the tower and whose duty it was to strike the hours with his hands on the tower bell. Not until 1360 was there installed in the tower a clock with a mechanism automatically striking the bell, thus saving the wages of a watchman (see Alfred Ungerer, *Les Horloges astronomiques et monumentales*, Strasbourg, 1931, p. 375). A similar evolution took place in Montpellier: the steeple-keeper of Notre Dame, who was supposed to strike the hours (*pulsator horarum noctibus et diebus*) was replaced in 1410 by a clock with jaquemart, an automatic instrument for striking the time, in order to save paying his high wages (see J. Renouvier et Ad. Ricard, *Des mâitres de pierre et des autres artistes gothiques de Montpellier*, Montpellier, 1884, p. 97). As it required five or six persons to ring the heavy bell in Montpellier (weight, with two clappers, 6800 pounds), the carpenter Pierre d'Ayric agreed by contract with the prior of Notre Dame de Tables in the year 1363 to construct for a price of sixty gold florins an engine that would need only one, or at the most two persons (see Renouvier and Ricard op. cit., p. 59, Latin contract, ibid., p. 164).

The same reasons also explain why at the very time that the ancient perpetuum mobile disappeared, the longing for an *artificial perpetuum mobile* arose and attempts to construct one were made.[149]

It is clear that with the advent of the new machine age great steps forward can be noted for the first time since antiquity in the field of *mechanical theory*, the intellectual reflection of the machine practice. We cannot elaborate on this issue but shall confine ourselves to observing that antiquity cultivated only statics as a theory of the balance and equilibrium and never realized that the task of the machine was to supply mechanical labor in order to save human labor; while now, in the latter part [70] or the thirteenth century, Jordanus Nemorarius and his school formulated some of the fundamental ideas of dynamics, the first recognition of the principle of *virtual displacements* which was unknown in antiquity, the demonstration of the *moment of force* and the solution of the problem of the *inclined plane*, the elementary basic mechanical form of every machine.[150]

The first sporadic constructions of machines in Western Europe go back as far as the middle of the twelfth century.[151] Then they multiplied without however becoming so numerous that they could be said to constitute an industrial revolution. There were a considerable number of machines *using water as their motive power*. In Villard de Honnecourt's *Sketchbook* there is a drawing of a *sawmill* driven by falling water; the saw moves by an ingenious arrangement vertically up and back, while

[149] The *Sketchbook* of the French architect Villard de Honnecourt (ca. 1235) contains the oldest known drawing of a *perpetuum mobile* (*Facsimile of the Sketchbook of Wilars de Honecort [Villard de Honnecourt], an Architect of the Thirteenth Century*, Trans. by Robert Willis, London, 1859, p. 35).

[150] P. Duhem, *Les Origines de la Statique* (Paris, 1905) I, 98–123, 147; E. Jouget, *Lectures de mécanique* (Paris, 1908) pp. 25–28; G. Sarton, op. cit., II, 22 and A. P. Usher, op. cit., 53–55. The beginning of a new orientation is noticeable already in the middle of the twelfth century in the striving for a new classification of the sciences. The growing importance of machines can be seen, e.g., in Dominicus Gundisalvus, who undertook a new division of the sciences in physics, mathematics und metaphysics treating machines (*De ingeniis*) as a subdivision of geometry. See his major work, *De divisione philosophiae*, written about 1150 (ed. L. *Bauer, Beiträge zur Geschichte der Philosophie des Mittelalters*, 1903, vol. IV. no. 2–3, pp. 122–124). There he distinguishes two methods, one a static discipline "de ponderibus ... secundum quod mensurantur vel secundum quod mensuratur cum eis et eis"; the other a dynamic discipline, which "considerat secundum quod moventur vel secundum ea, cum quibus moventur; et tunc est inquisitio de principiis instrumentorum, quibus elevantur graviora et super que permutantur de loco ad locum" (*De divisione philos.* pp. 121–122). The understanding begins to spread that the task of machines lies in practical use, that is, in saving labor and costs: "Sciencie ergo ingeniorum docent modos excogitandi et adinveniendi, qualiter secundum numerum corpora naturalia coaptentur per artificium aliquod ad hoc, ut usus, quem querimus proveniat ex eis ... " (ibid., p. 122).

[151] The study of this matter is made more difficult because many reports on the first application of machines are forgeries of later origin. Thus, the much-quoted date of 1105 as that of the erection of a windmill by the Benedictines is just as false as a Hamburg document dating from 1152 and containing a privilege for a cloth-cutter (see Franz M. Feldhaus, *Die Technik der Antike und des Mittelalters*, Potsdam, 1931, p. 261).

the horizontally placed log is slowly driven under the saw; the speed of this forward movement of the log has to be adjusted to the speed of the saw.

Wool-weaving industry supplies us with another example of the application of water power. This was chiefly practiced by women as a domestic industry. Fulling by foot could not be done by women, however, because it was too strenuous – it [71] was done by a special class of workers, the *fullers*. During the thirteenth century the *fulling process was mechanized*, the action of man's two feet was replaced by that of two wooden, water-driven hammers, alternately raised and dropped on the cloth. This meant a great saving of human labor because a great number of hammers could be set to work simultaneously with but one man standing by to watch the cloth.[152]

But it is exaggerated in this connection to speak of an "industrial revolution" in the thirteenth century, as Carus-Wilson does; the technical innovation of which we speak affected only one *small part* of the processing of textiles, the fulling process, and produced relatively slight social effects.[153] The same is true of the fourteenth century although at that time water power was applied to spinning[154] and in the following century, the mechanization of the textile industry made further [72] progress.[155]

Another invention that was completed by the middle of the thirteenth century is the *mechanical clock*;[156] of its various applications, the most important, since the

[152] E. M. Carus-Wilson, "An Industrial Revolution of the 13th century," *The Economic History Review 11* (1940) 43–44; and Aloys Schulte, *Geschichte des mittelalterlichen Handels und Verkehrs zwischen Deutschland und Italien*, Leipzig, 1900 I, 118.

[153] A fulling machine modernized in the sixteenth century, "machina per follar panni di lana," is reproduced in Zonca's *Novo Teatro di Machine*, 1607, pp. 42–43.

[154] "As early as 1341 there were in Bologna large spinning-mills driven by water, and they produced an output equal to that of four thousand spinners" (G. Libri, *Histoire des sciences mathématiques*, [Paris, 1841], II, 233). This Bolognese silk-spinning jenny, composed of several thousand component parts enjoyed great fame for centuries; it is also reproduced in Zonca's *Novo Teatro* of 1607. See *Filatoio da acqua*, pp. 68–69.

[155] Leonardo da Vinci (ca 1500) made a drawing of a complete spinning-machine (*Codice Atlantico*, folio 393 v) with an automatic thread distributor that uniformly distributed the threads over the spools – a device that was not reinvented until 1794 in England. Another model of a mechanically driven spinning mill ("il molinello ò filatoio") is given by Branca in his book on machines published in 1629, Fig. 20. Leonardo da Vinci also made dozens of drawings of various machines for the manufacture of cloth, weaving-looms, shuttles, et cetera. Particularly interesting is a cloth pressing and shearing machine (*Codice Atlantico*, folio 397) intended to replace shearing by hand. The cloth is driven under the shearing table over cylinders rotated by a shaft and for that reason is tightly stretched; as it comes on the table it passes through the scissors that rapidly open and close mechanically. These inventions of Leonardo, however, remained mostly on paper and had no social influence. Besson (op. cit., Fig. 59) gives the reproduction of another model of a "pressoir à presser drap."

[156] See Lynn Thorndike, "Invention of the Mechanical Clock about 1271 A.D." *Speculum, A Journal of Medieval Studies 16* (1941) 242. It is obvious that mankind received enormous new material for observation and for reflection upon the action of forces from the above mentioned revolutions of technique. Here, in the machinery, in the turning of the waterwheels of a mill or an iron works, in the raising of a stamping hammer in an iron foundery, in the motion of the wheels of a clock or in the observation of the trajectories of cannon projectiles, we have the simplest mechanical activities, those simple quantitative relations between the homogenous labor power of the water

fourteenth century, has been in the form of a planetarium.[157] Another interesting application is the use of the clock mechanism in flour mills.[158]

[73]

Although the tower clocks in the cities played a great part *in the regularization* of the urban life and work, they did not lead to any industrial revolution either – if only because the number of clockmakers was very small.

Yet these inventions while they did not bring about an industrial revolution left important traces on the thinking of entire generations; it is possible to speak here of an "intellectual revolution." While in the early Middle Ages, during the period of miracles and mysteries, all exact science was regarded as sorcery,[159] in the thirteenth century the first automata penetrated into an irrational world. Today it is difficult to visualize the intellectual revolutions connected with the discovery and improvement of the clock machines. In the *Paradiso*, the part of the *Divina Comedia* written between 1316 and 1321, Dante glorifies the hourstriking clocks of the Italian cities.[160] The great French historian of the fourteenth century, Froissart, wrote in 1368 a poem entitled "L'Horloge Amoureuse" in honor of the clock.[161] Philip de Maisières in his "Songe du Vieux Pélerin" (1389) wrote a dithyrambic praise of Jacopo de Dondi's mechanical planetarium built at Padua in 1344.[162] What fascinated contemporaries in the mechanical clock was not so much the fact that it indicated the time as the *automatism* of the clock-work,[163]

[74]

---

machines, clocks, cannons and their performances, that is, those relationships on which modern mechanics orients its basic concepts.

157 See Lynn Thorndike, *A History of Magic and Experimental Science*, vol. III (1934) Chapter XXIV: "John de Dondis and Scientific Measurement . . .," pp. 386–397.

158 In a Milan document of 1341 we read: "Adinvenerunt facere molendina quae non acqua aut vento circumferentur, sed per pondera contra pondera sicut solet fieri in Horologis." (Libri, op. cit. II, 232).

159 As late as in the second half of the fourteenth century, Langland says in his "Vision of Piers the Ploughman" through the mouth of Dame Study:

> But astronomy is a hard thing and evil *to know*; Geometry and geomancy are guileful of speech; . . . For *sorcery* is the sovereign book of that science. There are mechanical devices of many men's wits. . . . All these sciences I myself in sooth Have found among the first to deceive folk." (Lynn Thorndike, *A History of Magic*, vol. III, 1934, p. 422.)

160 *Del Paradiso*, Canto X, 139–144.

161 Pierre Dubois, *Horlogerie: description et iconographie des instruments horaires du XVIe siècle*, Paris: Didron, 1858, pp. 18–21.

162 Ibid., p. 18.

163 "Automata," says Ducange, "videntur esse instrumenta mechanica, *sponte moventia*" (*Glossarium mediae et infimae latinitatis Paris*, 1840, I, 508, verb. "Automata.") Dante too admired the wheels moving at various speeds. The hour wheel seems to be motionless while the second wheel flies:

> As wheels within a clock-work's harmony So turn that unto him who watches them The first seems quiet, and the last to fly. (*The Divine Comedy of Dante Alighieri*, Trans. by Jefferson B. Fletcher, New York, 1931, *Paradiso*, Canto XXIV, 13.)

which seemed something almost divine. The vertical motion of a slowly falling weight in a planetarium was transformed by a transmission device into the circular motion of the clock mechanism. One single weight released the motion of numerous wheels that turned at various speeds and described various courses, a fact that necessarily led men to ponder over the causes of this difference in the relations of time and space. The automatism of the rotations of the planetarium was adjusted to the velocities of individual heavenly bodies. This experimental imitation of celestial mechanics robbed it of every mystic veil and suggested the idea that the motions of the heavenly bodies obeyed principles similar to those of the mechanical planetarium.

How strongly the thinking even of the thirteenth century was influenced by contemporary machines can be seen in the well-known passage from Roger Bacon's "Epistola de secretis operibus artis et naturae" (cap. 4) in which he foresees the invention of automatically driven ships, carriages, flying machines, submarines, suspension bridges and gigantic magnetic machines.[164] Thorndike justly emphasized that such passages in the works of Bacon should not be regarded merely as dreams, but as indications that there was *an interest in mechanical devices* and that men were already beginning to struggle with the problems which have only recently been solved.[165] [75]

But all this by no means exhausts the influence of machines upon thought. The invention of the mechanical clock is very closely connected with scientific chronometry, that is to say, exact quantification of time, which is a prerequisite of exact observations in all fields. While sand and water clocks were not capable of accurately measuring small intervals of time,[166] the mechanical clock made such measurements possible for the first time. Thus Geoffrey of Meaux was able to [76]

---

[164] "Machines for navigation can be made without rowers so that the largest ships on rivers and seas will be moved by a single man in charge with greater velocity than if they were full of men. Also carts can be made so that without animals they will move with unbelievable rapidity ... Also flying machines can be constructed so that a man sits in the midst of the machine revolving some engine by which artificial wings are made to beat the air like a flying bird. Also a machine shall in size for raising or lowering enormous weights ... Also a machine can easily be made for walking in the seas and rivers, even to the bottom without danger ... And such things can be made almost without limit, for instance, bridges across rivers without piers or other supports, and mechanisms, and unheard-of engines" (Roger Bacon, *Opera hactenus inedita Rogeri Baconi: tractatus brevis et utilis ad declarandum quedam obscure dicta Fratris Rogeri*, ed. Robert Steele, Oxford: Clarendon, 1920, p. 533).

[165] Lynn Thorndike, op. cit., vol. II (1929) 654–655. Thomas Aquinas too wrote a work (that remained unfinished) on water conducting and mechanical engineering ("liber de acquarum conductibus et ingeniis erigendis"), as we learn from the correspondence carried on after his death (1274) between the University of Paris and the General Chapter of the Dominicans in Rome. See Lynn Thorndike, op. cit., II, 601.

[166] For measuring small intervals of time Galileo used the pulse in *Dialogo sopra i due massimi sistemi del mondo tolemaico e copernicano* (1632). See first day.

determine the exact duration of the lunar eclipse of March 18, 1345; according to him it lasted three hours, twenty-nine minutes and fifty-four seconds.[167]

The impulse to precise time measurement that proceeded from the mechanical clock discussed here increased with time in extent and strength and was condensed a century later in the dialog (ca. 1450) of Nicholas of Cusa, *Idiota de staticis experimentis*[168] to a program of systematic experimental research in all areas of physics. There, one participant in the dialogue, Idiota, or the man in the street, who usually is identified with Nicholas of Cusa himself, decries mere book learning and urges the importance of measurement of natural phenomena especially *weighing* by means of the balance (ascertaining the specific weight of all elements including air) and *timing* by means of water clocks. Since all phenomena take place in time, it is also important for science to measure all these phenomena in the course of time, for from the difference in time elapsed, e.g. in respiration or the pulse (e.g. of a healthy or sick person, of a young or old person) to measure the swiftness and force of the wind or of projectiles, just as from the difference in weight one can discover the "rerum secreta."[169]

*****

The invention of the clock produced a similar revolution in historical writing. Three periods can be distinguished[170] in the manner of indicating time used by the great chronicler Froissart in describing the events in France and Flanders between 1330 and 1400. Until 1379 the time is indicated inaccurately by the hours of prayer (*horae canonicae*). After 1370 public striking clocks became more frequent in the cities of Flanders and northern France, and Froissart began (after 1380, first in Ghent, occasionally in Paris) to cite the modern time indications in addition to the hours of prayer for cities that had striking clocks. In his description of French events between 1389 and 1400, Froissart uses only the modern hours.[171] The new method of indicating time expresses the fact that with the development of the cities and the urban [77] middle class, there occurred a transition from ecclesiastical to worldly computation of time, a process in which the clock played an important part.[172]

Another field of influence of machines upon thought is that of modern *mechanical* theory. Later we shall see that the fundamental ideas and propositions of mechanics were based upon observation and analysis of machines.

---

[167] Thorndike, op. cit., III (1934) p. 290.

[168] Nicholas of Cusa, *Opera omnia*, Academiae Litterarum Heidelbergensis, vol. 5., ed. Ludwig Baur, Leipzig, 1937.

[169] Ibid. pp. 221f, 233ff.

[170] *Chroniques de J. Froissart*, ed. Simeon Luce (Paris, 1869–1899) vol. 9, 10, and 11.

[171] Gustav Bilfinger, *Die mittelalterlichen Horen und die modernen Stunden* (Stuttgart, 1892) pp. 157, 224–226, and Alfred Ungerer, op. cit., p. 54.

[172] Froissart's reports on the events in Spain and Portugal, 1382–1386, do not contain indications of time by hours, except the midday meal of Count de Foix. One may conclude that there were no public striking clocks in Spain at that time and that only the Count had a castle clock. In fact, we know from other sources that the first public striking clock in Spain was received by the Seville cathedral in 1400.

Automatically driven machines were admired because their whole mechanism and the motions of their parts were the result of the preliminary detailed computations made by their designers. In order to make a machine work, its inventor had to elaborate in advance the intermediate and interlocking motions of the gears, levers, guide-pulleys and screws, and to discover the technique of transmitting these movements from one part to another, so that the desired goal would be achieved in the end. Thus the emergence of the automatic machine in the scholastic world of the irrational represented *the birth of rationality*, of the principle of a mechanism logically worked out to the last detail, binding all the dead component parts of the machine and all the partial motions transmitted from one part to another into a significant, intelligible and, so to speak, living whole. And in this mechanical [78] principle of automatically driven machines worked out to the last detail, binding all the component parts of the machine and all the practical motions transmitted from one part to another into a living moving whole – lies the difference from the "automats" of antiquity, which lacked this technique of *transmitting these movements* from one part to another and also lacked the subsequent return to the point of departure of the motion. The "automats" of antiquity were not absolute automats and for certain functions needed constant human assistance. Thus for instance H. Diels[173] showed that only a certain part of the functions occurred mechanically (automatically), for instance, in the clock of Gaza, twelve eagles holding wreaths in their claws tell the hours by bending down at the stroke of the clock and opening their claws thus letting the wreaths that they hold fall upon the heads of Herakles figures thus displaying the hours. But the wreaths could not automatically return to their original position, and the operator had to take these wreaths off the 12 Herakles figures by hand and put them back in their former places in the claws of the eagles. Similarly the operator had to light by hand the lamps that moved past the doors of the night. Thus the absolute automatism of motion was lacking.

In addition, the machine – and Descartes recognized this immediately and clearly – opened up new vistas for mankind; its historic task was "to diminish man's labor."[174] Hence it is clear that a direct line leads from the first machines we have observed here to Descartes' mechanistic conception of the world, which represents the last link in a long chain of an intellectual struggle of modern thinkers beginning with Gundisalvus, Jordanus Nemorarius, and his school.[175] [79]

*****

[173] "Ueber die von Prokop beschriebene Kunstuhr von Gaza," *Abhandlungen der Königlich Preussischen Akademie der Wissenschaften, Phil.-hist. Klasse*, 1917, Nr. 7., p. 19.

[174] See *Discourse on Method*, I (*Philos. Works*, I, 84).

[175] This historical role of the machine in the development of modern culture is overlooked by authors like Gina Lombroso, Berdyaev and Georges Duhamel, who from their humanistic point of view look upon "materialist" inventions with scorn and contempt and see in the "technical age" only a "despiritualization of life" or "spiritual turpitude," destined to bring the destruction of culture in its train (see Gina Lombroso, *La Rançon de Machinisme*, Paris, 1931; Nicolas Berdyaev, "Man and Machine," in *The Bourgeois Mind and Other Essays*, London, 1934; and Boyd G. Carter, "Georges Duhamel: Humanism versus the Machine," *The French Review 17*, No. 3, January1944).

This triumphant progress of the machine left its mark in another area too: man's relation to the universe. Since antiquity man had been regarded as a universe in miniature; in Plato's Timaeus the world is referred to as a *perfect animal unity* and an analogy is suggested between the structure of the human body and the spherical form of the universe. Likewise, Aristotle speaks of the animal as the microcosm and the universe as the macrocosm. Medieval philosophers frequently use the same analogy or, when speaking of the structure of the universe, use the general expressions "fabricatio mundi," "theatrum mundi," et cetera.[176]

The once accepted terms of "macrocosm," "microcosm," "theatrum mundi," "fabricatio mundi," continued to be used throughout the Middle Ages and even later; Grosseteste (died in 1253), for instance, wrote a fragment entitled "That man is a microcosm." But at the end of the twelfth and the beginning of the thirteenth centuries, characteristic new expressions emerge; now the analogy is no longer made between the universe and the human body, but between the *universe and the machine*. Thus Alanus de Insulis (Lille), one of the most prominent thinkers of the end of the twelfth century (1126–1202) says in his treatise "Distinctiones [80] dictionum theologicarum": "the abyss is called the earthly machine inasmuch as it was created in the beginning Abyssus dicitur *mundana machina*, prout primo fuit creata."[177]

If formerly philosophers had attempted to make the structure of the world more comprehensible by comparing it to the "perfect animal" (as Plato did), now the animal itself had become a problem and could be understood only when it

---

But when Lynn White Jr. says: "The chief glory of the later Middle Ages was not its cathedrals or its epics or its scholasticism: it was the building for the first time in history of a complex civilization which rested not on the backs of sweating slaves or coolies but primarily on non-human power. . . . The study of mediaeval technology . . . therefore . . . reveals a chapter in the conquest of freedom." (Lynn White Jr., op. cit., p. 156), he confuses potentialities and realities and forgets that the replacements of human by non-human labor spread only to a negligible section of production of that time. In the Merovingian era (7th and 8th c.) the domanial water mill (*farinarium*) became widespread and replaced the hand mill. The use of the seignoral mill was "obligatoire" for the population but not "gratuit"; it became an instrument of monoplistic exploitation of the peasant population (Ferdinand Lot, *Histoire Du Moyen Age. Tome I. Les Destinées de l'Empire en Occident de 395 à 888*, Paris: Presses Univ. de France, 1928. (*Histoire Générale,* ed. by Gustave Glotz) pp. 350–51, Tome II; Augustus Fliche, *L'Europe Occidentale de 888 à 1125*, Paris: Presses Univ. de France, 1930). In view of this fact, the well-known opinion of J. Stuart Mill: "It is questionable if all the mechanical inventions yet made have lightened the day's toil of any human being," seems more justified with regard to that period. See *Principles of Political Economy*, Bk. 4, Chapter 6, vol. 2, p. 312.

[176] Thus, Maslama Ibn Ahmed (died ca. 1007) of Madrid, wrote a commentary on Ptolemy's *Planisphaerium*, translated into Latin during the first half of the twelfth century: *Sphaerae atque astrorum coelestium ratio, natura, et motus: ad totius mundi fabricationis cognitionem fundamenta*.

[177] M. Baumgartner: *Die Philosophie des Alanus des Insulis, im Zusammenhange mit den Anschauungen des 12. Jahrhunderts*, Münster, 1896, p. 71 (*Beiträge zur Geschichte der Philosophie des Mittelalters 2,4*, ed. C. Bäumker).

was compared to a machine, such as a mechanical clock. Thus *Thomas Aquinas* denied free will to beasts and likened their natural functioning to that of a clock. He explained the sagacity of animals by interpreting them and their *functions as machines*, because animal motions are made without the use of reason and without any decision of the will (*electio*), in other words, purely mechanically.[178]

As machines developed in the course of the next centuries, such analogies multiplied. As we have mentioned before, in the second half of the fourteenth century numerous planetaria driven automatically on the model of clocks were constructed to illustrate the motions of the heavenly bodies. Gradually people became accustomed to consider the universe itself as a machine constructed according [81] to mechanical principles. In his work "De commensurabilitate motum celestium" *Nicolas Oresme* discussed the problem of the regularity of and rational numerical proportion between the velocities of the stars and the heavens as a basis for scientific predictions. Any irrational proportion, he thought, would be intolerable for the heavenly Intelligence that moves the orbs. For this intelligence was conceived as operating in a manner resembling that of the clockmaker who strives for the maximum of harmony when constructing a clock: "For if anyone should make a mechanical clock, would he not make all the wheels move as harmoniously as possible?"[179] In the first half of the fifteenth century, Giovanni da Fontana (1395–1455) again compared the universe to a mechanical clock.[180] At about the same time Nicholas of Cusa repeatedly referred to the universe as "machina mundi," "machina mundana."[181] More, from the idea of the world as a well-designed machine there develops logically from Cusa to Newton, Voltaire and Kant a new conception of God, who like a builder of machines maintains the "ordo universi" and its harmony.[182] [82]

By the end of the fifteenth century the influence of Leonardo da Vinci in the field discussed here marked a turning point. According to him, "the whole world, also the living one, *is ruled by the laws of mechanics; the earth is a machine, and man is one too*. He conceives the eye as a 'camera obscura'... he determines the point

178 "Et idem apparet in motibus horologiorum et omnium ingeniorum humanorum quae arte fiunt ... Et propter hoc etiam quaedam animalia dicuntur prudentia vel sagacia, non quod in eis sit aliqua ratio vel electio" (*Prima secundae summae theologicae*, Quaestio XIII, art. 2 ad fin. (Rome, 1882–1930) VI, 99–100).

179 L. Thorndike, op. cit., III, 1934, 405.

180 "O mirabilis sapientia divina quod tam mobile horologium aedificavit et mirabiliter moveri iussit" (ibid., IV, 1934, 169).

181 Nicolai de Cusa Cardinalis, *Opera*, Basileae, 1565. *De docta ignorantia*, T.I., lib. III, cap. XI, p. 38 and cap. XII, p. 39. In lib. II, cap. XIII, p. 42 we can read: "Mundi machinam perire non posse."

182 "Est igitur ordo universi, prima et praecisior imago, aeternae et incorruptibilis sapientiae, per quem tota mundi machina pulcherrime et pacifice persistit." (*De venatione sapientiae*, Chapter XXXII. De Cusa Cardinalis, *Opera*, op. cit., I, 324).

where the reflected rays intersect."[183] Leonardo's pioneering work in the field of comparative anatomy is based on the realization that the functioning of the animal body as well as the motions of its several parts are subordinated to the laws of mechanics. Thus, Leonardo thinks of muscles in terms of their relations to bones as levers.[184]

In the sixteenth century, Chancellor Thomas More, in his *Utopia* (1516) referred to God as the master workman who "in the manner of other artisans offered the machine of this world for the contemplation of man."[185] And a hundred years before Descartes, the physician François Rabelais, in Book III of the first edition of his *Pantagruel*, published in Paris in 1546, characterized the human body as a [83] machine that is carried on its feet and whose most important task is to produce blood.[186]

At the beginning of the sixteenth century, Georg Joachim Rheticus, a disciple of Copernicus, related in his *Narratio prima* the birth of the Copernician theory, and tried to explain it by the example of a mechanical clock. While the Ptolemaic system is compelled, in order to explain all phenomena, to assume a large number of intersecting circles, the Copernican assumption of a rotating earth simplifies the whole mechanism and makes the hypothesis of so many circles superfluous. The clockmakers too, says Rheticus, try to construct their mechanisms in the simplest possible way and to eliminate every wheel that is not absolutely indispensable; they achieve this aim by letting another wheel, often by a slight change in position, perform the function of the eliminated wheel. "Should we not attribute to God, the creator of nature, that skill which we observe in the common makers of clocks?"[187] And Copernicus himself in his famous preface to *De Revolutionibus* (1543), dedicated to [84] Pope Paul III, describes the universe as a machine.[188]

The same thing can be found in the work of the Spanish theologian Raymond Sébonde, *Theologia naturalis* (1569). In Chapter XVII the idea is developed that God, the master, not only created the universe once, but that he must constantly

---

183 M. Herzfeld, *Leonardo da Vinci* (Leipzig: Diederichs, 1904) CXV, and CXXII.

184 A. P. Usher, *A History of Mechanical Inventions* (New York, 1929) p. 61.

185 "...quem ceterorum more artificum arbitrantur, mundi hujus visendam machinam homini ... exposuisse spectandam" (Thomas Morus, *De Optimo Reipublicae Statu de que Nova Insula Utopia*. Libri duo, Londoni, 1777, Lib. II, p. 148).

186 "... *cheminent les pieds et portent toute ceste machine*." Cf. *Oeuvres de François Rabelais*, édition critique par Abel Lefranc, Paris, 1931, V, 54. In the English edition (*The Works of Mr. Francis Rabelais*, Philadelphia, 1912, I, 338) this important passage was rendered inaccurately: "the feet ... bear up to the whole bulk of the corporal man."

187 See the English translation of the *Narratio prima* by Edward Rosen in: *Three Copernican Treatises* (New York, 1939) p. 137.

188 "Hanc igitur incertitudinem mathematicarum traditionum de colligendis motibus sphaerarum orbis, cum diu mecum revolverem, coepit me taedere, quod nulla certior ratio motuum *machinae mundi*, qui propter nos, ab optimo et regularissimo omnium opifice, conditus esset ..." (*Nicolai Copernici Thorunensis de revolutionibus orbium coelestium*. Libri VI, Thoruni, 1873, p.5).

watch it and recreate it so that his created work, the machine produced by his will, will be preserved in the full perfection of its functioning.[189]

Finally, it is interesting to note that nearly a century before Thomas Hobbes, mechanical concepts were applied even to social phenomena. In England, John Hales, the presumptive author of the *Discourse of the Common Weal of this Realm of England* (1549; first printed 1581) looked upon society as similar to the physical body, as a *mechanism* which behaved in accordance with unalterable laws. In the third dialogue he developed a detailed philosophical theory of social and economic causation illustrated by the example of a clock: "As in a clock theare be many wheales, yet the first whele beinge stirred, it drives the next, and that the third, till the last moves the Instrumentes that strikes the clocke. So, in the makinge of a howse . . ."[190] [85]

Thirty years later in his *Essays* (1579) Montaigne compares the social structure of the time of the civil wars with a machine that seems to malfunction and to be breaking down.[191] How important the role of mechanistic ideas in the outlook of this period was can be seen from the fact that even authors like Cornelius Agrippa de Nettesheim, to whom the rationalist mode of thinking was alien and who was a representative of a sort of Platonism, a vitalistic view of nature, nevertheless could not rid himself of mechanistic ideas. According to his view, a soul permeates the universe; this *anima mundi* is an intelligent world soul which translates the divine dictates for lower forms of souls and is interpreted by Cornelius in his *De occulta philosophia* (1533) monistically, as a unity that binds the dispersed parts of the world into a unique *world machine*.[192]

Thus, from this list of writers which I believe to be representative although not all-inclusive, it is clear that the conception of the world as a machine, although not scientifically elaborated, was fairly widespread several centuries before Descartes.

*****

189 I quote here from Montaigne's French translation: "Ainsi ce grand maistre ouvrier ni endormy ni nonchalant, porte sans cesse, enferme et soustient en sa main, sans peine et par la seule volonté cette machine, son bel ouvrage." *Oeuvres Complètes de Michel de Montaigne*, Paris, 1924–32, IX (*La théologie naturelle de Raymond Sebon*) p. 34. Hamlet's last words in the letter to Ophelia refer to his body as a machine:

Thine evermore most dear lady, whilst
this machine is to him, HAMLET. (Act II. Sc. 2).

190 John Hales, *Discourse of the Common Weal of this Realm of England,* Elizabeth Lamond ed. Cambridge, 1893, p. 98. (See Eli F. Heckscher, *Mercantilism*, trans. M. Shapiro (London, 1935) II, 311–313.)

191 "A voir nos guerres civiles, qui ne crie que cette *machine* se bouleverse et que le jour du jugement nous prend au collet . . ." (I,§26, 116). See Fortunat Joseph Strowski, *Montaigne*, 2. éd. Paris, F. Alcan, 1931 p. 244.

192 "Anima mundi, vita quaedam unica, omnia replens, omnia perfundens, omnia colligans et connectens, ut unam reddat totius mundi machinam," Henricus Cornelius Agrippa ab Nettesheym, *De occulta philosophia*, Bk. II, Chapter LVII (*Opera* vol. 1, London; Bering, n.d., p. 241).

If the machines of the early period exerted such a powerful influence on human thinking in various scientific fields, this is true *a fortiori* for the machines of [86] the fifteenth and sixteenth centuries, when the motive power of water, gradually extended to ever new activities, was finally applied to the iron industry and mining (15th century), a fact which led to one of the greatest revolutions in the history of technique and *the industrial revolution of the fifteenth and sixteenth centuries*. The impulse for this development came from the immense *extension of the world market* as a result of the voyages of exploration of the Portuguese and the Spaniards, that required increased production.[193] As this increase could not be achieved by artisan methods, the need arose for a means of accelerating production, that is to say, for machines. And this need was satisfied.

The beginning of the fifteenth century saw the invention of blast furnaces generating high temperatures that could not be achieved with the primitive clay-ladles and air-"bloomeries" erected on high ground by peasants and small forest smiths. The technical point of departure of this innovation was the utilization of water as motive-power in the extraction and processing of iron ore in the operation of large bellows for the treatment of the ore, whereby the high temperatures indispensable for liquefying the iron were secured.[194] This technical innovation soon led to a social change, to the transfer of the centers of the iron industry from high mountains and forests to the river valleys, where in the place of the numerous small peas- [87] ant founderies and forest smiths there arose big capitalist enterprises characterized by mass production and the concentration in one spot of several productive processes: extraction of the ore, its processing and forging. Here, clustered together, were big stone-built blast furnaces with water-wheels, water-driven hammers and bellows, heavy stamping-mills for the crushing of the ore, houses and cabins for scores of workers, stables for horses used in hauling the ore, timber and coal, and several other machines and buildings, all of which required a large capital investment.[195]

It might also be mentioned that the famous Dutch inventor Cornelis Drebbel in the first quarter of the seventeenth century invented very cleverly constructed ovens and furnaces, in which the fire could be kept at the desired temperature.[196] In

[193] See John U. Nef, "A Comparison of Industrial Growth in France and England from 1540 to 1840," *Journal of Political Economy 44* (1936) 220.

[194] One of the earliest examples of the application of water power to a blast furnace in England was Bishop Langley's furnace in Durham, run by John Dalton, the records of which for the years 1408–09 are preserved (Julius Pratt, "Machinery in Sixteenth Century English Industry," *Journal of Political Economy 32* (1914) 783).

[195] The triumph of the new technique and the new large-scale form of enterprise quickly was achieved, for instance in Champagne (France), as early as the second half of the fifteenth century, and was glorified in *Nicolas Bourbon's* Latin "*Poem on the Iron Industry*" (Paris, 1517). The creator of modern metallurgy was *Vanoccio Biringuccio* whose book *De la pirotechnia, Libri X* (Venice, 1540) was the first scientific and systematic exposition of metallurgy and the most important book about the art of casting until the French Encyclopedia of 1785. (A French edition of this book, "*La pyrotechnie, ou l'art du feu*" was published in Paris in 1556).

[196] G. Tierie, *Cornelis Drebbel, 1572–1633,* Amsterdam: H. J. Paris, 1932, p. 42–43.

England great water hammers were utilized in the forges by which iron was brought into commercial shape. For instance the "great water-hammer" working in Ashdown Forest, Sussex in 1496 weighed seven or eight hundred pounds. By the close of the sixteenth century such hammers were very common.[197] [88]

The revolution in mining that started in the second half of the fifteenth century was also due to the use of water power. In the mines of the Middle Ages the depth of the workings seldom exceeded a few fathoms. When greater depths had to be reached, the primitive system of drainage in use until then proved inadequate: the ditches were flooded by underground waters.[198] Only the application of water-wheels as the motive power for powerful *pumps*, for transporting and raising machines, made real workings in depth possible: they made possible the construction of expensive adits or long tunnels, ventilation shafts, stamping-mills for crushing ores and above all the installation of drainage engines that pumped water from the mines night and day, automatically, without the use of additional labor power.[199]

In mining too the technical revolution led to a *far-reaching social change*. As the underground workings progressed, the growing need for capital for the construction of mines, shafts, drainage installations, et cetera, led to far-reaching transfers of property and a powerful concentration of capital. In Germany and the adjoining regions, from the fifteenth century on, the cooperative enterprises of small medieval owners as a result of lack of capital became dependent upon a few big backers who had large sums at their disposal, usually big ore merchants (for instance, the Fuggers in Augsburg) who gave them advances and acquired shares in their enterprises, while the original small owners were reduced to the status of wage workers.

This is not the place to follow the historical development of machinism in various branches of industry. We shall confine ourselves to making a few observations on the technical advances in France and Holland, the countries where Descartes lived and could constantly observe the technical achievements of his time. The first stamping and cutting mill is said to have been erected before 1532 in Nuremberg, Germany, but more specific information is not available.[200] [89]

In France, despite the raging religious wars, the *first stamping-mill* replacing the hand hammer for coinage purposes was erected as early as 1552, under the direction

---

[197] Pratt, op. cit., p. 784. In Goslar, for instance, a contact was signed on Sept. 18, 1478 with the famous engineer and citizen of Krakow, Johann Thurzo, concerning "the Waterpumps of the Ramelsberg" (see Clamor Neuburg, *Goslars Bergbau bis 1552. Ein Beitrag zur Wirtschafts- und Verfassungsgeschichte des Mittelalters*, Hannover: Hahn 1892, p. 244).

[198] C. Neuburg, op. cit., p. 190.

[199] C. Neuburg, op. cit., p. 218. Georg Agricola in his treatise on mining written in 1550, which is not a theoretical work but merely describes the inventions and technique that represented the *accumulation of generations of experience*, demonstrates in his text and illustrations, the numerous and complicated water-driven wheel-machines, chain and suction pumps, hauling machines, water-driven hammers, et cetera, then used in mining. (See Georg Agricola, *De Re Metallica*, Transl. from the first Latin edition of 1556 by Herbert C. Hoover and Lou H. Hoover, London, 1912).

[200] L. Beck op. cit. II, 527.

of Olivier; the heavy steel cylinders between which the gold and silver bars passed were driven by water-wheel and shaft. After several improvements, mechanical stamping was introduced in *all* French mints in 1639.[201]

By the end of the fifteenth century the first stamping mills for rolling iron and copper plates were introduced in France. "This is the first triumph of the machine."[202] Although the overwhelming mass of industrial organizations remained artisan in character, in limited fields of industry the artisans with their long professional training and virtuosity were replaced by *machines* operated by untrained workers. These performed the work better, faster and cheaper, while achieving an excellence and uniformity of the product that were unattainable by artisan methods. The new technique although still in its beginnings was sufficiently widespread to attract attention and be admired for its efficiency.

[90] Such an enthusiastic admiration for the new technique is revealed in the memorandum submitted in 1604 by the royal councilor, B. de Laffemas, Henry IV's "valet de chambre." In it he describes the great advantages of the stamping and rolling mill (*moulin de forge*) as compared with the former artisan technique. "... the iron is cut and split in as many pieces and as small and in a manner as one desires, something that formerly was done exclusively by hand by locksmiths and other such workers and only at great expense ... And copper and brass as well that is stamped and flattened by hand by the coppersmiths and other workers at great expense, are fashioned by the said mills in sheets as thin as one desires, and shaped as one desires, and more of them in one day than a coppersmith could make in a month, and at lesser cost."[203]

In 1609 two applicants were given the right to introduce the German method of *drawing wire* with water power in place of hand labor.[204]

[91] The machine technique, however, went beyond the field of metals. The above mentioned memorandum by Laffemas contains an impressive picture of a "new invention," a big modern spinning shop in France. This takes care of "great quantities of all sorts of woolens, bristles and cotton, linen, hemp, floss-silk and other similar materials using as laborers little children, blind people, old men, one-handed

---

[201] Ludwig Beck, *Die Geschichte des Eisens* (Braunschweig, 1895) II, 528, 945.

[202] Henri Hauser, *Les Débuts du capitalisme* (Paris: Alcan, 1927) p. 11. See also Henri Sée, *Histoire économique de la France* (Paris, 1939) 123.

[203] H. Hauser, op. cit., p. 11–12. Salomon de Caus, *Les Raisons de forces mouvantes. Livre troisième, traitant de la fabrique des orgues* (Frankfurt, 1615), gives us the earliest description and a reproduction of a hand-driven stamping mill for rolling leaden sheets for organ pipes. Even more interesting are the description and reproduction of a stamping mill for glaziers's lead, by Vittorio Zonca, 1607, who describes a machine taken over from industrial practice and constructed according to the principle of the lever, which automatically fashioned glazier's lead of the desired profile (See Vittorio Zonca, *Novo teatro di machine et edificii*, Padua, 1607, pp. 79–80. "Ruote da incavar il piombo per le finestre di vetro.")

[204] John U. Nef, *Industry and Government in France and England, 1540–1640* (Philadelphia, 1940) p. 85.

and simple-minded people who sit at their ease, *without toil and bodily pain*; it produces more in one day than can be done in three days by distaffs and with greater excellence."[205]

In 1589 came the invention of the knitting-loom by William Lee of Nottingham (England). "It is a machine exceedingly complex, consisting of two thousand parts, which, in a moment almost, can make two hundred meshes or loops without requiring much skill or labour in the workman."[206] At the invitation of the French government Lee came to Rouen in 1600 with nine workers and as many machines, because Henry IV had guaranteed him important privileges.

We shall pass over many other mechanical inventions of that era[207] in order briefly to dwell on the development of technique in Holland (and neighboring countries). As early as the twelfth century canals were built in Flanders; and beginning with the thirteenth century sluices were also constructed in order to maintain the waters in canals and rivers at a high level. Even then, powerful cranes existed; by means of big machines called *overdraege* (double-cranes) completely laden (albeit not very large) ships were raised, turned to one side and lifted to the upper level or lowered to the lower level.[208] In the second half of the sixteenth century Guicciardini described a gigantic lock built near the mouth of the Bruges canal, which by means of a powerful machine was locked during the low tide and opened during the high tide so that ships could sail directly out to sea.[209] In his description of Antwerp that for the most part was built on wooden pillars planted in swampy ground, Guicciardini mentions big mechanical water-driven hammers that served for driving long wooden posts into the water.[210] [92]

The first authentic windmill in Europe appears in Normandy about 1180. (In Persia they are found even in the tenth century). Within a generation the windmill had become – for geographical reasons – a typical part of the landscape on the plains of nothwestern Europe (Flanders, Netherlands);[211] for the fall of rivers was

205 H. Hauser, op. cit., pp. 12–13.

206 John Bekmann, *History of Inventions* (London, 1892) II, 368–369. [1814 IV, 311].

207 For instance a machine used with keeping the harbors clear of sand. In 1562 patent was granted to George Cobham and others for ten years for the importation of machines from abroad (Italy?) "to cleanse and carry away all shelves of sand, banks, etc. out of all rivers, creeks, or havens" (Julius Pratt, op. cit., p. 782).

208 Leopold A. Warnkönig, *Flandrische Staats- und Rechtsgeschichte bis zum Jahre 1305* (Tübingen, 1835) I, 322.

209 Lodovico Guicciardini, *Description de Tous le Pays-Bas*, Anvers, 1582, p. 371. [441].

210 Guicciardini, op. cit., p. 307. Such mechanical hammers are reproduced in Besson's and Branca's books on machines. See Jacques Besson, *Théâtre des instruments mathématiques et mécanique* (Lyon, 1578) Figs. 22 and 23; Giovanni Branca, *Le Machine*, Rome, 1629, Fig. 3.

211 In the thirteenth century there were 120 windmills in the vicinity of Ypres (see Lynn White Jr., op. cit., p. 156). Dante mentions "come quando ... par di lungi un molin che *il vento gira*" (*Inferno*, Canto xxxiv.6). In the 11th c. so-called tide mills (*aquimoli*) are mentioned (cited in a document from the year 1044), which turned for six hours in one direction and for six hours in the other direction. See Libri *op cit.* II 231–232. With regard to the extensive use of wind-mills in the sixteenth century England, see, J. Pratt op. cit., p. 776.

so gradual that expensive dams and mill-pounds often had to be constructed to run [93] water-driven mills. Windmills were much cheaper.

Of great technical interest is Guicciardini's description of the *water-supply system of Bruges*. Supplying the lower parts of the city presented no difficulties; aqueducts with a sinking level had been known since Roman times. But to supply the upper part of a city that had no springs was a different problem. By 1552 a building named the *Logis* (later *water huis*) was erected in the lower part of the city, as well as a water reservoir; a *water pump* driven by a horse-whim pumped the water to the upper part of the city in strong lead pipes laid under the surface of the earth with several smaller branch pipes distributing water to the houses and public fountains.[212]

Another machine that was widespread in Holland was the printing press. Printing shops began to spread in Holland with astounding rapidity as early as the last third of the fifteenth century; they were set up in Deventer, Delft, Gouda, Harlem, Leyden, Nimwegen, Leeuwarden, Zwolle, etc., chiefly under the impulse given to printing by the religious sects, especially the schools of the "Brothers in Common Life" (*Frères de la vie commune*) numerous in Holland. These sects, freed from the church censorship and from the educational monopoly of the clergy, for the first time put [94] printing at the disposal of lay education.[213]

*****

What conclusions can be drawn for our problem from this brief survey of technical development? To be sure, the new forms of enterprises and inventions comprised only a small fraction of the national production; by far the greater part of the production was still carried on by artisan methods. But the importance of this technical revolution consists in the fact, that the new forms of labor, because of their social effects, attracted great notice and deeply influenced contemporary thinking. – In his *Descriptio Britanniae* (1607) William Camden graphically shows us the victory of the new technique and big enterprise in the iron industry of Essex and Kent, where the most prominent lords and barons shared in the profitable iron industry and delivered weapons to the navy.[214] The new big enterprises in Sussex changed even the appearance of the landscape: in order to obtain more power from the waterfalls, the furnace owners often joined several small rivers into one strong waterfall, the sound

---

[212] Guicciardini, op. cit., p. 373 [442]. Thirty years later the new mechanical method of water-supply was introduced into England, probably from Holland. In 1579 a Dutchman, Peter Morris received from Mayor and Commonalty of London a five-hundred-year lease by which he was authorised to erect an engine under the first arch of London Bridge to supply city with water. By the year 1582 the new engine was in operation. (Julius Pratt, op. cit., p. 781).

[213] Henri Pirenne, *Histoire de Belgique* (Brussels, 1907) III, 289, sqs.

[214] W. Camden's *Britannia, newly translated into English* (London, 1695; ed. Edmund Gibson) p. 167. See also David and Gervase Mathew, "Iron Furnaces in South-Eastern England and English Ports, 1578," *The English Historical Review 48* (1933) 91–99.

of which, added to the "beating with hammer upon iron, fills the neighborhood round about, night and day, with their noise."

Guicciardini gives us similar details about Flanders, where blast-furnaces were set up at an early date near many tributaries of the Meuse River in the principality of Liège and the county of Namur.[215] [95]

These changes and their bearing upon the great profitability of the new enterprises aroused general interest in technique. "The newly awakened interest in mechanical improvements . . . spread among all classes in England from the nobility to the humblest artisan."[216]

Guicciardini reports a similar interest in inventions, stimulated by the profit motive, in the Netherlands. The Belgians, he thinks, have a special talent "for inventing all sorts of instruments including kitchen utensils to facilitate and shorten everything they undertake."[217] "There are always subtle minds that invent some unusual way of profiting and making new gains."[218] [96]

The theory and practice of *water-raising* by means of pumps or water wheels aroused great interest, for it was the fundamental prerequisite of industrial mechanization, water being at that time the most important motive power.[219] The Romans knew small hand pumps. In the Museo Arqueologico Nacional in Madrid [see below p. 230] there is a fire-engine dating from the Roman times, that was used in the mines of Huelva. Small water pumps were known in the Middle Ages and used, for instance, in architecture. There is a report dating from 1511 of the use of a pump during the reconstruction of the cathedral of *Bordeaux*; when subsoil water flooded the foundations, the workers demanded higher wages "since they draw out water by day and by night [quia extrahebant aquam de die et de nocte]." After a short time, when the raise of wages was cancelled, the workers destroyed the pump: "they broke the engine into pieces in the night [de nocte fregerunt ingenium]."[220]

---

215 Lodovico Guicciardini, op. cit., p. 464. As early as the sixteenth century Namur developed into a metallurgical center with numerous blast-furnaces, forges and water-driven hammers, where, to quote Guicciardini, iron was cast, hammered and forged amidst smoke, flames and noise as in the workshop of Vulcan. The emperor Charles V, who by the middle of the sixteenth century had the best artillery in Europe, had his cannon cast in Belgium (cf. L. Beck, *Die Geschichte des Eisens*, II, 850, 854; Henri Pirenne, *Histoire de Belgique*, op. cit., III, 246; A. Vierendeel, *Esquisse d'une histoire de la technique*, Louvain, 1921, p. 362).

216 J. U. Nef, "The Progress of Technology and the Growth of Large Scale Industry in Great Britain, 1540–1640," *The Economic History Review 5* (1934) 16.

217 Guiccardini, op. cit., p. 4.

218 Ibid., p. 465 [583].

219 We shall mention the works of Leonardo da Vinci, Guiseppe Ceredi, 1567; Bernard Palisay, 1580; Della Porta, 1601; Galileo, 1612; Salomon de Caus, 1615, and later Pascal, 1654. Galileo himself invented a machine for raising water using a horse-driven pump. See "Privilegio concesso a Galileo per l'invenzione d'una macchina da alzar acqua" 1593 in: *Le Opere di Galileo Galilei*, Edizione Nazionale, Firenze, vol. 19 (1938) 126.

220 See Julien Hayem, *Mémoires et documents pour servir à l'histoire du commerce et de l'industrie en France*. 1st. series, Paris, 1911, 78–79: Source: "un livre de comptes du trésorier de la cathédrale de Bordeaux conservé aux archives de la Gironde".

Pumping of water to very high levels was attended with special difficulties, because the waterproof joining of iron pipes was unknown; wooden or leaden pipes could not stand the greater pressure required for higher levels and burst.[221] In July 1565 a letter from William Humfrey to Sir William Cecil recommends "an Almain engineer, who can raise water one hundred fathoms high (= 500 or 550 feet), by a newly invented engine."[222] However further details are missing. The solution was to raise water by short stretches with the help of bucket wheels or pumps installed at every storey.[223]

[97]

The *pumps* were really very important for mining but only in connection with water power. Water power is responsible for the transformation of mining since the second half of the 15th century. In the middle ages many German mines were closed because the deeper one descended, the more there was contamination by bad air and water. It was only the application of waterwheels driven by flowing water as motors for strong pumps and strong lifting equipment that brought in fresh air and pumped

[221] As early as 1598 Cornelis Drebbel obtained a patent in Holland "to lead fresh water in great quantities through *leaden pipes* and to raise it, like a fountain, from low down upwards to the height of 30, 40, 50 or more feet." In praise of his work Constantin Huygens writes: "No one has made a cleverer contribution than Drebbel to the art of pumping up dead, as we call it, or left-over water from pools and of drawing it away – and no one ever will" (see G. Tierie, op. cit., p. 45–46). At the beginning of 1631 Descartes' friend the engineer de Villebressien came to Holland and lived with the philosopher in Onde Peins in the center of Amsterdam because he was interested in "l'Hydraulique où art d'èleves les Eaux" (see Gustave Cohen, *Écrivains français en Hollande dans la première moitié du XVIIe siècle*, Paris: Champion, 1920, p. 471.)

Such inventions for pumping water occur increasingly in the 17. c. Thus, for instance. Francis Potter (1594–1678), Commoner of Trinity College in Oxford, "made a notable improvement in an engine for raising water from a deep well at Kilmanton Parsonage." (R. T. Gunther, *Early Science in Oxford*, Oxford 1937, vol. XI, 229). Sir Eduard Ford (1605–1670) constructed similar machines; he was commissioned to improve the water-supply of London and constructed water-works (pumps) near Charing Cross and at Wapping (ibid. p. 230).

Finally, Sir Samuel Morland (died 1695), member of Magdalene College, succeeded in making efficient water-engines for raising water from wells to high places such as at the top of Windsor Castle. He was honoured by King Charles by the title of "Master of Mechanics" to the King, and by being sent to France to obtain further information on water-engines (R. T. Gunther, *Early Science in Cambridge*, Oxford, 1937, p. 74).

[222] Pratt, op. cit., p. 779.

[223] Jacques de Strada of Mantua (1512–1588) gives a reproduction of such a "triple pump to raise water from one basin to another up to the required height" (Jacques de Strada à Rosberg, *Desseins artificieux de toutes sortes de machines*, publié par Octave de Strada, neveu, Francfort sur le Main, I, 1617, II, 1618, Figs. 31, 39). Toward the beginning of the sixteenth century, Juanelo, Charles V's Italian mechanic, built the famous water machine ("artificio de Juanelo") near the bridge of Alcantara at Toledo (Spain), to raise the water from the Tagus to the citadel of the Alcazar, the highest point of the city. How strong the interest in such machines was at that time can be seen in contemporary writings that praise the water wheel as the "wonder of the world." People undertook great voyages to admire it; El Greco reproduced the machine in the background of one of his paintings (see Maurice Barrès, *Greco ou le Secret de Tolèdo*, Paris, 1912, p. 39). Water-wheels to raise water for irrigation purposes are of ancient origin. Famous is the waterwheel on the Crontes River (Syria), built in the first half of the thirteenth century by the engineer Qaysar ben abî al-Qasim for the prince of Hamah (Aldo Mieli, op. cit., p. 155).

out water and carried the ore with the rock to the surface that made deep mining and the excavation of deeper tunnels and shafts possible, just as the application of natural power (of water in stamping works, in mining etc.) made possible the use of a concentrated force, that far exceeded human power and therefore was independent of it and presented humankind with new tasks. It was the beginning of the technological age

The citizen of Krakow, Johann Thurzo, and his son knew how "with the aid of a machine driven by mechanical force or a so-called *Wasserkunst* to make flooded ditches passable again." Thurzo was well known in central German mining as an internationally active engineer who introduced rational procedures.[224] In order to avoid competition with the inventor, Jakob Fugger arranged to have his company join with Thurzo. Georg Thurzo, Johann's son, married the daughter of Jakob Fugger's brother (1497), and one can say that their capitalistic power in mining, the investment of massive capital, rested on the alliance with the technical knowledge of the Thurzos.

*****

The water machines, originally invented as driving-machines, "*machines mouvantes*" for industrial purposes, later, in the sixteenth and seventeenth centuries were also used for the purposes of domestic comfort, for the embellishment of palaces, gardens and grottoes of the wealthy and for moving complicated toy mechanisms set up there. These luxurious pleasure grounds contributed to popularizing the understanding of mechanical constructions and the mechanical conception of the world among wide classes of the population.[225] They played an important role for Descartes who saw them "in the grottoes and fountains that are in the gardens of our kings," to wit, on the grounds of the palais of Saint-Germain en Laye, and

---

224 *Max Jansen, Jakob Fugger der Reiche: Studien und Quellen*, Leipzig, 1910, p. 109.

225 The same can be said of the very complicated theatrical machinery employed in Italy since the sixteenth century. In the opera "The Battle of Apollo and the Serpent" performed in Florence in 1590, on the occasion of the wedding of Ferdinand I, Grand Duke of Tuscany, the serpent pierced by Apollo's arrows, extracted these arrows from its body and broke them with violent anger. "This was the triumph of theatrical machinery," says L. Celler (pseud. for Louis Leclerq) in *Les origines de L'opéra* (Paris, 1868) p. 332. The machines at that time played a more important part than the music: "dont les machines sont estrées celèbres ... et dans lequel les machines et la danse tiennent encore plus de place que la musique" (L. Celler op. cit. p. 328). In 1645 Mazarin, at the wish of Queen Anna, had the famous mechanic Torelli come from Parma with all his theater machines (ibid. 339). The universal proclivity towards theatrical machinery was so strong that even the great author of tragedies Corneille was forced to make concessions to popular taste and write "pieces à machines". One such "piece à machines" *Andromède* was performed with music by Assourcy in 1650 in the Theater *Petit Bourbon*. His second such piece was *Toison d'or.* Another popular machine was the flying dragon ("Draco volans") mentioned already in Della Porta's *Magia naturalis* (1561, 1st ed. 1558) lib. II, cap. 14, and later in Diego Ufano's book *Artillerie* (Anvers, 1621) p. 134, "How to make a flying dragon: the machine of the dragon being very subtle." The religious dramas (*les dramas sacrés*) and the urban spectacles of the fifteenth century already "contain the germ of the opera *with machines* that was later perfected by artists such as Brunelleschi or Leonardo da Vinci" (E. Borel, "L'Opéra ancien en Italie et en France," *La Revue critique des idées et des livres 23* (Paris, 1913) 674).

[98] described them in his "Traité de l'Homme" (1644) with a view to explaining the motions of the parts of the body, which result from the purely mechanical interplay of nerves and muscles by comparing them with the driving force of water in the above-mentioned toy mechanisms.[226]

De Fontenelle or "A Look behind the Scenes of a Theatre"

That our interpretation of Descartes' basic ideas is correct, that the mechanistic conception of Descartes is nothing but the reflection of the machines that existed in his time, seems to be confirmed not only by the Descartes texts just cited but also by the fact that contemporaries and prominent disciples understood him no differently.

Within French thought Bernard de Fontenelle marks the *highest point in the influence of the mechanistic ideas of Cartesianism*. With Fontenelle Cartesianism entered the Académie Française in 1691 and the Académie des Sciences in 1699. It should thus be especially interesting for us to see how Fontenelle understood Descartes. To this purpose we shall use his *Entretiens sur la Pluralitè des Mondes* (1686) in which he elegantly introduces the principles of Descartes' mechanical conception of the world.[227]

"Nature always appears to me in the same point of view as theatrical representations. In the situation you occupy at the opera you do not see the whole of its arrangements: the machinery and decorations are so disposed as to produce an agreeable effect at a distance, and at the same time the *weights and wheels are hidden by which every motion is effected*."

Nature appears to us in a similar way. "Nature so entirely conceals from us the means by which her scenery is produced, that for a long time we were unable to discover the causes of her most simple movements."

How should this be explained? To this purpose Fontenelle introduces "as spectators of an opera" Pythagoras, Plato, Aristotle, and finally Descartes sitting in an opera-box. They are "viewing the flight of Phaeton, rising on the wind; ignorant at the same time of the construction of the theatre," each of them interprets the phenomenon in a different way: one (Plato) says, "it is some *hidden virtue* in Phaeton which causes him to rise"; another (Pythagoras) says, "Phaeton is composed of certain numbers which produce his elevation"; a third (Aristotle) says, "Phaeton has a love for the top of the stage" and strives to reach it. Finally, Descartes appears in the box as he tells you "that Phaeton rises in consequence of being drawn by cords, fastened to a descending weight, which is heavier than himself. It is no longer believed

[226] Henryk Grossmann, op. cit., p. 208.

[227] I quote the *Conversations on the Plurality of Worlds* (London: Hurst 1803, pp. 9–10) in the translation by Elizabeth Gunning. The first three editions comprised only five evenings. The sixth evening appeared first in the fourth edition of 1698 and takes up "discoveries that have been lately made in the Heavens".

that a body ... can rise and descend without a counterbalancing weight; thus, whoever examines the mechanism of nature is only going behind the scenes of a theatre," where – as we know – are hidden the weights, wheels and the wires of the machinery.

One has hitherto considered the works of nature with more veneration than they deserve. In reality "the universe is but a *watch on a larger scale*; all its motions depending on determined laws and the mutual relation of its parts."[228]

Not only the great inventions that revolutionized the productive process and furthered the beginnings of large-scale capitalist industry had great importance and influence on contemporary thinking, but also the smaller inventions, for instance, the scientific instruments used by scientists in their professional work. These strongly impressed the imagination of such thinkers; they showed them in everyday practice that the hand equipped with an instrument was superior to the bare hand, even if trained. To this group belongs the *velo* invented by the Florentine architect L. B. Alberti and described by him in 1453 as well as the *pantograph* invented eighty years later by the Jesuit Christoph Scheiner; these inventions enabled anyone mechanically to enlarge or reduce all kinds of drawing to any scale, a task for which trained draftsmen were previously required.[229] Here also belong the "geometrical compass" with the help of which arithmetical, quadratic and cubic lines could be represented in the form of a graphic table, thus facilitating the work of scientists. [230] [99]

The end of the sixteenth century also saw the invention by Barozzi of a compass for the mechanical drawing of conic sections,[231] and earlier the invention of E. Radolt who until 1486 directed a famous printing shop in Venice. While formerly a special woodcut, made by trained draftsmen, was needed for every geometrical figure, Radolt, in his edition of Euclid published in 1482, for the first time composed these figures mechanically from lines and other parts just as words are composed of letters.[232] Francis Potter, commoner of Trinity College in Oxford, invented a beam-compass, an accurate instrument by which an inch could be automatically divided into thousand parts.[233] Finally one must list here the *improved portable handmill*, very popular at the time of Cardano and described by him, destined for the use of small households and serving to "sift and bolt flour"; a simple and ingenious device

---

[228] Bernard de Fontenelle, *Conversations*, pp. 9–10. – Leibniz attacks this last paragraph and accuses Fontenelle and those who think similarly of "confondant les choses naturelles avec les artificielles .... Ils conçoivent que la différence qu'il y a entre ses machines et les nôtres, n'est que du grand au petit." *Système Nouveau de la Nature* (1695) § 10. *Leibnitii Opera omnia*, ed. I. Ed. Erdmann, Besolini 1840, vol. I., p. 126.

[229] M. Cantor, op. cit., II, 268 and p. 635.

[230] Invented before the end of the sixteenth century by Michel Coignet from Antwerp, it was improved by Galileo and described in his "Compasso geometrico e militare" (1606) in *Le Opere di Galileo Galilei Edizione Nazionale*, vol. 2, pp. 335–424 and vol. 19, p. 222 (see Cantor, op. cit., II, 629–632).

[231] Ibid., II, 533, 634.

[232] Cantor, op. cit. 266

[233] R. T. Gunther, *Early Science in Oxford*, Oxford, 1937 vol. 11, 229.

enabled it to supply "two or three sorts of flour"; and this machine saved much labor, because "one man who turns the wheels ... does as much as three sifters and bolters." Thanks to this easily operated mill, families became independent of the millers.[234] Here belong also the above mentioned invention of the *circular slide rule* by R. Delamain (1630) and of the *calculating machine* by Pascal (1643).

In the sixteenth century technique reached a stage at which it was possible to adjust every type of machine to various purposes and environments. For instance, when Ramelli deals with bridges, he systematically mentions all the existing kinds of bridges: wooden, stone, iron, with arches, suspension, and finally "decomposable" bridges. [100] When dealing with mills he treats constructions that are driven by human force (treadmills) or by the wind or water, by clockwork mechanisms, i.e. by falling weights, mills built in the middle of a river, mills anchored on two pontoons, etc. Della Porta in his *Magia Naturalis* (1558) treats of all kinds of mirrors. Thus in Chapter V the author speaks "of flat mirrors that let the feet be seen above and the head below [de speculis planis ut caput deorsum, pedes sursum videantur]"; in Chapter VI he deals with "a mirror made of planes in which many images of one thing are made to appear [speculum e planis in quo unius rei imagines plures apparentur," etc.[235] Already Bacon had predicted that "*the art of discovery may advance as discoveries advance*."[236]

We shall briefly illustrate the strength and universality of the interest in technical inventions by two examples: the *exhibitions* of machines and the technical *literature*.

[101] The first official exhibition of machines was not organized in Paris until 1683, as we learn from a pamphlet published in that year, a catalogue describing various exhibitions of machines newly invented: derricks, pumps, mills of various sorts, clocks, levelling instruments, bridges, et cetera.[237] But long before the official exhibitions, various fairs, the centers of international trade, like the fairs in Lyon and Saint-Germain (Paris) and especially the Frankfurt fair, which was internationally

---

[234] Girolomo Cardano *De subtilitate (*1550) here quoted after the French edition: "De la Subtilité" (Paris, 1556) pp. 50–51.

[235] Op. cit., Book IV, Chapters 4–13. It is worth noticing that Della Porta in the 1589 edition of his book (p. 129) already mentions the idea of constructing telephones based on the the magnetic properties of the needle. One can see the impediments in the way of such ideas in the circumstance that Galileo, who is a major representative of the view that every non-visualizable effect and thus every action at a distance is to be considered absurd and unscientific, to be a *qualitas occulta* (*Dialogue*, 1632, First Day) alludes to this idea of Della Porta, whom he suspects of fraud.

The fact that the idea of Della Porta was not an isolated incident but rather was also considered by others can be seen in John Wilkins, *Mercury, or the Secret and Swift Messenger: Shewing how a Man May with Privacy and Speed Communicate his Thoughts to a Friend at Any Distance*, London, 1641.

[236] Bacon, *Novum Organum*, Book I, 130 (*Works* I, 223).

[237] *Explication des modèles des machines et forces mouvantes que l'on expose à Paris dans la rue de la Marpe,* Paris, 1683 (see Harcourt Brown, "The Utilitarian Motive in the Age of Descartes," *Annals of Science*, vol. I (1936) 190). Henry IV at the close of the 16th century had considered organizing a *museum of machines* and industrial models in the Louvre (see Gustave Fogniez, *L'Économie sociale de la France sous Henri IV, 1589–1610*, Paris, 1897, p. 102; and E. Levasseur, *Histoire des classes ouvrières et de l'industrie en France avant 1789*, Paris, 1901, II, p. 171).

famous, performed the same function as can be seen from a report on the Frankfurt fair dating from 1574.[238] "This fair," writes Etienne, "exhibits instruments with the help of which one man performs a work that without them would require the concourse of several workers." And Etienne stresses the rapidity amazing even at his time with which ever new inventions or improvements of old ones followed one another; as an illustration of this he cites "grinding wheels, a machine that gives the arms of a man all the power of a mill ... These wheels that, only a few years ago, aroused general admiration, as though they had fallen from heaven ... saw their importance diminished by inventions derived from them and constructed with incredible art and ingenuity ... machines worthy of Archimedes himself."[239]

The great influence of technology on contemporary thinking can be seen from the emergence (by the middle of the fifteenth century) and rapid growth of technical literature. From the great mass of this literature we shall cite here only a few characteristic examples. In addition to the great number of translations from classical authors,[240] one of the most characteristic books of the period was that [102] of Polydorus Vergilius of Urbino, *De Inventoribus Rerum* (Paris, 1505, the book of inventors), that in the sixteenth century alone had 39 editions. Guidobaldo del Monte's *Mechanicorum Liber* was published in 1577; two Italian translations of it were published in 1581 and 1615. Here belong the works on military machines: Roberto Valtino's *De re militari*, in 12 books written about the middle of the 15th century, published 1472 in Latin in Verona, 1476 in German and 1483 in an Italian translation in Verona; further, Justus Lipsius published his *Poliorceticon, sive de machinis, tormentis, telis, libri V*, in Antwerp, 1596; finally, Galileo's "Trattato Di Fortificazione," in *Le Opere*. Edizione Nazionale. Firenze 1932, Vol. II, 77, 146. The "*Théâtre des instruments mathématiques et méchaniques*" (Lyon, 1578) by Jacques Besson, successor of Leonardo da Vinci as "méchanicien du Roi," illustrated with 60 plates, was also published in two Latin versions (1582 and 1595), in German (1595), Italian (1582) and in a second French edition (1626). Faustus Verantius' *Machinae novae* (Venice, 1617) had engravings with captions in five languages (Latin, Spanish, French, German and Italian). Giovanni Batista Della Porta's *Pneumaticorum libri tres* (Naples, 1601) deals with hydraulic machines; he deals with other machines and instruments in his *Magia naturalis*, 1561 (French translation, 1612). Vittorio Zonca's *Novo teatro di machine* was published in Padua, in 1607. The printer Levinus Hulsius in Frankfurt am Main intended to publish a collection of 15 treatises describing all the machines and mechanical devices used at that time in German language, but as a result of his death in 1607 only 4 of these treatises were

---

238 Henry Estienne, *La foire de Francfort* (1574), Latin and French ed. by I. Liseux, Paris: Liseux, 1875. See also G. Fagniez, op. cit., p. 242.

239 Ibid., pp. 59–60.

240 Vitruvius was published in 1548 in the German translation of W. Rivius. The French translation, *Les dix livres d'architecture de Vitruve*, by Perrault, was published in 1673. Heron's *Automata* was published in an Italian translation by Bernardino Baldi in 1589: *Di Herone Alessandrino de gli automati: ouero machine semoventi, libri due* (Venice: G. Porro, 1589).

[103] published.[241] Even a book like Agostino Ramelli's *Le diverse et artificiose Machine* (Paris, 1588), was published in Italian, French and German – and all of these editions were richly illustrated. One must also mention here the book of Guido Pancirolli (1523–1599) of which the first volume, published toward the end of the sixteenth century deals with the *forgotten* inventions of the ancients[242] and the second volume with *modern* inventions unknown to the ancients.[243]

Finally, the Jesuit Gasparus Schott summarized and popularized all the technical knowledge of his time in his encyclopedia *Magia universalis naturae et artis*, 4 vols. (Würzburg, 1657–59). Volume 3 deals with mechanics and contains numerous [104] engravings of various machines.[244]

We have briefly outlined the most important stages of the revival of practical mechanics in western Europe from the thirteenth to the seventeenth centuries (although lack of space has obliged us to omit the important artillery and lifting engines). It should be mentioned only briefly that the builders of the gothic cathedrals with their heavy tower bells must have had instruments at their disposal to raise great loads to significant heights. In the *Sketchbook* of Villard de Honnecourt (1235) a powerful engine, a vertical screw for raising weights, with hand-spikes for turning[245] is pictured "one of the strongest machines for raising loads in existence," a judgment that even modern engineers subscribe to. In the machine book of Jacques Besson (plate 38a) there is a drawing of a *vertical screw* with handspikes as in Honnecourt's drawing.

One example that might be mentioned is the fact that the tower *della Magione* in Bologna along with its foundations was moved a considerable distance in 1455 without suffering any damage.[246] Parallel to practical mechanics *theoretical mechanics* was developed, particularly in Italy, France, and Holland. The striving for theoretical knowledge began in the practical field. Theoretical mechanics was born slowly, gropingly, from the struggle of man's reason with empirical matter and owed its great advances to the connection between the two, in contrast to antiquity when theory was separated from practice, because the latter was branded by the stigma

---

[241] See *Tractatus primus instrumentorum mechanicorum* (Frankfurt/M, 1605). Cf. also Cantor, op. cit., II, 630. Here belong the works on military machines: Roberto Valtino's *De re militari*, in XII books written about the middle of the 15th century, published in Latin 1472 in Verona, 1476 in German and 1483 in an Italian translation in Verona.

[242] Guido Pancirolli, *Pars Prior, Rerum Memorabilium sive Peperditarum*. There is an editio secunda (Ambergae, 1607/8).

[243] Guido Pancirolli, *Nova reperta, sive rerum memorabilium recens inventarum et veteribus incognitarum*. Ex Italico Latine reddita. (Frankfurt, 1631). There is also an English translation, *The History of Many Memorable Things Lost* (London, 1715) and another (London, 1727).

[244] In vol. III, Books II–VII (pp. 81–575) Schott deals successively with Magia Mechanica, Thaumaturgica, Statica, Hydrostatica, Hydrotechnica and Aerotechnica. Other machines as bellows, hydraulic organs and other musical instruments – are discussed in vol. II (Acoustica). See also Schott's *Mechanica Hydraulico-Pneumatica* (Würzburg, 1658) richly illustrated.

[245] Villard de Honnecourt, op. cit., p. 162.

[246] Libri op. cit. II. 217.

of manual labor and hence despised. But after the end of the Middle Ages, all those who during many centuries participated in the struggle for theoretical knowledge, from Jordanus Nemorarius, L. B. Alberti, Leonardo da Vinci, Guidobaldo del Monte, Niccolò Tartaglia and Girolamo Cardano up until Roberval, Galileo and Descartes, won their *mechanistic concepts* and theoretical theorems from the observation and analysis of machines and their functioning. Numerous texts by Leonardo da Vinci, Niccolò Tartaglia, Guidobaldo del Monte and others bear witness to the aspiration to study and theoretically to formulate the laws of motion above all on the basis of experience with the trajectories of cannon projectiles, with clocks and planetaria, water pumps and lifting engines.[247] [105]

The history of the origin of mechanics teaches us that the discovery of the law of falling bodies was closely linked to the history of firearms. The old Aristotelian milieu-theory of motion (which asserted that a hurled projectile was kept in motion by the air) was finally refuted by the repeated observations made on the projectiles fired from guns because the *inhibiting* effect of air resistance was recognized. With the rejection of the Aristotelian doctrine the path was free for new observations and new attempts at theoretical explanation. From Leonardo da Vinci by way of Tartaglia and Girolamo Cardano to Galileo there is an unbroken chain of scientific endeavor to develop a theory of the motion of falling bodies based on experience with firearms.

Experience with the Motion of Clockworks

It is difficult today to imagine the intellectual revolutions connected to the discovery and perfection of clock mechanisms. Scientific chronometry that is the exact quantification of time is the prerequisite of exact observation in all areas of knowledge; in the field of mechanics the clock is the first and most important machine with a uniform motion that occurs automatically by means of a system of

[247] Leonardo da Vinci's formulations of problems are characteristic: "If a bombard with four pounds of powder hurls a four-pound ball at its greatest power a distance of two miles, by how much must the charge of powder be increased to make it fire a distance of four miles? Does the power of the ball depend upon its initial velocity?" (Gabriel Séailles, *Léonard de Vinci, l'artiste et le savant*, Paris: Perrin, 1906, p. 353). Tartaglia's *Nova scientia* (1537) and *Quesiti et inventioni diverse* (1546) – his first book was devoted to the study of cannon projectiles – it had also, according to P. Duhem, a very great influence on the development of theoretical mechanics in the sixteenth century. (P. Duhem, *Les Origines de la Statique*, I, 197). Maurolycus of Messina (1494–1575), a theoretician of mechanics, wrote a *Treatise on Clocks* (G. Libri, op. cit., III, 108); likewise another theoretician of mechanics, Federico Commandino, wrote *De horologiorum descriptione* (Rome, 1562). Cardano too formulated his theoretical propositions on motion on the basis of clockmakers' experience (G. Cardano, *De rerum varietate*, 1557, Book IX "De Motibus"). Guidobaldo del Monte, in his *Mechanicorum liber* (1577) formulated the principle that weight and force are in inverse proportion to each other with regard to the distance they cover in the same time, on the basis of his experience with pulleys (M. Cantor, op. cit., II (1892) p. 524). And Galileo and Descartes similarly deduced their mechanical principles and concepts from the analysis of lifting engines, as we have shown above. (See above pp. 1–2.)

weights. At first the automatism of motion evoked more interest than time keeping. The public tower clocks of the cities of Italy and Flanders in the 13th and 14th centuries were massive geared machines in which the actual clock – that is, the time-measuring mechanism – was attached to the bell mechanism.[248] The planetaria that begin to appear in Italy in the 14th century were complex gear-works driven by weights that displayed the visible motions of the sun, the moon and the planets.

With this, a field of observation for scientific mechanics was created that had to have fruitful effects on the study of elementary laws of motion: the vertical motion of a slowly falling weight was transformed by a translation apparatus into the circular motion of the gear-mechanism. The automatism of the rotation of the planetarium had to be adapted – in accordance with astronomical motions – to the velocity of motion of individual celestial bodies. The motion of numerous gears and wheels that had to move with different speeds describing different paths was driven by a single weight – a circumstance that had to lead to systematic reflection on the causes of these differences with respect to time and space.

From the observation of machines the conviction was gained that a *perpetuum mobile* was impossible, because as Cardano explains, "in an extended stretch of time all natural bodies, the more they move, are worn away and consumed. [omnia [106] naturalia corpora longo temporis spatio, tum magis si moveantur, atteruntur atque consumuntur]" Thus every clock must be wound up if it is to move.[249] If labor is thus not obtainable without cost, the costs of labor can nevertheless be extraordinarily decreased, and the *fundamental principle of mechanics as well as the actual task of machines is to save human labor*. For, as Petrus Ramus says, God created the world after the model of a balance; hence one can lift great weights by the means of a small pulley.[250] Save human labor! This principle won through the study of machines was stressed by Conrad Dasypodius (1580), the professor of mathematics who constructed the astronomical clock of Strasbourg,[251] and the same principle is cited even in the [sub]title of the Italian edition of Guidobaldo del Monte's "Mechanics" (1581): " "Nellequali si contiene la vera Dottrina di tutti gli Istrumenti principali *da mover pesi grandissimi con picciola forza*" – a title that is reproduced literally in [107] Descartes' *Treatise on Mechanics* (1637).

[248] Pierre Dubois, *Horlogerie: description et iconographie des instruments horaires du XVIe siècle, précédée d'un abrégé historique: L'horlogerie au moyen age et pendant la renaissance, suivie de la bibliographie complète de l'art de mesurer le temps depuis l'antiquité jusqu'à nos jours*, Paris: Didron, 1858, p. 25.

[249] Giralomo Cardano, *De Subtilitate* (Basle, 1560) Liber XVII, 1082.

[250] "Deus universum libravit ... Magna igitur exiguae machinculae opera sunt" (Petrus Ramus, *Mathematicorum libri* (Basle, 1569) pp. 58, 61).

[251] "[Quae sit causas,] quod maxima pondera, minimis moveantur viribus: et quibusnam talis motus fiat machinis" (Conrad Dasypodius, *Heron Mechanicus* (Strasbourg, 1580) p. E2r.

VIII

Our historical survey of technological development has shown us how powerfully and universally the technical revolutions in the course of the sixteenth and seventeenth centuries made themselves felt in various fields of life and science. Descartes was not isolated from this technological trend. From his biographies we know that even in his Jesuit College at La Flèche he received technical training as an artillery officer and among other subjects studied "the art of fortifications and the employment of machines,"[252] and that throughout his life he preserved a strong interest in machines and technological problems. We know from the notes he took in his youth, *Cogitationes privatae*, 1619, that upon receiving the report of an ingenious invention he wondered whether without reading anything on the subject he, too, could not invent something like it;[253] that in his youth he experimented with self-made small automata.[254] We have already mentioned that from 1628 on, Descartes was engaged in work leading to the invention of his glass-cutting machine; we know that according to Baillet, he engaged as a volunteer in Maurice of Nassau's army in Holland (in 1618–1619) because he was attracted by the prince's fame as a mathematician and inventor of machines;[255] further, that in August 1628 Descartes took part as a volunteer in the siege of the Huguenot fortress of La Rochelle by Richelieu, because (as Baillet reports) this afforded him an opportunity to inspect the fortifications of the city built by the famous engineer Pompée Targon and the machines constructed on the seaside by the engineers Marillac and Du Plessis-Besançon and to discuss them with prominent engineers, particularly his friend Desargues, Richelieus's technical adviser.[256] It is further known that among his friends there was a representative of big industry, maréchal Fabert, who was connected with important foundries in Moyeuvre (Lorraine).[257] In addition, during his travels in Italy and Germany and during his stay in Holland, Descartes had abundant opportunities to become acquainted with the technological achievements of these countries.[258]

[108]

[109]

---

[252] P. Mouy, *Le Développement de la Physique Cartésienne* (Paris, 1934) p. 2.

[253] "Juvenis, oblatis ingeniosis inventis, quaerebam ipse per me possemne invenire, etiam non lecto auctore." (AT X, 214).

[254] AT X, 231, and Charles Adam, AT XII, 157.

[255] Adrien Baillet, *La Vie de Mr. Des-Cartes,* Paris, 1691, I, p. 41.

[256] AT I, 157. Gilson (op. cit., p. 286) thinks that it has been established with certainty that Descartes never took part in the siege of La Rochelle; according to Baillet's report Descartes returned from there to Paris on November 11, 1628. But Isaac Beeckman maintains in his *Journal* that at that time Descartes was in Holland. However, Baillet's report is too concrete and detailed to be an invention. Beeckman's entries in his diary do not disprove Baillet. Gilson himself says that Beeckman "is badly informed about Descartes' movements." In order to protect himself against curiosity, Descartes would not have hesitated to retain his old mailing address even though he had gone to France.

[257] Maxime Leroy, *Descartes social*, op. cit., p. 44.

[258] We know, for instance, that he saw the complicated machinery of the astronomical clock in Strasbourg, which in addition to the clock mechanism, moved numerous artificial heavenly bodies and figures and was regarded as a world marvel. (Descartes, Letter to Mersenne, Oct. 8, 1629, AT I, 25).

We have already seen how strongly he was impressed by the water machinery of the park of Saint Germain en Laye. In innumerable passages of all his principal works, Descartes extends his mechanical principles from physics to all phenomena of organic life: the human and animal bodies and their organs: the tongue, nose, eyes, lungs, heart, stomach, muscles and nerves are repeatedly referred to as machines[259] that function according to the same principles as a clock or other mechanical devices.[260] [110]

But the most striking illustration of the influence of mechanical devices on his thinking is perhaps his curious idea that could never have occurred to an ancient or mediaeval man and could only be the product of a "machine age," to wit, that when he looks down at the street from the upper storey of a house, what he sees passing by is perhaps not people, but only *automata* wearing hats and coats.[261] Even ideas as abstract as the famous proof of the existence of God from the reality of the idea of God are constructed, in the *Meditations* (1641) and in the answer to the first "objections" by Caterus, on the example of machines.[262]

The mechanical principles and laws thus gained from the study of machines were extended by Descartes to physics and biology, and finally to the universe;[263] and

---

[259] "I suppose that the body is nothing but a statue or machine" (AT XI, 120; see also pp. 125, 148, 145, 163, 173, etc.).

[260] Thus after expounding the automatism of the circulation of blood (Discourse, V) he says that all this "can happen apart from the direction of our free will. And this will not seem strange to those who, knowing how many different *automata* or *moving machines* can be made by the industry of man, without employing in so doing more than a very few parts." Further, he explains "that this movement which I have just explained follows as necessarily from the very disposition of the organs ... as does that of a clock from the power, the situation, and the form of its counterpoise and of its wheels." (Discourse V, *Philos. Works*, I, 115–116, 112). Descartes has similar things to say at the end of his *Traité de l'Homme* (1644) AT XI, 202. Likewise in the dialogue "Search after Truth" (1628) Eudoxe (=Descartes) speaks of parts of the body as the head, legs, nose, etc. "which constitute the *human machine*" (*Philos. Works*, I, 321).

[261] "I remember that, when looking from a window and saying I see men who pass in the street, I really do not see them, but infer that what I see is men ... And yet what I see from the window but hats and coats which may cover *automatic machines* [whose motions might be determined by springs]. (Second Meditation, *Philos. Works*, I, 155–156). The words in brackets are left out by Haldane and Ross, but can be found in the original text and in the English translation by John Veitch, *The Method, Meditations and Philosophy* (Washington, 1901).

[262] "This is illustrated in these Replies by the comparison of a very artificial machine, the idea of which is found in the mind of some workman. For as the engine corresponding to this idea must have some cause, i.e., the science of the workman... it is similarly impossible that the idea of God which is in us should not have God himself as its cause" (Summary of the third meditation, AT IX, 11, 83, 84; *Philos. Works*, I, 141–2).

[263] Already in his Discourse (V) Descartes says that "the laws of mechanics are identical with those of nature" (*Philos. Works*, I, 115). Likewise we read in Principles of Philosophy, IV, Art. 203: "And it is certain that there are no rules in mechanics which do not hold good in physics, of which mechanics forms a part or species, *so that all that is artificial is also natural*"(ibid., I, 299). In a letter to De Beaune (April 30, 1639, AT II, 542) Descartes also insists that "*all my physics is nothing but mechanics*." He makes a similar statement in his letters to Plempius (Oct. 3, 1637, AT I, 421 and Feb. 15, 1638, AT I, 524).

in his *Principia philosophiae*, Descartes explicitly states that he was inspired by machines when constructing his mechanistic system of the world.[264] [111]

Mechanics – to use an expression of Lewis Mumford – became the new religion and it gave to the world a new Messiah: the machine. Descartes was so dominated by mechanistic ideas that he could not think of the world or any of its parts without immediately comparing them with some machine. As Professor Adam says justly: "He constantly uses ... comparisons borrowed ... from mechanics .... These are not mere comparisons in his eyes but veritable assimilations, almost identifications."[265]

However, the most significant expression of Descartes' mechanistic ideas is not the many comparisons occurring in all his works between the human and animal body and various machines (clocks, sluice locks, mills, fountains, bells, water machines, organs, et cetera), but his cosmology, his thesis of the *formation of the universe* first advanced in the *Traité de la Lumière* (1632),[266] and later in Part III of the *Principles of Philosophy* (1644). In open contrast to the Biblical story of creation Descartes develops, in the form of a hypothesis, the theory that the sun, the earth and everything on it, the fixed stars, comets, the tide and ebb of the sea, et cetera, were not created ready-made, but were *formed automatically and mechanically* in the course of long periods of time from the simplest elements of matter and their motion; or rather, Descartes like an engineer creates before our eyes out of these most simple elements of matter and their motions a model of a second world, a world model that is identical to the real world, thus proving the necessity of its mechanical origin, that from these elements, in accordance with the established mechanical laws [112]

And Leibniz later says the same: "Je serais porté à allier ... le Physique avec le Méchanique ... Je crois que tout Physique dépend du Méchanique dans le fond, mais nous ne saurions encore arriver à ce fond-là" (Leibniz's letter to Grimarest, February 21, 1712 in *Leibnitii opera omnia,* vol. 5, *ed. Dutens*, Geneva, 1768, p. 63). The world is subject to mechanical laws although its ultimate grounds take us back to Metaphysics. "Quand je cherchait les dernières raisons du Méchanisme et des lois même du Mouvement, je fus tout surpris de voir qu'il était impossible de les trouvez dans les Mathématiques et qu'il fallait *retourner à la Métaphysique*" (Leibniz's letter to Remond de Montmort, January 10, 1714, in *Opera omnia*, vol. 5, *ed. Dutens*, p. 8–9).

Epistola de Rebus Philosophicis ad Fred. Hoffmann (1699): "Mihi videris de mechanismo naturae judicare rectissime, et mea quoque semper fuit scientia, omnia in corporibus fieri mechanice, etsi non semper distincte explicare possimus singulos mechanismos: ipsa vero principia mechanismi generalia ex altiore fonte profluere ..." (*Leibnitii Opera Philosophiae*, ed. Erdmann, Berlin, 1840, I. 161).

[264] "The example of certain bodies made by art was of service to me, for I can see no difference between these [machines made by artisans] and various natural bodies." (*Principles of Philosophy*, IV, art. 203, *Phil. Works*, I, 299). The bracketed words left out by translators run in French: "les machines que font les artisans"; not "bodies" made by artisans are in question but precisely "machines."

[265] Ch. Adam, *La vie et oeuvres de Descartes* (AT XII, 162).

[266] AT XI, 3–118: Descartes briefly summarizes this thesis in the Discours V (*Philos. Works*, I, 107–109).

of nature, *only this and no other world could arise. The solar system was evolved mechanically from chaos*, matter having in the course of time taken all possible forms and only those were retained which, according to the general laws of motion, offered adequate conditions of equilibrium and stability. This theory was the prelude [113] to those which Kant and Laplace advanced later.[267]

This was the intention of his first work (*Le Monde)*, which for well-known reasons was left unpublished and, except for the essay on light (*Traité de la Lumière*), has been lost. The essential content is preserved in the last two books of the *Principia* and one can assume that the publication of this work was then superfluous in terms of content.

*****

One problem still seems to require a separate discussion. Today, it is often admitted, as for instance by Professor G. N. Clark and R. C. Epstein, that the invention of machines and technical construction as a rule is not accidental but is conditioned by the times and is the result of a given social situation.[268]

A good illustration of the influence of social environment is Pascal's invention of his calculating machine (1643). This invention was intimately connected with the needs of a rising new social class – the centralized royal bureaucracy: Richelieu required large sums of money for his power politics. In 1639 he appointed Pascal's father tax commissioner for the province of Upper Normandy with headquarters at Rouen, and the latter made a great effort to satisfy the Cardinal's wishes and multiplied the vexatious taxes. Both the collection and imposition of taxes required a great deal of computation. Good calculators were rare, and the young Blaise Pascal was mobilized for this strenuous task. This gave the twenty-year old mathematician the incentive for inventing his automatic adding machine.[269] More accurately, Pascal's invention only adapted an old well-known mechanical device – the clock – to new requirements. For the clock, too, is nothing but an adding machine: it adds up time, seconds, minutes, hours. The complete revolution of a wheel of the lowest order

---

[267] Emile Boutroux, Descartes and Cartesianism, *The Cambridge Modern History*, New York, vol. IV (1907) p. 783). See Immanuel Kant, *Allgemeine Naturgeschichte und Theorie des Himmels oder von dem mechanischen Ursprung des ganzen Weltgebäudes,* 1755, and Laplace, *Exposition du système du monde*, 1796. And Kuno Fischer judges the philosophy of Descartes no differently from Boutroux: "The significance of his natural philosophical system consists in the attempt to explain the origin of the world system purely mechanically." Kuno Fischer, *Descartes und seine Schule*, Heidelberg, 1889 3rd ed. vol. I, Part 1, p. 356. This same thought is stressed by Christiaan Huygens in an epitaph after Descartes' death (1650) where he writes:

> Cette âme qui toujours en sagesse féconde Faisait voir aux esprits ce qui se cache aux yeux, après avoir *produit le modèle du monde*, S'informe désormais du mystère des cieux. (Foucher de Careil, *Oeuvres inédites de Descartes*, Paris: Durand, 1859, II, 236)

[268] "It remains the rule that accidental inventions are of no importance in the great social processes by which our technology has grown." (G. N. Clark, *Science and Social Welfare in the Age of Newton*, Oxford, 1937, p. 8. See also Ralph C. Epstein, "Industrial Invention: Heroic, or Systematic?" *Quarterly Journal of Economics 40* (1926) 232–272).

[269] Fortunat Strowski, *Pascal et son temps*, Paris, 1921, II, 50.

causes the partial revolution of the wheel of the higher order, and that is exactly what takes place in an adding machine.

The extent to which the calculating machine accorded to the needs of the time can be seen in the simultaneous multiplication of that invention. The Dutch Jesuit Johann Ciermans in his "Disciplinae mathematicae" (1640) mentions a calculating machine on wheels which he had invented that could carry out multiplications and divisions.[270]

Sir Samuel Morland invented an adding machine in 1666; Hooke's engine for Multiplying and Dividing was invented about 1670.[271] It is well known that Leibniz was occupied with work on calculating machines his entire life. During his stay in Paris in 1672 he displayed a calculating machine that could perform not only additions and subtraction but also large multiplactions and divisions and even draw square roots. The invention was presented to the scientific societies in Paris and London and brought Leibniz membership in the Royal Society.

However, it is less readily admitted that the same holds true for intellectual constructions and methods; they are often treated as if they were independent of the time and social environment: *spiritus flat ubi vult.* I think there is little justification for ascribing this special position to intellectual constructions, since there is no fundamental difference between inventing a technical and an intellectual construction. In the last analysis, every technical construction is also an intellectual one. Francis Bacon noted the social conditioning of ideas in general and of scientific methods in particular and he emphasized "that in the revolutions of ages and of the world there are certain floods and ebbs of the sciences, and that they grow and flourish at one time and fall at another."[272] Bacon admits that this is also true of his own contribution to scientific method: "And so this method is ... the offspring of time rather than of wit" ("et potius temporis partus quam ingenii").[273] This fact can be illustrated by many other examples of intellectual constructions: in the age of the first penetration of revolutionizing machine technique into industry and private life, the intellectual constructions too were strongly influenced by contemporary mechanistic ideas. In our own day we apply the term "machine" to nonmaterial things; for example, we speak of "bureaucratic machines," "political party machines," or "propaganda machines." A similar concept was current at the time of Descartes.[274]

[114]

[115]

Thus Galileo's discovery of the fundamental law of hydrostatics, concerning the equilibrium of two liquids in communicating vessels, was achieved by means of an intellectual construction that exactly follows the model of a machine (the balance).

[270] See M. Cantor, op. cit. Vol II, 1892.

[271] See R. T. Gunther, *Early Science in Oxford*, Oxford 1923, I, 129.

[272] *Novum Organum*, Book I, 92 (*Works* I, 198–199).

[273] Ibid., I, 122 (*Works* I, 216–217).

[274] The 16th century is the period in which the Theory of "Balance of Trade" develops in England; even the name points to the mechanical model of this phase. See W. H. Price, "The Origin of Phrase Balance of trade," *Quarterly Journal of Economics 20* (1906) 161.

In his "Discorso intorno alle cose che stanno in su l'acqua, o che in quella si mouvono" (1612),[275] Galileo explains how the equilibrium of two columns of water of unequal weight [equal height but different diameter] is established by imagining the two columns as *two different weights suspended on a balance with unequal arms.* The equilibrium is achieved by the compensation of the factor of velocity in one moving object and the factor of gravity in the other object. The very rapid ascension of the small quantity of water in the narrow pipe resists the very slow descent of the large quantity of water in the broad pipe. In brief, the liquids behave like solids on a balance with unequal arms. Galileo's consideration took the problem of the liquids on to the terrain of general mechanics. The hydrostatic phenomena that at first seemed so different from the behavior of solids and resisted being included in any rule finally, thanks to Galileo's intellectual construction, proved to be subordinated to the same mechanical law that is valid for solids. Later Pascal inferred [116] from this[276] that "a vessel full of water is a new principle of mechanics, and *a new machine* to multiply the forces ... a man by this means will be able to lift any weight."

Another illustration of a method constructed under the influence of contemporary mechanistic conceptions is offered by Hobbes who in "De Cive" (1642) declares that in his analysis of the state he follows a method in which "everything is best understood by its constitutive causes." He imagines the state and civil society as a big machine whose essence can be understood only when it is mentally decomposed into its elements deriving from human nature.[277]

Another example of an intellectual construction based on mechanical thinking, that is, formed according to the model of a machine is – as Sombart has shown – the system of exact book-keeping. It was first developed in Italy during the 13th and 14th centuries and found its highest expression in the first scientific system of double entry book-keeping in the *Treatise* of Fra Luca Pacioli (1494) which merely formulated theoretically the practice that had existed for more than 100 years.[278] As Sombart says, at the time in Italy "the general spirit was far advanced in the stage of rationalization and mechanization." Double entry book-keeping with its systematic arrangement of all economic credits and debits to an artful unity "can be called the

---

[275] *Le Opere*, Edizione Nazionale, vol. 4 (1932) p. 687, and 78.

[276] In his *Traité de l'equilibre des liqueurs* written in 1654, *Oeuvres de Blaise Pascal*, III, ed. Léon Brunschvicg and Pierre Boutroux, Paris: Hachette, 1908, p. 163.

[277] "For as in a watch, or some small such engine, the matter, figure, and motion of the wheels cannot well be known, except it be taken insunder and viewed in parts; so to make a more curious search into the rights of states and duties of subjects, it is necessary not to take them insunder but yet that they be so considered as if they were dissolved" (*The English Works of Thomas Hobbes*, ed. W. Molesworth, *Philosophical Rudiments concerning Government and Society*, London, 1841, vol. II, Preface, xiv). See also Frithiof Brandt, *Thomas Hobbes' Mechanical Conception of Nature* (London, 1928).

[278] See *An Original Translation of the Treatise on Double-Entry Book-Keeping by Frater Luca Paccioli*, transl. by Pietro Crivelli, London: Institute of Book-Keepers, 1924.

first cosmos erected on the *principle of mechanical thought*. In short *double entry bookkeeping was born of the same spirit as the systems of Galileo and Newton*."[279]

Must we not assume that what can be established with regard to the intellectual constructions of Bacon, Galileo, and Hobbes, is also true of Descartes' algebraic method?

Under the stimulus of the prevailing conditions a revolution in thinking took place toward the end of the sixteenth and the beginning of the seventeenth centuries; a *new ideal of science* arose: teleological explanations were definitively abandoned in favor of causal explanations, that is to say, of a notion that a valid understanding of things is achieved only when it has been shown that these things are constructed [117] upon the model of machines and operate mechanically, i.e., only when they have been explained in terms of their simplest elementary units and their motions. What is in question, Descartes says, is not to know how the causes operate in the last details. Often they are inaccessible to our senses and thus escape our real cognition. But the task of science is fulfilled when the causes are conceived as operating so mechanically that the visible phenomena can be *rationally* derived from them, even if the real processes take place in a manner different from the one we assume. *What is in question is only a definite type of rational, that is to say, mechanistic explanation* of the phenomena, and not one or another detail of such an explanation because "there is an infinity of different ways in which all things that we see could be formed." Descartes illustrates this point with the example of two differently made clocks keeping exactly the same time.[280] Robert Boyle (1627–1691) developed a similar idea. He reached the conclusion (1663) that machines taught man to regard the universe as a process in which the sequence of phenomena takes place in a purely mechanistic manner. Even when we cannot grasp this sequence and its causes in all [118] their details, we know that they are conditioned purely mechanically by material causes, to the exclusion of all supernatural factors.[281]

Under these circumstances, is it not natural and even necessary to assume that Descartes took the same mechanistic principles that he regarded as the only correct ones and that dominated all his thinking as a model for the construction of his algebraic method, an out of this method created *a kind of intellectual machine operating*

---

[279] See W. Sombart, "Die Entstehung der kapitalistischen Unternehmung," *Archiv für Sozialwissenschaft und Sozialpolitik 41* (1916) 325 and 318.

[280] "I believe that I have done all that was required of me if the causes I have assigned are such that they correspond to all the phenomena manifested by nature without determining whether it is by their means or by others that they are (actually) produced." (*The Principles of Philosophy*, Part IV, Art. 204, *Philos. Works*, I, 300).

[281] "As a man, that sees a screwed gun shot off, though he may not be able to describe the number, bigness, shape, and co-aptation of all the pieces of the lock, stock, and barrel, yet he may readily conceive, that the effects of the gun, how wonderful soever they may seem, may be performed by certain pieces of steel, of iron, and ... wood, [etc.] ... all fashioned and put together according to the exigency of the engine; and will not doubt, but that they are produced by the power of some such mechanical contrivance of things purely corporeal, without the assistance of spiritual or supernatural agents." (*The Works of Robert Boyle*, Birch edition, London, 1772, II, 47).

automatically, quickly and surely? This assumption is confirmed by the results of our analyses of the Algebra (see above, p. [46]); we were able to show that in his mathematical duels with Fermat and Roberval, Descartes claimed as advantages of his algebraic method all the features of machine work in its relation to the artisan technique: simplified manipulation, automatic operation and reduction of intellectual work to a minimum, and above all the rapidity and certainty of the results; we were able also to cite Descartes' own explicit statement emphasizing the mechanical applicability of the algebraic method by a comparison with mechanical work ("travail des mains") that the architect leaves to the manual laborers (carpenters and [119] masons).[282]

By so appraising the role of Algebra, Descartes, far from diminishing its significance, recognizes its real value. In the *automatism* of the algebraic method Descartes sees a *guarantee of its perfection*, because like every other automatism this method is independent of the subjective and individual factors of human nature, "just as a clock ... is able to tell the hours and measure the time more correctly than we can do with all our wisdom."[283] We also know that conversely Descartes inferred the automatic character of animal actions from their perfection.[284] In this contest we must again recall the great influence of the mechanical clock upon the development of our concepts. Man, in the words of Descartes, because of the subjective character of his nature, can measure time only incorrectly. Before the invention of the mechanical clock, in the Middle Ages, *the hours were unequal*; for the days were always divided into 12 day – and 12 night hours, although throughout the year the days and nights are of uneven duration; the relation between day and night changes steadily; similarly a slight journey from East to West alters the time of the sunrise by a certain number of minutes. In brief, the *empirical time* was [120] variable from day to day and from hour to hour. In contrast to this, the mechanical clock, its automatic and accurate timing, independent of the subjectivity of human nature, introduced regular, equal hours, an *abstract time* dissociated from variable empirical phenomena. Only now could the observed phenomena in various places be correlated exactly. Thus the mechanical clock "helped create the belief in an independent world of mathematically measurable sequences: the special world of science."[285]

The automatization of Descartes algebraic method was only a consequence of his theory of cognition. When he attempted to carry out his gigantic plan for a mechanistic explanation of the entire material world, the very first thing he tried to do was to get rid of the sensual qualities as subjective, whereas objective reality was conceived only as mechanical motion. Applied to the algebraic method this

[282] See above, p. 45, fn 101.

[283] Discourse V (*Philos. Works*, I, 117).

[284] Already in his *Cogitationes privatae* we read: "From the very perfection of animal actions we suspect that they do not have free will" (AT X, 219).

[285] Lewis Mumford, *Technics and Civilization*, New York, 1934, p. 15.

attempt implied that here, too, certain results could be achieved only if the subjective elements of human nature were eliminated or minimized and everything was reduced to the automatism of fixed mechanical rules of procedure.

Through this reduction of the algebraic method to fixed, mechanically applicable rules of procedure without much recourse to mental effort, Descartes' idea of a *Universal Science* in the above-mentioned double sense of the word could be realized. [121]

*****

In Descartes, science thus reveals a truly new character, popular in the profoundest sense of the term, and universally human; a character alien to antiquity, as well as to the Middle Ages and the Renaissance. Plato placed the philosophers high above the *masses* because of their intellectual superiority – according to him the masses are unable to philosophize: "Then the multitude cannot be philosophical ... And consequently the professors of philosophy are sure to be condemned by it."[286]

In contrast to Plato, Descartes anticipating and criticizing the danger of overspecialization, wanted to wrest science from the small circle of "specialists" and "virtuosi" and make it accessible to all intelligent people. Never before had such an idea been advanced. Thus Descartes established a *new ideal of science*: it was not to be the monopoly of an élite, but be created by the cooperation of all.

This belief in the intelligence of all people even though they have not enjoyed formal education expresses the well-known fact that since the Renaissance in Western Europe numerous men of the people who received no school training have distinguished themselves by the greatest achievements in the most varied fields of science [122] and art. At the time of Descartes everyone still remembered the case of Niccolò Tartaglia, one of the greatest mathematicians of the sixteenth century who discovered the general solution of equations of the third degree, although, because of his poverty, he had never graduated from a school.[287]

Descartes criticized the virtuosity of the learned specialists who out of vanity and the desire to be admired waste their energy and talents on ingenious games; he speaks ironically of "the solution of empty problems with which Logicians and Geometers have been wont to beguile their leisure."[288] "For really there is nothing more futile than to busy oneself with bare numbers and imaginary figures in such a way as to appear to rest content with such trifles, and to resort to those superficial

---

[286] *The Republic of Plato*, VI, 494. Trans. by John L. Davies and David J. Vaughan (London, 1923) p. 210.

[287] Descartes was driven to the same conclusion by his personal experience, too. "He educated his servant Gillot who became a professor in a School for Engineers in Leyden; and he made an astronomer out of Dirck Rembradtsz, a village cobbler." (On Rembradtsz, see AT V, 266–267; on Jean Gillot, AT I, 264–265, 325; AT II, 89 and AT XII, 262–263).

[288] *Rules for the Direction*, Rule IV, *Philos. Works*, I, 10. Boutroux too emphasizes the fact that in contrast to the surety and regularity of *method* that distinguishes modern science, ancient geometry was only "an enclosed arena where the *virtuosi* of demonstration alone could move ... Descartes explicitly proposes to break with the Greek tradition, and thereby profoundly differs from Fermat." (P. Boutroux, *L'Idéal scientifique*, op. cit., p. 105).

demonstrations, which are discovered more frequently by chance than by [scientific] skill."[289]

[123]

But even the great scientists of the past, says Descartes, such as Pappus, Diophantus and others who discovered several important truths, failed to make their discoveries accessible to others because they failed to show the method by which they reached them out of fear that they would thus lose their prestige.[290]

Descartes directed his criticism at the ancients but he had in mind his contemporaries too. For the scientists of the time of Louis XIII and Richelieu did not behave differently; as Brunschvicg aptly formulated it, they behaved like duelling musketeers: "They conceal the discoveries they publish behind enigmas ... they reserve them for an occasion when they will be able to confound a rival."[291]

Real scientific duels were fought among the scientists of that time, such as the duels between Descartes and Fermat and Roberval, and the opponents even appointed "seconds."[292] If in the duel between Descartes and Fermat every blow that was dealt marked an advance in mathematics,[293] on other occasions science was injured by such relationships. In 1654 when Pascal began to exchange ideas with Fermat concerning the calculus of probability the results achieved by his correspondent eighteen years earlier were unknown to him. And Pascal just as little suspected in 1658 when he submitted six problems concerning the cycloids to a public contest that four of them had been solved before – solved by Roberval, his father's and his own intimate friend.[294]

[124]

Only this contemporary background enables us fully to understand Descartes' attitude, his opposition to petty specialization. Only a *science universelle*, he felt, could eliminate the dangers threatening modern culture as a result of overspecialization. Descartes appeals to a new kind of audience the great masses of all intelligent people and is convinced that every individual is equipped with a "complete" reason, that is to say, with a reason that does not lack any essential component necessary for its correct functioning.[295] He is convinced that the "natural light" *in so*

---

289 *Rules for Direction*, Rule IV. *Philos. Works*, I, 11. The important word put here in brackets was omitted in the English translation.

290 "They acted just as many inventors are known to have done in the case of their discoveries, i.e., they feared that their method being so easy and simple would become cheapened on being divulged, and they preferred to exhibit in its place certain barren truths, deductively demonstrated with show enough of ingenuity, as the results of their art, in order to win us our admiration for these achievements, rather than to disclose to us that method itself which would have wholly annulled the admiration accorded" (Rule IV, *Philos. Works*, I, 12).

291 L. Brunschvicg, *Le Génie de Pascal* (Paris: Hachette, 1924) p. 3.

292 AT XII, 261–270.

293 AT XII, 261.

294 L. Brunschvicg, op. cit., p. 3.

295 "For as to reason or sense . . . I would fain believe that it is to be found complete in each individual, and in this I follow the common opinion of the philosophers" (*Discourse*, I, *Philos. Works*, I, 82). The widespread extent of such views at the time can be seen, for instance, in the book of the Spanish Jesuit Juan de Mariana, *De rege et regis institutione libri III* (Toledo, 1599) [Reprint Aalen: Scientia 1969], in which he bases the justification of his radical theory of popular resistance against monarchy which degenerates into tyranny (including even the people's right to tyrannicide) on the *common sense of mankind in distinguishing the honesty from the turpitude*. "Et

far as it has not been spoiled by false education is more fit for the cognition of truth than a science of the guilds [*Zunftwissenschaft*] degenerated by over-specialization. "This is confirmed by experience; for how often do we not see that those who have never taken to letters, give a sounder and clearer decision about obvious matters than those who have spent all their time in the schools."[296] And in the unforgettable [125] words of his treatise on the "Search after Truth" he rejects the scholastic wisdom of the learned in favor of a science that makes use of reason and natural light.[297]

For that reason Descartes' attitude toward his method is quite different from that of the learned specialists mentioned above. "I mean not to employ it to cover up and conceal my method for the purpose of warding off the *vulgar*; rather I hope so to clothe and embellish it that I may make it more suitable for presentation to the human mind."[298] [126]

---

est communis sensus quasi quaedam naturae vox mentibus nostris indita, auribus insonans lex, qua a turpi honestum secernimus" (*De rege*, Book I, Chapter VI, p. 74).

[296] *Rules for Direction*, Rule IV, *Philos. Works*, I, 9. In his letter to Mersenne about Herbert de Cherbury (Oct. 16, 1639, AT II, 598) Descartes also expresses the view that "*all men have the same natural light*." "quod etiam experientia comprobatur, cum saepissime videamus illos, qui litteris operam numquam navarunt, longe solidius et clarius de obviis rebus judicare, quam qui perpetuo in scholis sunt versati." (Reg. IV. AT X, 371) Similarly Descartes writes in the preface to the *Principes*: "Je suis seulement obligé de dire, pour la consolation de ceux qui n'ont point étudié, que . . . lorsqu'on a de mauvais Principes, d'autant qu'on les cultive davantage, . . . d 'autant s'éloigne-t'on davantage de la connaissance de la verité et de la sagesse. D'où il faut conclure que ceux qui ont le moins appris de tout ce qui a été nommé jusques ici philosophie, sont les plus capables d'apprendre la vraie." (AT IX.2, 8)

Similarly P. Charron (1601): "*Elle (la Science) ne sert point à la vie;* combien des gens riches et pauvres, grands et petits vivent plaisamment et heuresement sans avoir ouy parler de science? . . . Regardons un peux ceux qui . . . viennent des écoles et universities et ont la teste toute pleine d'Aristote, de Ciceron. Y-a-t-il gens au monde plus ineptes et plus sots et plus mal propres à toutes choses?" (*De la Sagesse*, Bk. 3, Chapter 14. p. 526) "Prenes un de ces *scavantaux*, memès le moy au conseil de ville en une assemblée . . . S'il se mesle de parler, ce seront de longs discours, des definitions, division d'Aristote; . . . Escutés en ce mesme conceil *un marchand, un bourgeois, qui n'a jamais on y parler d'Aristote; il opinera mieux,* donnera de meilleurs advis et expédiens que les scavants" (ibid.).

We find the echo of these views 150 years later with Adam Smith, who shared the belief that men were born equal. The differences and inequalities between them arise after the birth, not from nature, but from habits and different education (Adam Smith, *An Inquiry into the Nature and Causes of the Wealth of Nations* (London: Dent, 1920–21) I.ii.4).

Search after Truth, the student "ne se laissera pas tromper par les 'grandeurs de fortune,';. . . il 'sondera la portée d'un chacun'; il trouvera profit à la conversation 'd'un bouviez, d'un maçon, d'un marchand'." (Strowski, *Montaigne*, pp. 242–243)

[297] "A good man has no need to have read every book, nor to have carefully learned all that which is taught in the schools; it would even be a defect in his education were he to have devoted too much of his time to the study of letters. There are many other things to do in life, and he has to direct that life in such a manner that the greater part of it shall remain to him for the performance of good actions *which his own reason ought to teach him*, even supposing that he were to receive his lesson from it alone." ("The Search after Truth by the Light of Nature," *Philos. Works*, I, 305).

[298] Ibid., I, 11. One should compare this ideal of Descartes who wants to make knowledge accessible to all with the views of Spengler, who attacks the Western European countries ("the Faustean culture") for their "decisive mistakes": "instead of keeping strictly to themselves the technical knowledge that constituted their greatest asset," they "complacently offered it to the world." Thus began what Spengler calls "treason to technique": they began the export of secrets, processes,

Here, with Descartes, the last remnants of class distinction vanished in the sphere of human thinking. He values the "vulgar," the common man as much as the professor or dignitary. Similarly Descartes frees himself from the prejudice against women: he hopes that "even women" will be able to understand his *Discourse.*[299] While earlier philosophies emphasized the *differences* among people, resulting from the variety of religions, races, levels of civilization and social positions, Descartes emphasizes the *common universal elements of human nature* which join all humanity into one whole and which result from the identical character of human reason: thanks to reason all men, because they are thinking beings, are fundamentally equal. A century and a half before the French Revolution Descartes proclaimed the fundamental equality of human reason in impressive words. The opening sentences of the *Discourse* state with emphasis: "Good sense is of all things in the world the most equally distributed ... The power of forming a good judgment and of distinguishing the true from the false, which is properly speaking what is called [127] good sense or reason, *is by nature equal in all men*,"[300] whether he be prince or peasant.

The picture outlined here of Descartes differs from that handed down to us by tradition, which shows him as a cold scientist locked in a study, remote from life. We know, however, that he demanded of science not only abstract truth, but also practical usefulness for the greater well-being of mankind; he wished to become a "master and possessor of nature" in order to alleviate the lot of all those who labor. The same philosopher reveals that he believes in the common man, in the equality of reason in all men, and therefore, that he had an "enthusiasm for science"[301] that was no longer to be the "secret art" of a *few initiates* but the common result of the intellectual *cooperation of all intelligent people*.[302]

---

methods, etc., instead of keeping these as a monopoly for themselves. Thus the privileges of the leaders of civilization were betrayed: "The others have caught up with their instructors ... The exploited world is beginning to take its revenge on its lords" (Oswald Spengler, *Man and Technics*, New York, 1932, pp. 99–102). There are two different intellectual worlds that stand opposed here. The world of Descartes, convinced of the basic equality of all men, and the other world of Spengler, who believes in a Faustian master-race that has the right to rule over all other peoples.

[299] Letter to Father Vatier, Feb. 22, 1638 (AT I, 560).

[300] *Discourse*, I, *Philos. Works*, I, 81. Several years ago, in a discussion carried on in France, some attributed an *ironical* sense to these words of Descartes. Perhaps! But is the meaning of all the other passages we have cited that also emphasize the equality of human reason in all men ironical too? This objection only shows that its authors prefer to take the path of least resistance, that they strive to get rid of the *great problem raised by Descartes*, by a witty remark. No particular perspicacity is required to see that Descartes certainly did not assume the absolute mathematically exact equality of reason in all men since after all he tried to construct a method for the less talented! Equal or "complete" reason in each individual thus must mean something different, to wit, that everyone even endowed with only average gifts, possesses the necessary minimum required for the cognition of truth. In other words, every individual has a reason that does not lack any components essential for its functioning, hence that all men have a reason that fundamentally *operates in the same way*.

[301] Charles Adam, AT XII, 230.

[302] See above p. [55].

The last sentences of the *Discourse* constitute an open challenge to the specialists. Descartes addresses his work not to them but to the broad intelligent public, to every man with good sense, and is convinced that these men are better able to [128] appraise his work than the specialists. Hence, "although he wrote on philosophical and mathematical subjects with greater ease in Latin than in French,"[303] he wrote the *Discourse* not in Latin, the language of the professors, but in French, the language of the people: "If I write in French which is the language of my country, rather than in Latin which is that of my teachers, that is because I hope that those who avail themselves only of their natural reason in its purity may be better judges of my opinions than those who believe only in the writings of the ancients."[304]

[303] J. Sirven, op. cit., p. 28, note.
[304] Discourse, VI, *Philos. Works*, I, 129–30.

Additional Texts on Mechanism

Henryk Grossmann

1 From a Letter to Friedrich Pollock and Max Horkheimer, 23 Aug. 1935 (from Valencia, Spain)

But let me come back, to the actual purpose of my letter, in connection with my last piece of work. The problem still concerns me since it is still not yet completely solved so long as one *side* of it is not explicated. It is not enough to show – positively – why mechanics, for instance, arises around 1,500; one has to prove negatively, why, for instance, it could *not* arise in *antiquity*.

Therefore I am not satisfied with my paper in the *Zeitschrift*, but it was determined by polemical critical aspects, and under these circumstances I could not systematically enough develop the actual ideas.

Now, to the question, why no scientific mechanics could arise in antiquity, the answer has been given since Marx, that in a society based on slavery there was no occasion for saving human labor and replacing it by machines. Even Henri Hauser (*Les Debuts de Capitalisme modern*) agrees with this view. Now Marx's answer is on the whole right but not *concrete* enough. Marx was a visionary, he grasped the result correctly but without showing us the way by which he arrived at this result! I too advocated this view in the essay "The Capitalism of the Renaissance," which I gave you, dear Pollock, before your trip to America; but I don't find this answer completely satisfactory. Now the trip to Spain allowed me to answer this important question much more concretely, thanks to a genuine *trouvaille* which I made in the National Archeological Museum in Madrid! . . .

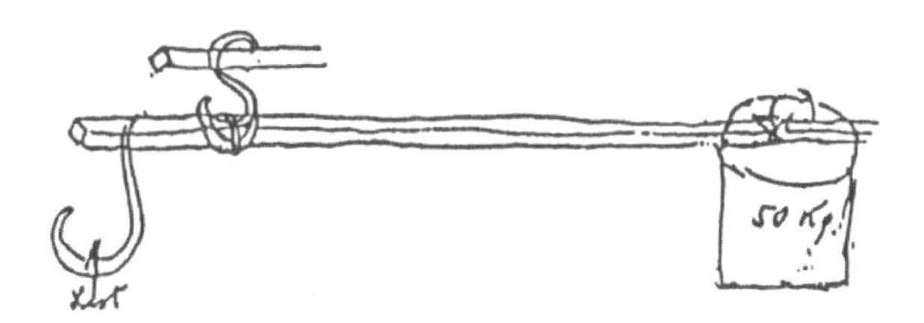

G. Freudenthal, P. McLaughlin (eds.), *The Social and Economic Roots of the Scientific Revolution,* Boston Studies in the Philosophy of Science 278,
DOI 10.1007/978-1-4020-9604-4_5, © Springer Science+Business Media B.V. 2009

Now to say that mechanism could not arise in antiquity because of slavery is not quite precise; there was mechanism in antiquity. You surely remember (1) the giant *catapults* in the Roman settlement near Frankfurt (whose name I have forgotten). Huge stone balls weighing 25–30 kg could be thrown great distances with impact. (2) The Romans also possessed the *balance* with unequal arms, and in French language areas "romaine" ("romana") is the name of the unequal armed balance. The Romans were familiar with the most important relationships of equilibrium on the lever with unequal arms. They also knew that the longer the arm, the smaller must the weight be to keep the load hanging on the other arm in equilibrium, – (3) And finally I found a real *water pump* in the museum!! A suction and discharge pump that could spray water great distances – a hand pump as a matter of fact. The pump is a complicated machine, and thus one must ascertain that in antiquity machines – (1) catapults, (2) levers, (3) pumps – were known.

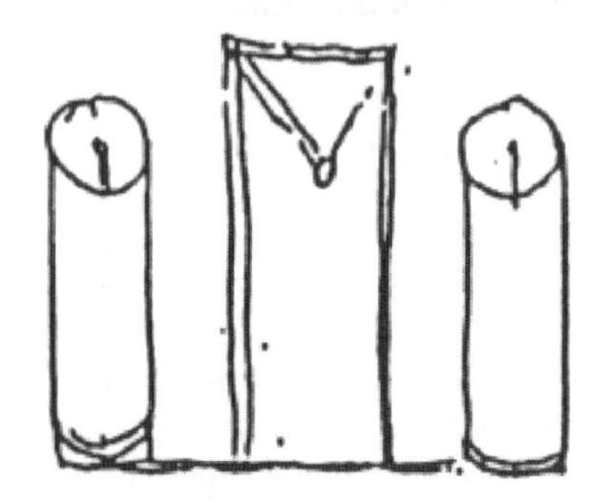

People like Tillich and like thinkers would thus say that it was merely an *accident* that mechanics arose in the 15th century and that it could just as well have arisen with the Romans.

Now the closer inspection of these machines allows me to show why theoretical mechanics could *not* arise in antiquity.

(1) The *catapults* did not have the purpose of replacing slave labor, but rather of hurling heavy stone balls over enemy embattlements into the enemy camp in order to cause damage. Slave labor could not help here; it was not being replaced.
(2) The balance in Madrid for instance is from a *Saliteria*, a saltpeter reduction plant (*Salpetersiederei*) in Temblegue (near Toledo). Here, too, the point was not the replacement of slave labor, but apparently *exact quantification* in the sale of saltpeter.
(3) Finally the details of where the pump was found are interesting. Namely, 80 m under ground in the iron pyrite *mines* at Valverde (Province of Huelva) in southern Spain near Gibralter. It was a *fire spray* in the mines. Here, too, it was not the purpose to replace slave labor; because of the great heat and danger of explosion, humans could not approach the source of the fire. It had to be dealt with at a distance.

In all three cases mentioned we have to do not with the replacement of human labor but with *special aspects* which provided occasion for the construction of the machines. Thus the fundamental principle of mechanics could not arise – the idea that with the machines we are dealing with the *replacement* and saving of *human labor*. I believe that the three cases I have adduced clearly demonstrate this and thus provide an answer to the question posed that is much more concrete than any given so far in the literature.

From Max Horkheimer *Gesammelte Schriften*, vol. 15 Frankfurt/Main: Fischer, 1995, pp. 392–396
Drawings by Grossmann.

2 Review of G.N. Clark, *Science and Social Welfare in the Age of Newton* (New York and London: Oxford University Press, 1937) and George Sarton, *The History of Science and the New Humanism* (Cambridge, Mass.: Harvard University Press, 1931) *Zeitschrift für Sozialforschung* 7 (1938) 233–237

Clark's little book, the fruit of unusual erudition and vast knowledge is a valuable contribution to the problem of the social determination of inventions and science, the attitude of the author with regard to the two being fundamentally different.

The introductory essay "Science and Technology" describes the general social prerequisites of scientific research in England and France during the 17th century. The major interest of the time was directed towards technology and mechanical inventions. Many wealthy aristocrats had their own laboratories, technical literature was widespread, and in both England and France scientific societies such as the *Royal Society* (1662) and some years later the *Academie Royal des Sciences* were founded, which provided a forum for organized technological studies often in cooperation with industry. On this background Clark sketches in a further essay ("Economic Incentives to Invention") in a masterly fashion the importance of economic incentives for invention or for the greater application of already available but not yet applied inventions. Even more important are the impulses to inventions, that derive from the social form of organization of labor and especially from the striving for profits. Inventions, Clark believes, are the answers of entrepreneurs to economic depressions in which profitability suffers through the fall of commodity prices. Improved mechanization was supposed to lower labor costs and thus restore profitability. The relative backwardness of mechanization in agriculture is not, according to Clark, due to the lack of appropriate inventions; it is rather determined by the fact that the machines are not needed in consequence of the surfeit of cheap labor: small farmers had available the cheap labor of their family members. Wherever there existed large scale firms producing for export, which employed expensive wage labor, machines were invented and introduced early on or changes in the social form of the organization of labor were carried out, such as turning to the employment of cheap slave labor in the sugar and tobacco plantations. In strongly depopulated Spain, where wage labor was especially expensive, the Sembador, a combined plowing, sowing and harrowing machine, was used as early as the middle of the 17th century.

However, Clark goes a step farther and attempts to show the regularity of the uneven technical development of individual historical periods. He attempts in analogy to the economic cycle to adduce the great historical boom and slump periods to explain the uneven character of technological development in these periods. The long period of monetary devaluation after the discovery of America meant the devaluation of all fixed rents, thus harming the creditors and benefiting the debtors, especially industrial debtors; this naturally encouraged the rise of the new capitalist entrepreneurs and their export. The noticeable scarcity of wage labor and the rising prices gave the impetus to introducing many inventions, especially in export industries, which were engaged in sharp competition on the world market. The great majority of inventions consisted in "labor saving" machines.

But the impulse to introduce technical improvements became even stronger in the succeeding period of decline, when from the middle of the 17th century the change in prices occurred in the opposite direction. The fall in prices sharpened competition and drove towards protectionism and to the wars for sales markets that were characteristic of the mercantilist era. This is the period of practical and intellectual Colbertism, in which technological activity and the number of machine inventions increased significantly.

The fourth essay ("Social Control of Technological Improvement") describes on the basis of numerous examples the impediments placed in the way of the introduction of technical improvements during the 16th and the 17th century, whether by laborers threatened with unemployment or by the guild organizations. The latter defended their monopolistic domination of the market, excluding all competition and in this way contributed much to slowing down the pace of technical progress and capitalistic development.

In spite of the masterful presentation some weak points of the book can be seen in the fact that although Clark factually presents the connection between technology and social history, he has not mastered it theoretically. He asserts, for instance, that we are only very poorly informed about the class structure in early capitalism. Thus he believes that the introduction of machines occasioned the transition to mass production of cheap and necessary everyday commodities, in contrast to production in early manufacturing, which satisfied only luxury needs. Against this it should be emphasized that mass production is not the simple result of the increased technical capacity as a consequence of the introduction of machines, but rather was only made possible by the rise of broad strata of the affluent bourgeoisie in the 16th century, as opposed to earlier periods, where production was restricted to luxury needs of the small circle of nobility and higher clergy.[1]

---

[1] See my work cited on the next page [p. 233].

The weaknesses of the point of view taken by the author come out much more strongly in the main essay of the book, "Social and Economic Aspects of Science." While Clark admits the social and economic determination of technical invention, his attitude towards science is completely different. Science is placed outside the sphere of influence of economics, and the saying *spiritus flat ubi vult* ["the spirit flows where it will"] applies to it. The external impulses to scientific research are said not to be exclusively economic; they come to a certain extent from four other sources: art, religion, medicine, and war, which are radically independent of economic influence. The development of medicine, for instance, is not determined by any social, in particular economic, conditions but rather is the expression of a general human tendency toward increasing the length of life. But above all, the disinterested striving for knowledge independent of all utilitarian intent is at work here. For this reason Clark criticizes the essay of the Russian professor B. Hessen, who undertook to explain the specific character of Newtonian mechanics from the social and economic conditions of his time.[2] In spite of Clark's assurances that he wants only to discuss Hessen's main thesis, he in fact gets lost in polemical details and, as will be shown below, does not even mention Hessen's main thesis at all. If science is independent of economic and social influence and is to be explained merely through a striving towards truth, then it is hard to understand why classical mechanics only developed in the two centuries from Leonardo da Vinci to Descartes, Galileo, and Newton and did not already arise in classical Greece in view of the high state of development of philosophy and mathematics at that time. Clark raises the objection against Hessen, that in support of the asserted relationship between the scientific view of Newtonian physics and mechanics and the practical problems of mining, transportation, and the war industries he has only cited the names of many important authors but has provided nothing more than general illustration. To this it should be replied that the demonstration of this relationship was not in fact the focus of Hessen's deliberations. As for the relationship of the mechanistic conception of such authors as Descartes and Galileo to the development of industry in the sixteenth and seventeenth centuries, allow me to refer to my own work, where a more detailed argument is attempted.[3] Hessen's main thesis, however, goes much farther, a fact that Clark seems to have missed. Hessen attempts, namely, to understand the general character of classical mechanics and physics in distinction to the later development of these sciences.

In nature we encounter various forms of motion of matter (mechanical, thermal, electromagnetic) which are interconnected and are transformed one into the other, and one never finds in nature pure forms of motion isolated as such. Nonetheless, it is characteristic of the entire classical mechanics from Leonardo da Vinci to its completion with Descartes, Galileo, and Newton that it is only the theory of a single

---

[2] B. Hessen, "The Social and Economic Roots of Newton's *Principia*," In: *Science at the Cross Roads*, London 1931. [Grossmann cites the separately paginated offprint].

[3] Henryk Grossmann, "Die gesellschaftlichen Grundlagen der mechanistischen Philosophie und die Manufaktur," *Zeitschrift für Sozialforschung 4* (1935) 200–215 [this volume, pp. 141–144].

form of motion, the mechanical form, and never deals with the transformation of mechanical motion into other forms of energy.[4]

Hessen explains this state of affairs through the fact that during the period of classical mechanics the machines employed in industry (lifting and water machines) were used only for the transmission of this one, mechanical form of motion. A change occurred only with the development of large industry immediately after Newton, in particular the invention and extended application of steam engines, which gave impetus to the study of new, thermal forms of motion of matter (Watt's investigations on the thermodynamic properties of steam) and led to the establishment of thermodynamics as a particular sector of physics.[5] The further development of thermodynamics through Carnot (1824) is very closely connected with the study of steam engines. Whereas in the machines of classical mechanics one form of mechanical motion was transformed into another form of the same mechanical motion (e.g. a rectilinear motion into a circular motion), the essence of the steam engine lies in the transformation of a thermal form of motion into a quite different, namely mechanical, form. In this manner and conditioned by new technical developments, there arose the entirely new problem of energy transformation, not a trace of which can be found in classical mechanics.[6] With the invention of electromagnetic machines – and based on experience with them – electromagnetism was elaborated as a special section of physics alongside the already mentioned sections of mechanics and thermodynamics, and the general doctrine of energy transformation and its special forms was perfected. Thus it can be seen that science is not at all independent of social development, but rather that it is precisely the economic-technical development of bourgeois society that can not only make understandable in terms of content the development of individual sectors of science but also allows us to comprehend the historical order of the individual stages of this development.

Sarton's book collects contributions of diverse content and uneven quality. The essay "The New Humanism" is a program of reform for the teaching of history. In the introductory methodological chapter Sarton presents, from the standpoint of his idealistic conception of history, a critical discussion of individualistic historiography, not because it is individualistic but because it singles out the "wrong" heroes. Sarton would like to replace the kings and great dignitaries who waged wars with the real "creators" of history, the creative architects, craftsmen, thinkers, social reformers and saints. Sarton denies the economic determination of science and art. On the other hand, he makes concessions to the opposing view when he admits that "a certain determinism in the order of discoveries" exists and when he emphasizes the great significance of practice. The most valuable part of the book is the essay "East and West" which summarizes briefly the results of the author's research. As opposed to many authors – e.g., Max Weber – who glorify modern science as the exclusive product of the western-European spirit, Sarton shows the

---

[4] Hessen, op. cit. 38, 43 [188], [194].

[5] Op. cit. 49 [199].

[6] Op. cit. 44 [194].

strong and significant influence of the Orient, the great scientific achievements of the ancient Egyptians, Babylonians, Persians, Indians and Phoenicians, later the Jews and Arabs. One can speak not only of a Greek "miracle" but also of an Arab miracle. The superiority for instance of Arab culture vis à vis western European culture up to the middle of the 13th century cannot be disputed. Western-European science of the last five hundred years was only possible on the basis of the five thousand years of scientific exertion of the Orient.

Henryk Grossman: A Biographical Sketch

Rick Kuhn

Henryk Grossman is best known as an economic theorist. This sketch outlines his life, with an emphasis on his commitment to Marxism, which underpinned not only his contributions to the theory of economic crisis, the labor theory of value, imperialism and Marx's method in *Capital*, but also his work in economic history, the history of economic thought and the history of science. It draws on *Henryk Grossman and the recovery of Marxism*, the full biography of Grossman, and only provides citations to sources not referenced there.[1]

Grossman's parents, Herz Grossman and Sara Kurz, had an unconventional relationship. Herz was much older than Sara and they had five children between 1876 and 1884, before getting married in 1887. Herz's business activities were very successful. The family was Jewish, but increasingly assimilated to the Polish high culture of Galician government, big business and art. For the constitutional reform of the Austro-Hungarian Empire in 1867 had given effective local control of its economically backward Polish province and its thoroughly undemocratic parliament to the upper reaches of the Polish nobility. On 14 April 1881, Herz and Sara's son, Chaskel, was born in Kraków, the administrative capital of western Galicia and the cultural capital of partitioned Poland. Chaskel is a Yiddish name (derived from Ezekiel) but he was probably known, like his father, as Henryk from an early age, and formally adopted this Polish name in early 1915.

Henryk and his surviving brother went to a twelve year, academic high school. The state languages of Galicia and the Empire, Polish and German, were on the curriculum every year, as were religion, geography, math and science. His overall results were above average, but not outstanding. During the semester following his father's death, in June 1896, his grades, particularly in mathematics, fell.

R. Kuhn (✉)
Australian National University, Canberra, Australia
e-mail: Rick.Kuhn@anu.edu.au

[1] In German speaking countries Grossman was generally known as Henryk Grossmann. Rick Kuhn, *Henryk Grossman and the Recovery of Marxism,* University of Illinois Press, Urbana and Chicago, 2007.

G. Freudenthal, P. McLaughlin (eds.), *The Social and Economic Roots of the Scientific Revolution,* Boston Studies in the Philosophy of Science 278,
DOI 10.1007/978-1-4020-9604-4_6,

Once he came of age Henryk did not, to his mother's disappointment, take on major responsibilities in the family's business enterprises, coal mining, warehousing of imports and exports and landed property. But young Grossman had certainly absorbed some ruling class tastes and even personality traits. The theatre, the great works of Polish literature, art exhibitions and especially classical music gave him pleasure. He shared the fashionable, slightly bohemian enthusiasm for skiing, mountain air, climbing and rambling in the Tatras. Standards of personal behavior in Galicia were set by the Polish nobility. In this spirit, Henryk developed considerable self-confidence and a strong sense of personal honor.

While still at school, however, Henryk had become involved in the socialist movement. By the end of his first year at the Jagiellonian University, he was taking an interest in the Jewish question and was critical of the nationalism of the Polish Social Democratic Party (PPSD), the Galician component of the federal Austrian Social Democratic Party. He agitated in favor of rights for Ukrainian students at the university in Lwów (now L'viv, in the Ukraine) and was soon a leading figure in 'Movement', the main organization of radical and socialist university students in Kraków. His political activities included support for Marxist parties in the Russian-occupied 'Congress Kingdom of Poland'. Grossman arranged the smuggling of literature and participated in a welfare body associated with the Social Democratic Party of the Kingdom of Poland and Lithuania, the Party of the influential Marxist theorist Rosa Luxemburg, who was by then living in Germany.

From around 1902, Grossman increasingly devoted his political efforts to revitalizing the Jewish labor movement in Kraków. He established and became the secretary of a new general Jewish workers' association in Kraków, 'Progress', which started out with twelve members. In 1903, when he was the secretary of 'Movement', the group's membership peaked at 110, while 'Progress' already had 130 adherents.

In the face of PPSD neglect of the Jewish working class and the escalating recruiting efforts of Labor Zionists, Grossman, many hundreds of workers and other Jewish university students split from the Polish social democrats in Galicia. They announced the foundation of the Jewish Social Democratic Party of Galicia on May Day 1905, in four Galician cities. As well as writing a pamphlet on the Jewish question before the split, Grossman was the principal author of the JSDP's founding manifesto, and probably the key document in the bulletin which preceded its first congress. He was elected the organization's first secretary.

Although never admitted to the federal Austrian Social Democratic Party, the JSDP quickly organized about 2,000 workers, the bulk of the PPSD's Jewish membership. The new party played a very active role promoting the upsurge in workers' struggles in Austria-Hungary, which followed the outbreak of revolution in Russia in January 1905. Conducting most of its agitation in Yiddish, it came to dominate the Jewish labor movement in Galicia. The JDSP's militancy in economic and political struggles helped it outpace the Labor Zionists. By leading strikes and boycotts, it was able to organize new sections of the working class into local general unions and branches of the Vienna based industry unions under its influence. From the second half of 1905 through to 1907, the campaign for universal male suffrage in Austria was one of the JSDP's most important activities, as it was for the whole Austrian social democratic movement. A crucial element in this and the Party's other work was its newspaper, the *Social democrat*, published weekly from October 1905.

At the JSDP's 1906 Congress, Grossman retired from the position of secretary, with its now routine responsibilities. But he remained on the Party Executive and in the day to day leadership of the Party until late 1908, apart from a period of study in Vienna, during winter semester 1906–1907. When the Party adopted its own national program at the 1906 Congress, it was Grossman who introduced the discussion. Previously the JSDP had simply accepted the federal Austrian Party's formal position on the national question, which favored transforming Austria into a federation of national territories. Now the Jewish social democrats demanded 'national cultural autonomy', the position of the Jewish Workers Union of Lithuania, Poland and Russia (known as the Bund), in the Russian empire.

By 1907, in the wake of the defeat of the revolution in Russia and as the economy moved into recession, the Austrian labor movement was on the defensive. During the second half of the year, Grossman wrote *Bundism in Galicia*, a history of Jewish workers' efforts to establish socialist organizations in Galicia. This pamphlet provided a sophisticated historical and theoretical justification for the existence of the JSDP. The pamphlet made a similar case about the nature of an effective socialist party to that of Vladimir Ilyich Lenin and anticipated György Lukács's explanation, in the early 1920s, that the working class was both an object and the potential subject of history. Grossman's nuanced materialist history of ideas offered a sophisticated and dialectical analysis of the relationships amongst political organization and consciousness, national oppression and the routine struggles of the Jewish working class.

Neither Jewish bourgeois nationalism in the form of Zionism, Grossman pointed out, nor Polish nationalism served workers' interests. Nor did the distant prospect of socialism offered by the PPSD address the immediate problems of Jewish workers as both an oppressed and an exploited group. Both the PPSD and the Zionists contributed to the passivity of the Jewish masses. In contrast to their approaches, 'analysis of *all the practical interests* of the Jewish workers' movement and all the important phenomena of Jewish social life' was a precondition for making socialism relevant to Jewish workers and winning them from rival ideologies. According to Grossman, only a Jewish working class party could do these things.

While remaining active in the Party, Grossman also took the opportunity of a quieter political period to get on with his academic career. He completed his doctorate (a first degree) in late 1908. Although his comrades knew that Grossman was about to leave Kraków for a long stay in Vienna, he was still the most prominent figure in the discussions at the JSDP's third Congress, in October 1908.

On 1 December 1908, soon after arriving in Vienna, Henryk married Janina Reicher, a painter and daughter of a wealthy commercial agent in Russian Poland. Socialist friends published congratulatory messages in the *Social democrat*. Not only was Grossman re-elected to the Party Executive, in 1908, but also in 1910 at the fourth Party Congress, which he did not attend. There was no room for him on the Executive after a special Congress in 1911, when the rump of the PPSD's Jewish organization fused with the JSDP.

In Vienna Grossman devoted most of his time to his new family—'Janka' gave birth to Jean Henri on 16 October 1910 and Stanislaus on May Day 1914—and academic research. Under the supervision of Carl Grünberg, the first professor with Marxist inclinations at a German speaking university, Grossman began a major study of the trade policy that the reforming Habsburg monarchs, Maria Theresia and her son Josef II, implemented in Galicia, between 1772 and 1790. Such a thesis was the basis for a higher doctorate, a prerequisite for a university post. The project involved extensive archival research in Vienna, Paris, Kraków and Lwów, in the course of which Grossman not only gleaned material for his thesis but also developed expertise in the history of official statistics in Austria and made a series of professional contacts.

The substantial *Austria's trade policy with regard to Galicia during the reform period of 1772–1790* by 'Henryk Grossmann' appeared days before the outbreak of World War I. It incorporated material and arguments from his earlier work on Galician politics and economic history, and the development of Austrian statistical collections, including academic articles in German and Polish published between 1910 and 1913. The book spelt out the details of Habsburg mercantilist policies during the short period of enlightened rule and their goal of increasing the province's value to the Empire, by promoting Galicia's economic development and trade. There were no references to Marx or any other Marxists. These might have raised questions about Grossman's suitability as a university professor. Much later, he asserted that the study was 'written from the standpoint of historical materialism'—a fair claim for this refutation of the nationalist orthodoxies of mainstream Polish and Austrian historians.

The War prevented Grossman from pursuing either an academic or a legal career. He did not share the patriotic fervor of July–August 1914 and only entered the army in February 1915, as a conscript. For a period, his military duties cannot have been very onerous as, in mid 1916, he was able to publish an account of the origins of official statistics in Austria and later a defense of it against criticism.

Grossman only spent a short time in the field, initially as a non-commissioned training officer, then fighting with his unit on the Russian front in Volhynia. Soon he was appointed as the representative of the recently established Scientific Committee for the War Economy of the War Ministry to the General Government in Lublin, the administration of the Austrian occupied sector of the Congress Kingdom of Poland. The Committee was essentially a high powered think tank on the economic management of the war. Its staff included the cream of the younger generation of professional economists in Austria, from across the theoretical spectrum. While working for the Committee, Grossman was commissioned as a lieutenant.

In February 1917, Grossman wrote a critique of economic statistics produced by the Statistical Office of the General Government, which included his alternative calculation of the wealth of the Kingdom of Poland before the War. In the winter of 1917–1918, he gave a lecture on the theory behind his estimates, at the prestigious Polish Academy of Science in Kraków. He also contributed an essay on the organization of credit to a book on the Kingdom of Poland before the War, published in mid 1917. During 1916–1917, Grossman may also have taken an interest in the social history of physics, particularly Descartes' understanding of science as a means to reduce human labor.

After the Bolshevik revolution in Russia, Grossman was recalled to the War Ministry in Vienna. He became a consultant (*Referent*) on economic aspects of peace negotiations in the War Economy Section of the War Ministry.

When, in early November 1918, the revolutionary actions of soldiers, sailors, workers and, in Austria-Hungary, nationalist mobilizations brought an end to the War and the German and Habsburg monarchies, Grossman seemed to have excellent prospects in peace-time Vienna. But the limited Austrian revolution meant that he could not pursue a career as a senior official of the Austrian Statistical Commission. Such a post required citizenship of the new German-Austrian Republic. And precisely this was denied to large numbers of Galician Jews living in Vienna, like Henryk Grossman, by the racist policy of the new Republic's coalition government, in which the Social Democrats were the largest party.

Instead, he accepted a responsible position with the Polish Central Statistical Office (GUS) in Warsaw. But, before he took up that job, he delivered a paper on the Marxist theory of economic crises at the Academy of Sciences in Kraków. The address expressed a number of the fundamental themes of his later work in economic theory. These included Marx's method in *Capital*, the inevitability of economic crises under capitalism, and the nature of capitalist production, as the contradictory unity of a labor process, creating useful things (use values) and a valorization process that creates value (that can take the form of money) and underpins profits.

At GUS, Grossman was entrusted with organizing independent Poland's first population census. This was a huge operation, employing at least 60,000 people. His work with GUS was interrupted by the brief Soviet-Polish War of June to September 1920. Grossman was conscripted into the army, as an officer. Fighting against the Red Army did not accord with his convictions. The authorities, suspicious about his behavior, stripped him of his military responsibilities and then had the police watch him. Back at GUS, while still in charge of preparations for the now delayed census, he wrote a report on rail freight statistics.

Grossman did not supervise the completion of the census. In the middle of 1921 he resigned from GUS. Carl Grünberg later explained that, 'as he was not prepared to accept the fudging ("Frisierung") of the census results in favor of the Polish majority and against the interests of the minorities, he left his post at the Statistical Commission and devoted himself exclusively to research and teaching'. How could a veteran of the Jewish working class's struggles against national oppression by the Austrian and Galician authorities be expected to go along with similar maneuvers by the newly independent Polish state? At GUS, there seems to have been sympathy for Grossman's stand. Well after his resignation, the Commission's journal published two articles by him: his wartime study of Russian Poland before the outbreak of hostilities and an account and analysis of the censes of the Napoleonic Duchy of Warsaw in 1808 and 1810.

The Free University of Poland (WWP), where Grossman taught from 1921 and took up a full professorship in economic policy in 1922, provided a less politicized atmosphere and a job that allowed him to pursue his own research interests. The WWP was not a state institution and its staff included many leftists from the Polish Communist Workers' Party (KPRP), which Henryk Grossman joined in 1920, and the Polish Socialist Party (PPS).

One of his projects resulted in a monograph on Simonde de Sismondi's contribution to political economy. It paid particular attention to the relationship between Sismondi's method and analysis of economic crises and Marx's approach. Beyond Sismondi and even 18th century physiocracy and mercantilism, Grossman's interest in the history of political economy extended back to the 15th century and even antiquity. He collected materials on the economic ideas of Copernicus and investigated the history of slavery in Christendom. But the largest of his projects at the WWP was on the foundations and nature of Marx's theory of economic crises.

Given that the Communist Party was subject to considerable police repression, cultural and educational front organizations were particularly important for the KPRP. The largest such organization was the People's University. It offered popular and specialist courses and collaborated with the trade unions' Workers' School. Other bodies close to the KPRP, like the 'Workers' Culture' association and the 'Book' publishing cooperative, used the facilities of the People's University. At the start of 1922, Henryk Grossman was the secretary of the People's University and soon became its chairperson, continuing in this role until 1925. He contributed to the journal *Workers' culture*; he translated, introduced and annotated several important works by Marx, which had not been published in Polish before.

Like other Communists, Grossman was subjected to harassment and arrest. Between 1922 and 1925 he was arrested and held in custody five times, for up to eight months. He was, however, never convicted. It seems that an unofficial arrangement with the Polish authorities for a kind of qualified exile was involved in his departure from Poland in 1925. He would leave the country but could visit for two weeks a year, so long as he only saw his family and did not engage in politics. Even before he departed, Henryk and Janka's marriage had apparently broken down, as a result of pressure from her family while he was in prison. They were never divorced and remained on good terms.

Carl Grünberg, now the Director of the Institute for Social Research in Frankfurt am Main, offered his former student a post as his assistant. The Institute was a remarkable body, associated with the University of Frankfurt and devoted to Marxist research. It provided Grossman with a good income, stable employment and a very favorable environment for his studies.

Grossman did not join the Communist Party of Germany (KPD), as a condition of his residence in Germany, and probably also of his arrangement with the Polish authorities. While still a sympathizer of the Communist International, he was no longer subject to Party discipline, especially in his research which was very much shaped by his earlier experience as an activist in the Jewish workers' and Communist movements. Like many Communists, out of loyalty to the Russian revolution and commitment to militant working class struggle, Grossman supported the Soviet Union, the KPRP, the KPD, and the Communist International despite all their political zigzags during the 1920s and, by 1930, the eradication of the last vestiges of workers' power in Russia.

The German police, concerned about Grossman's political associations, delayed the award of his higher doctorate. But he eventually received it, in March 1927. *Austria's trade policy* was accepted as his *Habilitationsschrift* and he delivered a trial lecture on 'Sismondi and classical political economy'. His inaugural lecture as a *Privatdozent* at the University of Frankfurt, was 'Oresmius and Copernicus as monetary theorists'. In March 1930 Grossman was appointed to an extraordinary professorship in the University's Faculty of Economics and Social Sciences.

Grossman's first publication in Germany appeared in 1927. It was a long critique of Fritz Sternberg's *Imperialism*, a large study of contemporary capitalism, which had received financial support from the Institute. The book dealt with issues Grossman was working on for a major publication. Against Sternberg's views, Grossman presented his own account of the scientific method used by Marx in *Capital*, foreshadowed in his 1919 lecture on economic crises. He argued that the simplifying assumptions Marx made early in *Capital*, in order to grasp fundamental processes, were progressively lifted as he introduced complicating factors, step by step. In this way, the analysis successively approached the empirical reality of capitalism. In contrast to Sternberg's voluntarist understanding of the process of revolution, in which its timing was determined only by the consciousness raising efforts of a socialist party, Grossman identified the Marxist position by quoting 'a specialist in revolutionary matters and at the same time a Marxist'. This expert, Lenin, had emphasized that the impossibility of the ruling class maintaining its domination in the old way and an acute deterioration in the living conditions of subordinate classes are objective preconditions for revolution.

The law of accumulation and breakdown of the capitalist system (being also a theory of crises), Grossman's best known work, appeared in 1929. The book employed Marx's method in *Capital*, moving from the abstract to the concrete. In their competitive drive to make profits, by increasing the productivity of their workers, individual capitalists invest a larger and larger proportion of their resources in machinery and equipment (constant capital) and a smaller proportion in wages (variable capital). But living labor is the source of new value and hence profit. Although more commodities are produced, the rate of profit, that is, the ratio of newly created value to capitalists' total outlays, tends to decline; in response, many will eventually stop investing and there will be a crisis. At a more concrete level of analysis, Grossman drew attention to 'counter-tendencies' to this process, significantly expanding Marx's discussion of these.

The book drew a great deal of attention and was widely reviewed. But, despite his precautions—noting the correctness of Lenin's comment that, for capitalism, 'there are no absolutely hopeless situations' and implicitly endorsing Lenin's approach to the politics of revolution—Grossman's argument was criticized for being a 'mechanical' theory of capitalist breakdown, by orthodox Communist and social democratic reviewers alike.

Shortly after *The law of accumulation* appeared, Grossman published a 'small *programmatic* work'. 'The change to the original plan of Marx's *Capital* and its causes' was concerned with Marx's method in *Capital* and its implications for the Marxist understanding of capitalism. Two further essays on economic theory were published in 1932. One developed Grossman's critique of Rosa Luxemburg's economic analysis, her treatment of gold production in particular. The other applied Grossman's insights about Marx's method to the question of the relationship between the value of commodities, reflecting the amount of socially necessary labor time involved in their production, and their prices.

In addition to his theoretical work and teaching, between 1930 and 1932 Grossman wrote and edited the entries on the socialist movement for the fourth edition of Ludwig Elster's three volume *Dictionary of economics*. His most significant essay in the dictionary, a survey of Marxism after Marx, was also published separately to mark the fiftieth anniversary of Marx's death. The survey made Grossman's Communist sympathies very clear and included an account of his own work on Marxist economics. This tacitly responded to reviewers' misrepresentations of *The law of accumulation* by insisting that 'a fatalistic policy of waiting for the 'automatic' collapse, that is without actively intervening, can not be in question for the proletariat'.

After Grünberg retired in 1930, Grossman was by far the best known member of the Institute for Social Research. He had substantial academic reputations as a researcher in the areas of economic history, the history of economic thought and economic theory. The items he wrote for Elster's dictionary gave him prominence as an authority on the history and theory of socialism before a much wider, lay audience. But the core intellectual concerns of the Institute began to shift away from Grossman's areas of interest under the new director, Max Horkheimer.

After the Nazi takeover in Germany, Grossman moved from Frankfurt to Paris, in March 1933. The failure of the German labor movement to stop the Nazis led him to reassess his political position and to adopt a very critical attitude to the official Communist movement under Stalin's leadership, while reaffirming his commitment to Marxism. He associated with other dissident Marxists in the German exile community in Paris.

Although separated from the Institute's head office, initially located in Geneva and then in New York, Grossman remained one of its members and worked on projects formulated in correspondence with Horkheimer and his associates. While in Paris, he wrote two reviews for the Institute's *Journal of social research* (*ZfS*) which touched on crisis theory and prepared the entry on Sismondi for the *Encyclopedia of the social sciences*.

The Institute had funded a project undertaken by Franz Borkenau, on the origins of the modern scientific world view. Horkheimer, drawing on advice from Theodor Wiesengrund Adorno, who had no expertise in the area, agreed to publish an article by Borkenau, summarizing his conclusions, in the *ZfS* in 1932. The appearance of Borkenau's book-length study, *The transition from the feudal to the bourgeois world view*, was delayed because there were concerns that Borkenau's analysis was neither Marxist nor accurate. In response to criticisms from the Institute, inspired by Grossman, Borkenau wrote a new first chapter that related intellectual developments to the emergence of manufacturing. But the Institute still had major reservations. Horkheimer's perfunctory preface disowned the book's analysis.

To increase the distance between the Institute and Borkenau, Horkheimer initially lined up Walter Benjamin to write a review for the *ZfS*.[2] Benjamin failed to come up with the goods, so Horkheimer turned to Grossman, who had previously undertaken historical research into the development of ideas, their relationship to changes in the mode of production and, at least as far back as 1917, had been interested in the social history of Descartes' contribution to science.

Grossman threw himself into the critique of Borkenau and transformed the project from a mere defense of the Institute's credibility into the presentation of a systematic alternative account of the emergence of the modern mechanical world view. He became immersed in the project. 'The problem of the origins of mechanistic thought has so gripped me and taken up all of my efforts' he told Paul Mattick 'that I have spent almost all of my time for months in the Bibliothèque Nationale in the literature of the 16th and 17th Centuries'. By July 1934 he was thinking of writing not only a review article dealing with Borkenau's arguments, but also a book, 'Cartesianism and manufacture' which would settle accounts more with Weber's position in *Protestant ethic and the spirit of capitalism* than Borkenau.[3]

Grossman wrote three essays which demolished the central arguments in Borkenau's book. Borkenau contended that modern mechanics, the precondition for the emergence of a mechanistic world view, dated from the middle of the 17th Century. The development of mechanics was in turn, based on a metaphor between physics and economic practice: the division of labor in the stage of capitalist development, known as 'manufacture', that preceded industrialization.

In 'The capitalism of the Renaissance period', Grossman demonstrated that modern mechanics was elaborated under the influence of capitalist development around 1500. On the other hand, in 'Manufacture of the 16th–18th centuries', he showed that a systematic division of labor did not become characteristic of manufacture until the second half of the 18th century. In other words, modern mechanics existed *before* the middle of the 17th century, when Borkenau asserted it arose. And a

[2] Letters from Walter Benjamin to Gershom Scholem 15 January 1933 and 17 October 1934, Walter Benjamin *Walter Benjamin Briefe 2 1929–1940* Suhrkamp, Frankfurt am Main 1978 pp. 561, 624.

[3] Letter from Grossman to Friedrich Pollock, 13 August 1934, Leo-Löwenthal-Archiv, Universitäts- und Stadtarchiv Frankfurt am Main A 325 p. 86. Grossman outlined the structure of the book on a contents page, headed 'Der Kartesianismus und die Manufaktur' 23 July 1934, Max-Horkheimer-Archiv, Universitäts- und Stadtarchiv Frankfurt am Main (MHA) VI 9 p. 409.

systematic division of labor, which, according to Borkenau explained the formulation of mechanics, only emerged well *after* that period. 'The beginnings of the capitalism and the new mass morality' responded to Borkenau's uncritical reliance on Max Weber's analysis, by demonstrating that capitalism predated Protestantism. This essay was only published in 2006.

Initially Horkheimer agreed to publish the first two of Grossman's essays. Then, concerned that they would give too much publicity to Borkenau's book, he changed his mind a couple of weeks later and demanded a single review 'to distance us from his work'.[4] Faced with Horkheimer's 'painful' decision, Grossman was disappointed. He agreed to proceed as Horkheimer wished but pointed out that the publication of a straightforward critique, separately from new research results, could damage the Institute. 'The more significant the mistakes, indeed the *factual mistakes* that I have to insist on, the more people will ask why the Institute published such a book'.[5] Then Horkheimer requested that the article should be about 32 printed pages long, raising the possibility of a later, pamphlet-length supplement to the journal.[6] Grossman tried but failed to fit and trim his overabundant material into a compact article before the end of the year.[7] Eventually he gave up on this effort and, in early January, sent a manuscript three to four times the length requested in early January, along with two letters justifying his work.[8] 'You set me an unachievable task,' he complained.

The core of Grossman's positive argument in the article drew on his understanding of Marx's scientific method. This, with the assistance of Marx's comments about machinery in *Capital*, he applied to the history of science, demonstrating that constructing, working with, and observing machines had made it possible to set aside some of the concrete appearances of physical phenomena—complicated by different kinds of motion and friction, for example—to identify abstract mechanical work. Grossman supported his case with references to the discoveries of Leonardo da Vinci and René Descartes amongst others. The material foundations of mechanics and the mechanistic world view lay in the impetus given by capitalism to the invention of new machines.

Horkheimer's response to the draft was expansive praise: the piece was 'entirely excellent'.[9] In view of the significance and quality of the work no sections were to be excised. References to Marx were, however, edited out. Grossman's response indicated his continuing political engagement,

---

[4] Letter from Horkheimer to Grossman 25 September 1934, MHA IV 9 p. 404, referred to in MHGS15 p. 225; letter from Horkheimer to Grossman 8 October 1934 Max Horkheimer *Max Horkheimer, Gesammelte Schiften, Band 15: Briefwechsel 1913–1936* Frankfurt am Main: Fischer, 1995 (MHGS15) pp. 236–237.

[5] Letter from Grossman to Horkheimer 26 October 1934 in MHGS15 p. 254.

[6] Letter from Horkheimer to Grossman 7 November 1934 MHGS15 pp. 257–258.

[7] Letter from Grossman to Horkheimer 26 October 1934 MHGS15 p. 254.

[8] Letters from Grossman to Horkheimer 4 January 1935 MHGS15 pp. 293–296, 5 January 1935 MHGS15 pp. 297–299.

[9] Letter from Horkheimer to Grossman 26 January 1935 in MHGS15 p. 301.

> After all, we all fight for the great proletarian cause. But since the destruction of the labor movement, a satisfaction which every fighter felt earlier, before the world war, from recognition from within his movement is no longer possible. So one is happy with finding satisfaction in the narrow circle of ties and gets encouragement to further work from it.[10]

Grossman had expressed the hope that the study would be 'a nice contribution to the materialist conception of history and not through general prattle, à la Bukharin, but through *concrete* research of historical material'.

The history of science was on Grossman's mind during a visit to Spain in the summer of 1935. In Madrid he saw some examples of machines used in antiquity, at the National Archaeological Museum. From Valencia, he wrote a long letter, including diagrams, about them to Friedrich Pollock and Horkheimer. Given the prevalence of cheap slave labor, Grossman argued, these were not designed to save labor but to perform functions which were otherwise impossible. They did not, therefore, prompt the formulation of a theory of mechanics, which only emerged during the 15th century, in association with the application of machines to labor saving purposes (this volume pp. 231–233).

Around the start of 1936, Grossman moved to England. The political situation in Europe and the prospects of war between France and Germany were one consideration in the move, the unsatisfactory research conditions in Paris another. He again devoted most of his time to economic research. In September Horkheimer asked for an article to be published in the *ZfS* in 1937, suggesting that he might use the methodological part of his work on crises. Instead, Grossman, proposed a long and original piece to mark the 70th anniversary of the publication of the first volume of *Capital*.

Divergence in the outlooks of Grossman and Horkheimer became clearer after Grossman moved to New York in October 1937. Horkheimer's shift away from Marxism was one element in the widening gap. Another was a radical change in Grossman's political views in 1935–1936. The Spanish Civil War and his perception that the Soviet Union was playing a progressive role in the conflict, seem to have had a major impact on him. He became and remained an uncritical supporter of the USSR under Stalin. The Hitler-Stalin Pact and the outbreak of World War II led Horkheimer to make more open and sweeping criticisms of the Soviet Union, while Grossman's attitude to Russia did not change. A political gulf therefore opened up between Grossman and the group around Horkheimer in addition to growing theoretical differences. Nevertheless, relations remained friendly for a period and Grossman happily participated in Institute activities and seminars. Between 1938 and 1942, he wrote nine book reviews for the Institute's journal. One was his summary of Marx and Engels on the civil war in the United States. Six were related to crisis theory. Two dealt with issues in the history of science.

One of the reviews concerning the history of science examined books by George Clark and George Sarton (this volume pp. 233–237), the other works by Lynn

[10] Letter from Grossman to Horkheimer and Friedrich Pollock 7 February 1935 in MHGS15 p. 316.

Thorndike. In 1937 Grossman praised Clark's account, in *Science and social welfare in the age of Newton*, of the development of technology under the pressure of socio-economic circumstances during the 17th Century. But he was critical the way Clark underplayed these factors in explaining scientific advances and engaged in a sustained defense of Boris Hessen's analysis of Newtonian physics. Sarton's chapter on the influence of 'oriental' culture on European science came in for particular praise, as an antidote to the orthodoxy, shared by Max Weber, that exaggerated the 'western European spirit'. Arabic was superior to western European culture until around the middle of the 13th Century. Grossman regarded volumes 5 and 6 of Thorndike's *A History of Magic and Experimental Science* as valuable compendiums of information about the thought of the 16th Century. But Thorndike failed to account for the rise of rationalist thinking in the 12th and 13th Centuries, its decline and then revival in the 16th Century. This was one of the concerns of the reviewer's own current research.[11]

Eventually the stresses caused by political and theoretical differences, a salary cut and financial insecurity led to a crisis in relations between Grossman and the Institute. Their association was subsequently very limited, although he continued to draw his salary and remained on good terms with the office staff.

After this break and prolonged delays, just before Christmas 1941, the Institute issued a tiny duplicated edition of the essay on Marx's economics that Grossman had been working on since 1937. *Marx, classical political economy and the problem of dynamics* identified two aspects of Marx's original contribution. The first was the importance of the distinction between use value and value. The second was Marx's conception of capitalism as a dynamic system, which Grossman contrasted with the static approach of contemporary neo-classical economics. The essay was and remains one of the most impressive critiques of the methodological underpinnings of what is now simply called 'economics' in most universities and the media.

In 1943 Grossman published the two part 'The evolutionist revolt against classical economics', in the *Journal of political economy*. It, like the study of dynamics, cleared a path to identifying Marx's original contribution to social theory. The argument challenged two related, false conceptions: that Marx was the first to introduce an historical perspective into economics; and that this was due to the influence of Hegel on Marx. Much of the essay dealt with the question of 'how dynamic or evolutionary thinking actually entered the field of economics' and laid the basis for Marx's approach.

During the 1940s Grossman continued to work on the question of crises under simple reproduction, a study of the early history of modern science, with a focus on Descartes' contribution and the history of economic thought. These studies grew out of several of Grossman's ongoing research projects. He also dealt with 'The social history of mechanics from the 12th to the middle of the 17th century' and 'Universal science against the specialists: a new attempt to interpret Descartes'

[11] See the brief outlines of Grossman's research project, 5b 'Capitalism in the 13th Century'; 5c 'Kapitalismus im 13 Jahrhundert' 'Mitarbeiter 1939', MHA IX 58.

philosophy' in separate manuscripts.[12] (See above p. 157.) Grossman had written these works in German and did 'not have the money to have them translated into English. I cannot pay a translator from my meager income and therefore cannot prepare my manuscripts for publication'. The Institute may have provided some funds, for his essay on Playfair as an early theorist of imperialism was worked into English by Norbert Guterman, a professional translator and 'freelance associate of the Institute',[13] and 'Descartes' new ideal of science: universal science vs. science of an elite' was also translated.

Having sent the Playfair piece off to Guterman, in the spring of 1947, Grossman wrote to Christina Stead and Bill Blake that 'I am not decided what to do now: finish Descartes? or the book on Marx's simple reproduction (which I regard as my chief contribution to Marxist theory)'.[14] This may have been a matter of tinkering with the Descartes manuscript more than anything else. A year later he wrote of visiting Paris in order to find a publisher for it.[15]

After his break with the Institute, Grossman's standard of living dropped dramatically because his salary was frozen despite rapid inflation, and his health deteriorated. He was, however, far from isolated. Amongst the refugees in the United States were old acquaintances from Kraków, Vienna and Germany. He struck up new friendships, amongst Poles and Germans, particularly Communists, and became very close to the Australian novelist Christina Stead and her partner, the American writer and economist, Bill Blake. Already thinking about leaving the United States for Europe in September 1946, Grossman's desire to cross the Atlantic increased as the Cold War intensified. His most fruitful contacts were in East Germany. He was one of the most prominent Marxist economic theorists still living and his transition from New York to the Soviet Zone of Germany would be a coup for its emerging administration.

The University of Leipzig invited him to take up a chair in political economy. On his arrival, in March 1949, he received a very friendly welcome. Grossman lived in a conveniently located and well-appointed apartment, paid a low rent and received a salary higher than his income in New York. In Leipzig, he found a number of friends and acquaintances from Germany before the War and exile in New York.

The new professor of political economy threw himself into teaching, social interaction and political life, joining the Socialist Unity (i.e. Communist) Party in June. But he was unable to sustain these activities for long, due to ill health. He was in poor physical shape, with arthritis, Parkinson's disease, weakened kidneys, already a problem in New York, and a heart condition. By March 1950 he was in hospital,

---

[12] 'Aufzählung der nicht veröffentlichten und nicht registrierten Arbeiten von Prof. Dr. Henryk Grossman', Henryk Grossmann', PA 40, Universitätsarchiv Leipzig p. 68.

[13] Rolf Wiggershaus *The Frankfurt School: Its History, Theories and Political Significance* Polity, Cambridge, Mass. 1995 p. 386.

[14] Letter from Grossman to Stead and Blake 4 and 5 May 1947, Box 17 Folder 125, Stead Collection, National Library of Australia.

[15] Letter from Grossman to Schreiner, 14 April 1948, Albert Schreiner Nachlaß, Bundesarchiv, Berlin Ny 41 98/70 p. 103.

had suffered a small stroke and undergone an operation for a prostate growth. The doctors concluded that he had less than a year to live and his cousin, Oskar Kurz who had also been an exile in New York, had already come from Vienna. As an important public figure, he received privileged treatment in Leipzig's premier hospital, the Polyclinic.

Grossman was able to return home and continue his research for a period. He tried to have his work published in East Germany. But during 1950–1951 the regime conducted a campaign against people who had returned from exile in the west after the fall of the Nazism. Grossman, extremely ill and with an international reputation, was not regarded as a threat. He remained a supporter of the authorities in Russia and its German ally. Yet he continued to propound his original contributions to economic theory, which had been inspired and informed by working class struggles and Marx's own conception of socialism as working class self-emancipation, although Stalinist mouthpieces had 'refuted' them. Grossman died on November 24, 1950. None of his work was ever published in the German Democratic Republic or the Soviet Union.

Grossman's contribution Marxist economics was largely dormant until the late 1960s, when a new generation of young Marxists, initially activists in the West German new left, rediscovered them. The recovery of his study of 'The social foundations of mechanical philosophy' and his manuscript on Descartes took even longer.

Boris Hessen: In Lieu of a Biography

Gideon Freudenthal and Peter McLaughlin

Hessen has not yet found his biographer. The following compilation by the editors collects the most essential verifiable information available from various publications. We hope this volume will encourage further research into Hessen's life and work.[1]

Boris Mikhailovich Hessen (Gessen) was born on August 28, 1893 (new style) in Elisavetgrad, a small Ukranian manufacturing city (now called Kirovograd in Russian, and Kirovohrad in Ukranian). Elisavetgrad was the seat of the provincial government of Cherson, the region now called Kirovohrad. Half the working population of the town was employed by Ellworthy Ltd., a wholly British owned company that produced a large part of the agricultural machinery used in Russia. Boris' father, Mikhail Borisovich, seems to have been a fairly prominent member of the city's Jewish middle class, serving on the board of directors of one of Elizavetgrad's banks and as a member of the local Jewish poor relief agency.

The populace was on the whole Ukranian, literacy was low. Boris attended the local "gymnasium" for eight years where the language of instruction was presumably Russian. Hessen reported on a later application to the college of Red Professors that he had German, French, English, and Latin. Instruction at such a school would have included Latin and French. It is likely that English, too, was offered, since Hessen knew enough English to study in Scotland after finishing secondary school. Hessen was also fluent enough in German to attend conferences in Germany, to apply successfully for leave to study in Germany, and to serve on the board of a German-language physics journal (*Physikalische Zeitschrift der Sowjetunion*).

In 1913 Boris went to study at the University of Edinburgh with his classmate Igor Evgenevich Tamm, who was later to win the Nobel Prize for physics in 1958. The two shared an interest in physics and reportedly also in subversive

G. Freudenthal
Tel Aviv University, Tel Aviv, Israel
e-mail: frgidon@post.tau.ac.il

[1] This compilation is primarily based on Hessen's handwritten CV of 1924 published and translated into German by R.-L. Winkler 2007 and on Winkler's biographical sketches in this and other papers in the bibliography. We have extensively used Josephson 1991 and have also drawn on Gorelik 1995 and Hall 1999.

G. Freudenthal, P. McLaughlin (eds.), *The Social and Economic Roots of the Scientific Revolution,* Boston Studies in the Philosophy of Science 278,
DOI 10.1007/978-1-4020-9604-4_7, © Springer Science+Business Media B.V. 2009

politics. From the matriculation records at Edinburgh we know that Hessen studied primarily physics and mathematics, taking calculus with E.T. Whittaker. The year at Edinburgh was not to be repeated. Before Hessen could return for the second year, war broke out in Europe, and travel to Edinburgh was not possible. Since in Czarist Russia there were impediments for Jewish students at the university, Hessen was unable to continue in physics; he enrolled at the Polytechnic Institute in Petersburg from 1914 till 1917, where he studied mathematics and economics, auditing courses in mathematics at the University. He began privately to study philosophy and the history of mathematics.

Before the outbreak of revolution Hessen had returned to Elisavetgrad and pursued revolutionary activities. In 1917 he was a member of the Elisavetgrad Organization of Internationalists and after the October Revolution a member of the Elisavetgrad Soviet. He joined the Bolshevik Party in 1919. The story is told that he also helped to expropriate his father's bank. He saw action with the Red Army in the campaign that ended occupation of Elisavetgrad by the insurgent A.N. Grigoriev.

From 1919 to 1921 he was an instructor of political administration. From 1921 to 1924 he was instructor for political economy at the Communist Sverdlov University in Moscow. In 1924 Hessen entered the College of the Red Professoriate, a training center for communists destined to teach at the University. He remained on at the college as a lecturer and as deputy director of the section for natural sciences. From 1926 on Hessen also taught "methodology" at Moscow University in the Faculty of Mathematics and Physics, which included history and philosophy of science.

In late 1930 Hessen became director of the expanding Institute for Physics at Moscow University, and in 1931, when Physics was upgraded from an Institute to a Faculty, Hessen became the first Dean. In 1933 he was made corresponding Member of the Academy of Sciences. In 1934 when the Academy moved from Leningrad to Moscow, Hessen gave up the deanship and became Deputy Director of the FIAN (Physics Institute of the Academy of Sciences). He retained this position until 1936, the year of his arrest. Hessen was also on the editorial board of *Physikalische Zeitschrift der Sowjetunion* and of *Progress in Physics* (UFN).

From the beginning of his university career Hessen was involved in projects on the history and philosophy of science. He belonged to the "Circle of Physicist-Mathematician-Materialists" established in 1927. He was among the general editors of a series in *Classics in Science*, introduced to provide materials for Marxist discussions of science. He lectured on history and philosophy of science at the Institute of Philosophy and at the Institute of History of Science and Technology of Moscow University. A national handbook of scientists in the Soviet Union from 1930 lists Hessen's area in the Physics Department as "History and Philosophy of the Natural Sciences"; his area of work is said to be "physics, methodology of the exact natural sciences, foundations of statistical mechanics and of relativity theory." Among the fruits of this professional activity are the famous paper of 1931 "The Social and Economic Roots of Newton's 'Principia'" and a 700-page collection of translated sources and expositions by Hessen intended to serve as a textbook for courses on the history of physics. This book was in proof when Hessen was arrested and has remained unpublished.

In 1928 Hessen published an elementary introduction to relativity theory under the title *Osnvnye idei theorii otnositel'nosti*. He also published a defense of quantum mechanics as an introduction to the Russian translation of Arthur E. Haas's *Materiewellen und Quantenmechanik*. Although Hessen published some technical papers in statistics, most of his publications are either in popular science or in philosophy and history of science. In 1928 Hessen applied for a four month study-leave to work on statistics with Richard von Mises in Berlin and to attend a summer course for foreign physicists conducted by Max Planck and Arthur Eddington. He also attended the 6th German Physicists' and Mathematicians' Conference in Königsberg 1930 at which the Vienna Circle held its Second Conference for the Epistemology of the Exact Sciences. Hessen there contradicted Hans Reichenbach's characterization of history of science as recreational reading, arguing on the contrary that it provides a key to understanding contemporary science.

Hessen's activities from his first major publications in 1928 to his arrest in 1936 can best be characterized by two main concerns: the defense of the relative autonomy of physics against interference in the content of physical investigation from without and the positive elaboration of the compatibility and affinity of Marxist materialism with the newest developments in physics as opposed to classical physics. He repeatedly defended the Theory of Relativity against charges of idealism, even at times when he was the only Marxist at the Communist Academy sessions to take this position.

Hessen's Marxism was based on serious study of the sources. He was familiar not only with the main writings of Marx and Engels but also drew on unpublished work known only to scholars at the time.

On August 21, 1936 Hessen was arrested and charged with involvement in the terrorist activities of a Trotskyist-Zinovievist conspiracy. At the trial on December 20, 1936 Hessen and two alleged co-conspirators were convicted of complicity in the 1934 murder of S. M. Kirov and of planning terror attacks on leading Soviet officials. All three were condemned together. Hessen and one of the others (Arkady Ossipovich Apirin) were executed by firing squad that same day. The third accused (Arkady Mikhailovich Reisen) was sentenced to ten years penitentiary and died in prison.

In 1955 Hessen was rehabilitated.

Bibliography

Gorelik, Gennady (1995) *Meine antisowjetische Tätigkeit ...: Russische Physiker unter Stalin* (transl. by Helmut Rotter), Braunschweig: Vieweg.

Hall, Karl Philip (1999) *Purely Practical Revolutionaries: A History of Stalinist Theoretical Physics* (Dissertation, Harvard University).

Josephson, Paul R. (1991) *Physics and Politics in Revolutionary Russia*, Berkeley: University of California Press.

Winkler, Rose-Luise (1987/88) "B.M. Hessen" in *Porträts russischer und sowjetischer Soziologen*. Sonderheft Soziologie und Sozialpolitik. Beiträge aus der Forschung. Berlin and Moskau: Akademie der Wissenschaften, 208–210.

Index